T0319402

Cellular V2X for Connected Automated Driving

Cellular V2X for Connected Automated Driving

Edited by

Mikael Fallgren
Ericsson
Ericsson Research
Stockholm, Sweden

Markus Dillinger
Huawei Technologies
Huawei German Research Center
Munich, Germany

Toktam Mahmoodi
King's College London
Department of Engineering
London, UK

Tommy Svensson
Chalmers University of Technology
Department of Electrical Engineering
Gothenburg, Sweden

This edition first published 2021

Registered Offices
John Wiley & Sons, Inc., 111 River Street, Hoboken, NJ 07030, USA
John Wiley & Sons Ltd, The Atrium, Southern Gate, Chichester, West Sussex, PO19 8SQ, UK

Editorial Office
The Atrium, Southern Gate, Chichester, West Sussex, PO19 8SQ, UK

For details of our global editorial offices, customer services, and more information about Wiley products visit us at www.wiley.com.

Wiley also publishes its books in a variety of electronic formats and by print-on-demand. Some content that appears in standard print versions of this book may not be available in other formats.

Library of Congress Cataloging-in-Publication Data

Names: Fallgren, Mikael, editor. | Dillinger, Markus, editor. | Mahmoodi, Toktam, editor. | Svensson, Tommy, editor.
Title: Cellular V2X for connected automated driving / [edited by] Mikael Fallgren, Markus Dillinger, Toktam Mahmoodi, Tommy Svensson.
Description: Hoboken, NJ, USA : John Wiley & Sons, Inc., [2021] | Includes bibliographical references and index.
Identifiers: LCCN 2020020433 (print) | LCCN 2020020434 (ebook) | ISBN 9781119692645 (hardback) | ISBN 9781119692638 (adobe pdf) | ISBN 9781119692652 (epub)
Subjects: LCSH: Vehicular ad hoc networks (Computer networks)
Classification: LCC TE228.37 .C45 2021 (print) | LCC TE228.37 (ebook) | DDC 388.3/1240285–dc23
LC record available at https://lccn.loc.gov/2020020433
LC ebook record available at https://lccn.loc.gov/2020020434

Cover Design: Wiley
Cover Images: © kalawin jongpo/Getty Images, © lushik/Getty Images, © FingerMedium/Getty Images, © darekm101/Getty Images, © olaser/Getty Images, © oonal/Getty Images, © Scharfsinn/Shutterstock

Set in 9.5/12.5pt STIXTwoText by SPi Global, Chennai, India
Printed and bound by CPI Group (UK) Ltd, Croydon, CR0 4YY

C9781119692645_070421

Contents

List of Contributors

Taimoor Abbas
Huawei Technologies
Lund, Sweden

Jesus Alonso-Zarate
Centre Tecnològic de Telecomunicacions de Catalunya (CTTC/CERCA), M2M Communications Department
Barcelona, Spain

David Armand
Orange Labs Services, Department of Security
Châtillon, France

Mate Boban
Huawei Technologies, Huawei German Research Center
Munich, Germany

Hellward Broszio
VISCODA GmbH
Hannover, Germany

Jose Leon Calvo
Ericsson, Ericsson Research
Herzogenrath, Germany

Hanwen Cao
Huawei Technologies, Huawei German Research Center
Munich, Germany

Massimo Condoluci
Ericsson, Ericsson Research
Stockholm, Sweden

Kai Cordes
VISCODA GmbH
Hannover, Germany

Markus Dillinger
Huawei Technologies, Huawei German Research Center
Munich, Germany

Hieu Do
Ericsson, Ericsson Research
Stockholm, Sweden

Mikael Fallgren
Ericsson, Ericsson Research
Stockholm, Sweden

Antonio Eduardo Fernandez Barciela
PSA Group
Vigo, Spain

Gabor Fodor
Ericsson, Ericsson Research
Stockholm, Sweden

and

KTH Royal Institute of Technology
Stockholm, Sweden

Nil Garcia
Chalmers University of Technology
Department of Electrical Engineering
Gothenburg, Sweden

Charalampos Kalalas
Centre Tecnològic de Telecomunicacions de Catalunya (CTTC/CERCA), M2M Communications Department
Barcelona, Spain

Efstathios Katranaras
Sequans Communications
Paris, France

Hyowon Kim
Hanyang University
Seoul, South Korea

Apostolos Kousaridas
Huawei Technologies, Huawei German Research Center
Munich, Germany

Marc Lacoste
Orange Labs Services, Department of Security
Châtillon, France

Andres Laya
Ericsson, Ericsson Research
Stockholm, Sweden

YunXi Li
Ericsson
Stockholm, Sweden

Zexian Li
Nokia Bell Labs
Espoo, Finland

Per Lindberg
Volvo Cars
Gothenburg, Sweden

Jian Luo
Huawei Technologies, Huawei German Research Center
Munich, Germany

Maliheh Mahlouji
King's College London, Department of Engineering
London, UK

Toktam Mahmoodi
King's College London, Department of Engineering
London, UK

Konstantinos Manolakis
Huawei Technologies, Huawei German Research Center
Munich, Germany

Laurent Mussot
Orange Labs Networks
Châtillon, France

Keerthi Kumar Nagalapur
Ericsson, Ericsson Research
Gothenburg, Sweden

Mikael Nilsson
Volvo Cars
Gothenburg, Sweden

Yvan Rafflé
Orange Labs Services, Department of Security
Caen, France

Stephan Saur
Nokia Bell Labs
Stuttgart, Germany

Malte Schellmann
Huawei Technologies, Huawei German Research Center
Munich, Germany

Panagiotis Spapis
Huawei Technologies, Huawei German Research Center
Munich, Germany

Erik Ström
Chalmers University of Technology
Department of Electrical Engineering
Gothenburg, Sweden

Tommy Svensson
Chalmers University of Technology
Department of Electrical Engineering
Gothenburg, Sweden

Rémi Theillaud
Marben Products
Suresnes, France

Ricard Vilalta
Centre Tecnològic de Telecomunicacions de Catalunya (CTTC/CERCA), Optical Networks and Systems Department
Barcelona, Spain

Fuxi Wen
Chalmers University of Technology
Department of Electrical Engineering
Gothenburg, Sweden

Henk Wymeersch
Chalmers University of Technology
Department of Electrical Engineering
Gothenburg, Sweden

Yunpeng Zang
Ericsson, Ericsson Research
Herzogenrath, Germany

Foreword by Christian Micas

In 2020, after nearly a decade of research and innovation efforts in cellular technologies applied to connected automated driving (CAD), the first commercial launches of cars equipped with cellular vehicle-to-everything (C-V2X) were announced by vehicle manufacturers. These developments will establish 2020 as a landmark for the uptake of CAD on roads, paving the way to fully automated driving. C-V2X-enabled CAD has the potential to provide great societal and environmental benefits by significantly reducing the risk of road fatalities, lowering carbon emissions through enhanced traffic conditions, and overall decreasing the cost of transportation of people and goods.

5G is expected to be the power engine of this transformation, thanks to its unique features in terms of very high throughput, low latency, redundancy, and ultra-high reliability. It will allow vehicles, road infrastructure, and pedestrians to exchange real-time data supporting advanced use cases in the field of mobility and transport.

The European Commission contributed to the definition of the 5G vision as early as in 2013, with the launch of the 5G infrastructure Public Private Partnership (5G PPP) – a joint initiative in the field of research and innovation with the information and communication technologies (ICT) industry – to deliver technologies, architectures, solutions, and standards for the fifth generation of mobile communications. This European initiative has an overall budget of €700 million over the 2014–2020 period. It is funded under Horizon 2020, the European Union's R&I programme (2014–2020) and is matched by an industrial investment of at least three to five times that amount.

In September 2016, the Commission adopted the "connectivity package," an ambitious strategy for moving Europe into the gigabit connectivity era. It includes a reform of the telecom regulatory framework, with specific spectrum and investment-friendly measures; it was adopted in December 2018, and a 5G Action Plan was developed to put in place the right framework conditions for the launch of 5G in Europe in 2020. These actions include a calendar for the assignment of 5G pioneering bands across the Member States, as well as the setting of targets for uninterrupted provision of 5G connectivity along major European transport paths by 2025.

As part of 5G PPP Phase II, 2017–2019, vertical experiments, trials, and demos have contributed significantly to the definition and validation of use cases for CAD, in particular thanks to the 5GCAR project. This two-year Horizon 2020-funded project proved the validity of C-V2X for advanced CAD use cases and made significant contributions to the standardization of C-V2X technical specifications and the definition of business models that will enable it.

Phase II of the 5G PPP initiative has been followed by a third and last phase of large-scale trials and pre-deployment projects, launched in November 2018 and September 2020. During this phase, experiments are examining use cases for CAD over motorways, rail, inland waterways, and maritime paths, with a focus on cross-border sections of the so-called 5G corridors. This extended pan-European footprint of large-scale trials will test 5G connectivity solutions for CAD and related business models in a wide variety of geographic situations and road conditions, including multi-modal mobility and transport use cases; this is all thanks to the versatile nature of 5G technology, which offers multipurpose, multi-service network platform features.

These large-scale trials are gathering a broad range of stakeholders from different industries, in particular mobile network operators, telecom vendors, tower companies, public and private road operators, vehicle manufacturers, and their OEMs. In addition to testing, demos, and validation of 5G solutions for CAD, they will also define and assess the various business models required for the provision of dedicated 5G connectivity services for CAD and contribute to the definition of innovative 5G ecosystems in the field of mobility and transport. The outcome of these 5G cross-border corridor trials will contribute to a pipeline of projects toward the large-scale deployment of 5G corridors in the European Union, with the financial support of the proposed Connecting Europe Facility 2 (CEF2) Digital, with a budget of €1.8 billion over 2021–2027 and additional funding possibilities under the Recovery and Resilience Facility.

For the next multi-annual financial framework covering the 2021–2027 period, building on the 5G PPP, the Commission has proposed launching a new partnership covering beyond 5G research as well as deployment activities with an enlarged scope. The target is addressing a complete value chain covering connectivity, device aspects in an IoT context, and service aspects in the context of distributed computing moving to the network edge. Several technologies, such as THz and LiFi communication in ultra-dense networks, are moving toward cloud-native architectures; edge computing and networking have been identified as future R&I topics, including artificial intelligence and blockchain supporting both network management and user applications. The European partnership is also expected to play a strategic role in coordinating the CEF2 Digital deployment program, in particular 5G corridors. Combined, they will have the capacity to contribute to the development of the next generation of CAD in the present and next decade. Moreover, energy efficiency, cybersecurity, and trust will receive particular attention.

This book addresses in a remarkable way the different dimensions of CAD and the complex set of related technical and business challenges; it provides, in a very intelligible and clear manner, a comprehensive picture of the various aspects of C-V2X technologies for CAD, ranging from potential applications and business models to technical descriptions of radio, systems, and network design. As a reference book on C-V2X, it benefits from key findings of the two-year EC Horizon 2020 5GCAR project, in which its editors and authors have played key roles.

Christian Micas
Senior Policy Officer, Future Connectivity Systems
Unit, DG CNECT, European Commission

Foreword by Maxime Flament

Beyond the age of the internet, when each and every new thing is interconnected, there are still dramatic connectivity gaps in our daily lives, especially when talking about mobility: in our cars, trains, and planes. Electronics have invaded our vehicles to make them safer, but those vehicles are still oddly isolated from the rest of the world. By design, vehicles gather information from a plethora of sensors. They intelligently fuse data to create a high-definition, real-time, self-conscious image of themselves and their surroundings. This is done carefully, in a fail-safe manner, to avoid any possible catastrophic malfunction. Today, even lower-end vehicles have their share of radar, cameras, and other sensors that enable life-saving features such as emergency braking. All this to say that sensor-based advanced driver-assistance systems (ADAS) is finally a success story.

When it comes to connectivity, today's reality is that 130 million vehicles are connected to the internet via mobile networks, i.e. vehicle-to-network (V2N); not all of them gather useful data for intelligent transport system and services, but this is a pretty good basis. The trend can only increase, as most vehicle manufacturers have announced their intention to connect their entire fleets to cloud or backend services. At the same time, the arrival of 5G opens the prospects of guaranteed quality of service together with, when required, very low latency. Past the hype, and with some market consolidation, this will bring huge added value to the operation of vehicles during their lifetime: mobility, insurance, comfort, traffic, logistics, maintenance, software updates, product improvements, and entertainment. 5G connectivity is, in the process, decoupling from delivered services to become truly multipurpose and respond to varying requirements.

Right now, the 5G Automotive Association (5GAA) is making sure all players are pushing in the same direction: faster and timely commercial deployment of 3GPP standards, starting with LTE-V2X including both mobile networks and short-range communications. Then, with the addition of 5G mobile networks, new radio (NR) and 5G NR short range, the industry will have new tools to seriously complement ADAS and automated driving. Once vehicle actuation is involved, the complexity reaches another order of magnitude in the domain of functional safety. At the moment, it is highly unlikely that manufacturers, owners, users, insurance companies, and society will accept that automated vehicles depend on data or computing performed outside the vehicles. A change of paradigm is needed. Further work on edge computing and ultra-reliable low-latency communication will provide more insights into the operation of automated vehicles relying on external data and computing. Until the reliability of the network connection and the availability of edge computing can be guaranteed, long debates will continue about which communication channel should be used for automated driving use cases. There is a difficult balance between the computing needed in the vehicle and computing that can be outsourced. For the time being, services running outside the vehicle will be able to support and improve automation and safety but will not be used alone to make decisions.

This being said, there are also substantial non-technical challenges. The first is organizational: who does what to deploy life-saving features based on commercial mobile networks? Second is user acceptance: how to achieve consumer trust for new services dealing with private data. This does not stop at cybersecurity but also requires user-centric design and direct perceived value. In addition, connectivity comes with new kinds of liabilities: the more the manufacturer and owner-user know about the vehicle, the more they have the duty to fix it. This may be one of the most challenging issues for manufacturers: their responsibility vis-à-vis defects, which may be detected faster with sufficient analysis of the collected data, and how quickly actions are launched to mitigate potential safety risks.

5G, including C-V2X, is a new door to create a real internet of vehicles. It comes with great opportunities but also with many challenges. The 5GCAR project began a new wave of European-funded activities, including cooperative work where communication researchers innovate together with vehicle electronics engineers: an exchange resulting in concrete connected car deployment ambitions built on the use of multi-billion-dollar mobile network investment. This is the kind of cooperation that makes European industries stronger.

I am happy to be linked to the important results outlined in this book and hope there will be even more progress in the future. Most important, we need more win-win cooperation between key market players in the connected vehicles business: the automotive and telecommunication sectors working together to unlock new services and open new market opportunities. Stay tuned!

Dr. Maxime Flament
CTO, 5G Automotive Association (5GAA)

Foreword by El Khamis Kadiri

Since April 2018, the connected vehicle has become a must, given the eCall regulatory framework. At PSA, the eCall service has already been available for many years. But there is no doubt that in recent years the connected vehicle has both been enhanced technically and seen a boost in the market.

Connectivity is an enabler for three main pillars in automotive use cases: infotainment, original equipment manufacturer (OEM) use cases, and cooperative intelligent transport system (ITS). The infotainment giants of the digital era are squarely on board; and mirroring solutions, voice recognition engines, and personal assistants are now widely available. This convergence is also apparent in the adoption of operating systems: a relevant part of the industry is migrating from a Linux OS to an Android OS. This is also affecting onboard capabilities, which are evolving from connectivity provided by the customer to built-in connectivity. Embedded SIM cards (eSIMs) and split billing are crucial enablers. And the final pending element is an evolution of the business approach as mobile network operators (MNOs) migrate from cost per volume to another business model.

OEM use cases are based on an extended-vehicle paradigm. Thanks to connectivity, the car is linked to the OEM cloud and can exchange data in both directions for software updates, maintenance, onboard sensors, and so on. The OEM cloud provides computational power for managing data and acts as a hub for defining connected services to the market, such as remote operations, fleet management, etc.

Finally, cooperative ITS is a field where connectivity can improve the capabilities of advanced driver-assistance systems (ADAS). The benefits of connectivity enable a new sensor to complement the limitations of onboard sensors (lidar, camera, and radar), thus improving ADAS possibilities at all levels of autonomous driving.

5G is the first mobile generation for which the verticals have been taken into account from the beginning. This is the case for the automotive sector: at PSA, we have mobilized significant resources to participate in defining this new technology. The company's 5G program is addressing multiple vectors; R&D activities for several projects have been launched based on open innovation and with 5GCAR playing a major role. PSA is teaming up with MNOs, telco vendors, and technology providers to build technical components adapted to automotive needs, while paying attention to onboard limitations and the industry's cost sensitivity. And this process is not over – it will continue in the coming years with new 3GPP releases.

Another relevant vector is participation in international forums where 5GAA has played a major role in coordinating OEM voices in the connectivity-standardization bodies where, historically, the automotive sector was not relevant. A vital fact during the lifetime of 5GCAR has been the European Commission's rejection of the Delegated Act. The adoption of 5G in the automotive sector is not an easy path. The higher cost of the onboard modem must be justified from both technical and business perspectives: the infrastructure roll-out is a must to boost adoption, but it has encountered significant difficulties in terms of cost and delays. Sidelink is a promising approach that can fill some of the infrastructure gaps, but the Delegated Act proposed a technology that would have moved in the opposite direction of successful 5G deployment. Fortunately, due to the act's rejection, Europe will stay competitive compared to other regions like North America and China with respect to 5G connected vehicles.

5G opens a complete new era for connectivity in the automotive sector. It will no longer be a commodity but an enabler for new driving functions and OEM use cases. There are still many challenges to be faced like the arrival of the sidelink New Radio technology, the deployment and adoption of the millimetre waves and a new paradigm in terms of the relationship between the devices and the infrastructure. Some of these topics evoke what is already called Beyond 5G. This book gives the perfect base to understand the new role of the connectivity in the mobility sector covering use cases, radio interfaces and the core architecture evolution and with a vision of the business challenges and impacts.

El Khamis Kadiri
Head of Innovation Connected Vehicle
PSA Group, France

Foreword by Magnus Eek

Volvo Cars has had a focus on safety since the company was founded in 1927 by Gustaf Larsson and Assar Gabrielsson. In 1958, Nils Bohlin became Volvo Cars' first dedicated safety engineer; a year later, he designed the three-point safety belt. A few years after that, the Swedish company waived its patent rights to enable all automakers to use the safety belt. It has been estimated that this single act may have saved over a million lives globally.

Ninety years down the road, we Volvo employees are still constantly challenging ourselves to follow our safety vision; it's in our DNA. Today's vehicles are equipped with various safety sensors that have outstanding capabilities for detecting objects, following objects, and detecting dangerous scenarios in line-of-sight (LOS) using cameras, radar, and lidar (light detection and ranging). In addition, Volvo Cars has offered a Connected Safety program since 2001, starting with eCall (which sends SOS information about accidents, the vehicle's position, etc. to alarm central). Connected Safety features launched in 2016 inform other vehicles about hazards and slippery conditions ahead on the road. Since 2020, Volvo on Call (convenience services) has been offered on all vehicle models.

As an addition to today's safety technologies, research into V2X (vehicle-to-everything) wireless technologies is ongoing to evaluate how to further improve safety, traffic efficiency, and driving comfort on our roads. The V2X sensor is a candidate for use as a complementary sensor to cover both LOS (more than ≈400 meters) and non-line-of-sight (NLOS) scenarios and enables collaboration between V2X sensors. To make full use of V2X technology, an ecosystem must be deployed that provides a worldwide/regional/market infrastructure. As an original equipment manufacturer (OEM), Volvo Cars requires system reliability, interoperability, high performance, and security; and vehicles must be equipped with V2X sensors to gain the advantage of vehicle-to-vehicle (V2V) and vehicle-to-infrastructure (V2I) communication.

For over a decade, the European Union has been contributing substantial funding to enable research and deployment of cooperative intelligent transport system (C-ITS). Through groups such as the CAR 2 CAR Communication Consortium (C2C-CC), contributions to harmonize C-ITS services have been delivered. In Europe, services have been defined in Day 1 C-ITS and Day 1.5 C-ITS services lists that cover hazardous location notifications and signage applications.

In recent years, developments have enabled multiple V2X access layers on the market: IEEE 802.11p, 3GPP LTE-V, and latest 3GPP 5G NR-V2X. In Europe, debates are ongoing about which technology to use, and investigations are under way to see if they can co-exist at the same dedicated wireless frequency spectrum.

As an OEM, one key advantage we see with 5G NR-V2X technology is that it enables a combination of 5G short-range communication between wireless sensors, meanwhile supporting traditional uplinks/downlinks with the same hardware. Additionally, 5G NR-V2X offers edge computing to reduce latency and secure bandwidth, which are some of the key performance indicators (KPIs) we require for C-ITS and connected services.

In the current decade, our vehicles will transform from stand-alone products to objects in a system of systems: C-ITS. 5G features like mobile edge computing (MEC) and network slicing may be key enablers of C-ITS.

On the commercial side, market challenges exist: for example, we need to understand the commercial landscape and new supply chains, such as who will implement MEC and how network slicing will affect business models. Customers on the network need a seamless approach to network subscription, independent of where they drive (reliable, safe, and at a low cost). Vehicles drive across mobile network operator (MNO) networks, regions (urban and rural areas), and countries. To utilize the full benefits of 5G, vehicles expect full service at any position and at all times. Worldwide harmonization of the wireless spectrum to enable V2X technologies would be beneficial to simplify and keep component cost lower.

Positioning is a challenge for many C-ITS services that use global positioning system (GPS) as a reference, since the current accuracy of positioning needs to be further improved. In C-V2X in-coverage modes, the 5G NR network can provide complementary positioning information (for relevant V2X sensors within a zone) and notifications about potentially dangerous scenarios. In the absence of infrastructure (in C-V2X out-of-coverage modes), how do we handle scenarios that require precision in positioning when only GPS is available (assuming bad GPS accuracy)?

For use cases where a vehicle needs to transfer (upload or download) a large amount of data in a very short time, or as a complementary technology for positioning, 5G mmWave with a higher-frequency band may be more suitable for urban areas than rural areas, due to limitations on communication range. For both areas, challenges remain in the time it takes to deploy infrastructure and the technical challenge of the communication distance between the NG-RAN (next generation of radio access network) and the 5G user equipment (e.g. vehicles implementing 3GPP Release 15 or later).

Magnus Eek
Product Owner – Wired and Wireless Communication Technologies
Volvo Cars, Sweden

Preface

The mobile communications industry is on a path to using wireless connectivity to connect all kinds of vehicles and road users. The automotive industry and various transportation systems are part of this journey, with vehicles becoming increasingly aware of their immediate surrounding from various types of integrated onboard and external sensors. The knowledge acquired by vehicles can be shared locally by different types of short-range communication enablers, while long-range communication solutions can provide additional information with added value. With relevant information from both nearby and further away, a vehicle can adapt its behavior based on what lies ahead and thereby make more informed decisions.

Connected vehicles are among the primary enablers of safe, efficient automated driving both during the early stages of automation and in more advanced automation stages. There is, hence, a strong technology trend in which the mobile communications industry and the automotive industry are becoming interwoven to enable new functionalities and capabilities for future automated driving. In addition, there is strong, steady, increasing need for high-capacity mobile broadband to provide automotive cloud connectivity for onboard users. However, this transformation in the two industries needs to take place in tandem with other stakeholders and academic research to enable advanced solutions for traffic safety and increased driving comfort.

Globally, and for many years, stakeholders such as the telecom industry, vehicle manufacturers, traffic authorities, smart cities, and others related to transportation have recognized the value of cooperation through communication to increase safety and traffic efficiency and reduce energy consumption and pollution. In the coming decade, *cellular vehicle-to-everything (C-V2X)* is seen as an essential enabler of progress toward these societal and economic targets. In addition, various industry associations and standard-settings organizations are working jointly to facilitate fifth generation (5G) mobile network assisted driving and automation.

The 5G Communication Automotive Research and innovation (5GCAR) project, running from June 2017 to July 2019, played a pioneering role in bringing these two industry sectors together with substantial contributions to drive the joint vision forward. Discussions about working on a book mainly based on the 5GCAR project began at the launch of the project. We felt it was time to collect all the good work we were planning to do and disseminate it in a coherent and accessible way, while also reaching out to a broader audience than those who typically read our project deliverables and scientific publications. At that time, we also discussed the idea of potential publishers, but it wasn't until 2018 that we started more hands-on planning for the book. During IEEE Globecom 2018, we began to talk with Sandra Grayson at Wiley, and we felt that we had the same vision for a book on connected and automated driving.

You may also be interested to hear how the 5GCAR project got started, since there has been a lot of work behind it. It goes back to the early days of European-funded research toward 5G, initiated by European Commission Framework Program 7 (FP7) with the Mobile and Wireless Communications Enablers for the Twenty-Twenty Information Society (METIS) project, which started in November 2012. Six months later, METIS use cases like traffic efficiency and safety, traffic jams, blind spots, and real-time remote computing for mobile terminals were released. Back then, the telecom industry realized both the need for continued growth in the telecom sector and the potential for connecting various kind of machines for advanced information and communications technologies (ICT) solutions toward a smarter society. At this stage, several areas were identified as particularly promising, such as industrial production systems (Industry 4.0), smart grids, smart cities, safer and more efficient transportation systems, agriculture, and the use of ICT for health (eHealth). Thus, mobile communications were seen as having significant potential to act as key enablers for sustainability in the broad sense via digitalization. Some of these ideas and early requirements for 5G were summarized in four white papers by the 5G infrastructure Public Private Partnership (5G PPP) and one paper by the Next-Generation Mobile Network (NGMN) alliance in early 2015, and since then many 5G publications are now available.[1] Back then, connected vehicles, smart grids, and smart manufacturing systems were identified as the most promising areas for early uptake. eHealth and smart grids have not yet taken off in a broad sense. Both intelligent transportation systems (ITS) to enable safe and efficient transport, and Industry 4.0 for more efficient and agile manufacturing, had good momentum. The strongest interest in 5G at this point turns out to come from the manufacturing and automotive industries. For these reasons, work planning within the 5G PPP took off to coordinate proposals to work closely with identified vertical key industries. In collaboration with the automotive sector, the identified key areas are cooperative ITS (C-ITS), *connected automated driving (CAD)*, and connected road user (CRU) services. For instance, mobile networks and broadband-connected vehicles are already in many cars on the market. As a result, and in parallel, the 5G Automotive Association (5GAA) was launched in September 2016. The 5GAA has played an important role in the convergence of the telecom and automotive industries, and the establishment of the 5GCAR project was one of the early successful outcomes.

Through this engagement, the automotive industry came to realize that huge challenges lie ahead when it comes to digitalizing cars and relying on external industry partners for offloading of on-board processing. Connectivity and data storage and processing seem to be promising way forward. We believe the 5GCAR project has played an important role in the convergence of the telecom and automotive sectors to find common solutions, by creating a research environment in which telecom and automotive researchers and engineers have worked closely. We sincerely believe that such co-creation is the way for true transformation to happen by bringing people together to solve problems. El Khamis Kadiri from the PSA Group concludes that "5GCAR has been a success story of how different sectors can work

1 For references and additional material, please visit the book's web page at Wiley.

together to build new solutions and face the enormous challenges in the mobility domain to be faced in the coming years. Connectivity and 5G will be crucial tools"; and Magnus Eek at Volvo Cars adds that "The H2020 5GCAR final demonstration showed the benefit of sharing V2X sensor data between the 5G network system and connected vehicles to help to predict various dangerous scenarios and avoid them."

The material in this book originates primarily from such close collaborations in 5GCAR. The book's content is provided by researchers from partner institutions in the project. Hence, authors are from the telecom industry (Ericsson, Huawei, Nokia, and Orange), the automotive industry (PSA Group and Volvo Cars), an industrial equipment provider (Bosch), academia (King's College London and Chalmers University of Technology), research institutes (CTTC and CTAG), and small to medium-size enterprises (Sequans, Marben, and VISCODA). In June 2019, this team of researchers, in the form of the 5GCAR consortium, demonstrated cooperative maneuvers to enable and achieve a coordinated vehicle lane merge on a highway, cooperative perception in terms of see-through and long-range sensor sharing, as well as protection of vulnerable road users through cooperative safety. We have posted a few videos of these demos on the book's web page at Wiley. There you can also download background and supplementary material for this book, in the form of 5GCAR project deliverables, publications, tutorials, and presentations. Please have a look.

As a last note, we wish to thank all of the 5GCAR project partners who ensured the successful completion of the project. A special thanks to all of you who contributed as authors or editors to this book. We have enjoyed working with all of you in the 5GCAR project and toward this book!

1. Target Audience and Reader's Guide

The objective of this book is to promote recent joint telecom-automotive research on C-V2X communications solutions, to support their standardization, and to accelerate their commercial availability and global market penetration. The vision is to address society's connected mobility and road-safety needs with regard to applications such as assisted and autonomous driving, ubiquitous access to services, and integration into intelligent transportation. This book is designed to offer both introductory and in-depth knowledge of how mobile connectivity can pave the way to automated vehicles. Toward this end, it addresses – in addition to academia from the telecommunication and automotive sectors – experts and managers in industry, spectrum regulators, and road traffic authorities. We believe that mobile network operators, telecommunication suppliers, automakers and their suppliers, and, in general, all vehicle manufacturers, including the motorcycle and bicycle industries, will adopt advanced 5G connected automated driving solutions for the new decade. Thus, we expect a large market penetration of cellular-supported road use by the end of this decade. We hope that this book will attract a large audience and inspire engineers who wish to develop joint innovation and development projects leading to integrated solutions, interoperability testing, large-scale pilots, and trial deployments in coming years.

The book covers both nontechnical and in-depth technical topics ranging from business models and spectrum considerations to radio, networking, and security and privacy considerations for C-V2X. The book summarizes the current status of the field, gives some recommendations for further activities, and concludes with a future outlook. If you would like to read just specific parts of the book, here are our suggested reading orders based on your perspective:

- *Engineers, researchers, and students*: Chapter 1, 4–9, 2, 3, and 10
- *Policymakers, marketing, business, and management*: Chapters 1, 2, 3, 10, 4–9.

Finally, please consider visiting the book's web page at Wiley for background and supplementary material.

We hope you will enjoy reading this book as much as we have enjoyed writing it!

Mikael Fallgren, Markus Dillinger, Toktam Mahmoodi, and Tommy Svensson

List of Abbreviations

1G	first generation
2D	two-dimensional
2G	second generation
3D	three-dimensional
3G	third generation
3GP-DASH	3GPP defined the progressive download and dynamic adaptive streaming over hypertext transfer protocol
3GPP	3rd Generation Partnership Project
4G	fourth generation
5G	fifth generation
5G PPP	5G infrastructure Public Private Partnership
5GAA	5G Automotive Association
5GCAR	5G Communication Automotive Research and innovation
5GMF	Fifth Generation Mobile Communication Promotion Forum
6G	sixth generation
AA	authorization authority
AAA	authentication, authorization, and accounting
ABG	alpha-beta-gamma
ACC	adaptive cruise control
ACEA	European Automobile Manufacturers Association
ACI	adjacent channel interference
ACIR	ACI ratio
ACK	acknowledgment
ACMA	Australian Communications and Media Authority
AD	autonomous drive
ADAS	advanced driver-assistance systems
ADC	analog to digital converter
AF	application function
AGC	automatic gain control
AI	artificial intelligence
ALP	average localization precision
AMF	access and mobility management function
AOA	angle of arrival

AOD	angle of departure
API	application programming interface
APT	Asia-Pacific Telecommunity
AR	augmented reality
ARCEP	Autorité de Régulation des Communications Electroniques et de la Postes (i.e. French Telecommunications Regulatory Authority)
ARIB	Association of Radio Industries and Businesses
ARQ	automatic repeat request
AS	automotive supplier
ASIL	automotive safety integrity level
AT	authorization ticket
ATIS	Alliance for Telecommunications Industry Solutions
AUSF	authentication server function
AV	autonomous vehicles
AWGN	additive white Gaussian noise
BAM	broadcast announce message
BEREC	Body of European Regulators and Electronic Communications
BLER	block error rate
BM	beamformed multicast
BOF	beginning of frame
BOS	beginning of symbol
bpcu	bit-per-channel-use
BS	base station
BSM	basic safety message
BSS	business support systems
BTP	bidirectional transport protocol
C-ITS	cooperative ITS
C-V2X	cellular V2X
CA	certificate authority
CACC	cooperative ACC adaptive cruise control
CAD	connected automated driving
CAFÉ	clean air for Europe
CAM	cooperative awareness message
CAN	controller area network
CAPEX	capital expenditures
CAV	connected and autonomous vehicles
CBTC	communication-based train control
CCAM	cooperative, connected, and automated mobility
CCI	co-channel interference
CCSA	China Communications Standards Association
CD	code-division
CDD	cyclic delay diversity
CDF	cumulative distribution function
CE	control element
CEN	European Committee for Standardization

CEPT European Conference of Postal and Telecommunications Administrations
CES Consumer Electronics Show
CHF charging function
CNMC Comisión Nacional de los Mercados y la Competencia (i.e. National Commission of Markets and Competition)
COTS commercial off the shelf
CP control-plane
CPE common phase error
CPM collective perception message
CRB Cramér-Rao bound
CRL certification revocation list
CRS cell-specific reference signal
CRU connected road user
CS conservative scenario
CSI channel state information
CSI-RS CSI reference signal
CSIR CSI at the receiver
CSIT CSI at the transmitter
CSMA carrier sense multiple access
CSMA/CA carrier sensing multiple access with collision avoidance
CSP communication service provider
CT core network and terminals TSG of 3GPP
CTL certificate trust list
CTS clear to send
CU cooperative user
D2D device-to-device
DANE DASH-aware network element
DCC decentralized congestion control
DEB direction error bound
DENM decentralized environmental notification message
DL downlink
DMRS demodulation reference signal
DOA difference of arrival
DS delay spread
DSM digital single market
DSRC dedicated short-range communications
DTT digital terrestrial television
E evolution
E-UTRA evolved UTRA
E2E end-to-end
EA enrollment authority
EATA European Automotive Telecom Alliance
EC enrollment certificate
ECA enrollment certificate authority
eCall emergency call

ECC	Electronic Communications Committee
ECDSA	elliptic curve digital signature algorithm
ECIES	elliptic curve integrated encryption scheme
ECU	electric control unit
eD2D	enhanced D2D
EDCA	enhanced distributed channel access
EE	end entity
EIRP	effective isotropic radiated power
EM	element manager
EN	European Norm
eMBB	enhanced mobile broadband
eNB	evolved node B
eSIM	embedded SIM
eUICC	enhanced UICC
ETC	electronic toll collection
ETSI	European Telecommunications Standards Institute
eURLLC	enhanced URLLC
EV	electric vehicles
eV2X	enhanced V2X
FAD	fully automated drive
FCC	Federal Communications Commission
FD	frequency-division
FDD	frequency-division duplex
feD2D	further enhanced D2D
FF	fast fading
FoV	field of view
FR	frequency range
FSPL	free space path-loss
FSS	fixed satellite service
GA	genetic algorithm
GBR	guaranteed bit rate
GBS	geometry-based stochastic
GDOP	geometric dilution of precision
GDPR	General Data Protection Regulation
GFBR	guaranteed flow bit rate
GI	guard interval
GN6ASL	geonetworking to IPv6 adaptation sub-layer
gNB	next generation node B
GNSS	global navigation satellite system
GOSPA	generalized optimal sub-pattern assignment
GP	guard period
GPRS	general packet radio services
GPS	global positioning system
GPU	graphics processing unit
GSA	General Services Administration

GSM	Global System for Mobile Communications
GSMA	GSM Association
GTP	GPRS tunneling protocol
HAD	highly automated driving
HARQ	hybrid automatic repeat request
HD	high definition
HPBW	half power beamwidth
HPLMN	home public land mobile network
HSM	hardware security module
IA	initial access
ICI	inter-carrier interference
ICT	information and communication technology
ICV	intelligent and connected vehicle
ID	identity document
IEEE	Institute of Electrical and Electronics Engineers
IEEE-SA	IEEE Standards Association
IMT	International Mobile Telecommunications
IoT	Internet-of-Things
IP	Internet Protocol
IPR	intellectual property rights
ISG	Industry Specification Group
ISM	Industrial, Scientific and Medical
ISO	International Organization for Standardization
ITS	intelligent transport system
ITS-AP	ITS application provider
ITS-S	ITS station
ITU	International Telecommunication Union
ITU-R	ITU Radiocommunication Sector
ITU-T	ITU Telecommunication Standardization Sector
iUICC	integrated UICC
KPI	key performance indicator
L	level
LCCF	local certificate chain file
LDPC	low-density parity check
LEDBAT	low extra delay background transport
LEK	Electronic Communications Act
LMF	location management function
LNA	low-noise amplifier
LOS	line-of-sight
LPF	local policy file
LS	least square
LSF	large-scale fading
LTCA	long-term certificate authority
LTE	long term evolution
LTE-V	LTE-vehicular (the V2X part of LTE in 3GPP)

M-MIMO	massive MIMO
MU-MIMO	multi-user MIMO
MA	misbehavior authority
MAC	medium access control
MAE	mean absolute error
MANO	NFV management and orchestration
MAP	map data
MBMS	multimedia broadcast multicast service
MC	multi-cell
MCD	multimedia content dissemination
MCS	modulation and coding scheme
MEC	multi-access edge computing (formerly mobile edge computing)
MIC	Ministry of Internal Affairs and Communications
MIIT	Ministry of Industry and Information Technology
MIMO	multiple input multiple output
MMSE	minimum MSE
mMTC	massive machine type communication
MNO	mobile network operator
MP	message passing
MPC	multi-path component
MSE	mean squared error
MSP	mobility service provider
multi-RAT	multiple RAT
NACK	negative-acknowledgment
NAS	non-access stratum
NB-IoT	narrowband IoT
NCU	non-cooperative user
NCS	non-conservative scenario
NEF	network exposure function
NF	network function
NFV	network function virtualization
NFV-I	NFV infrastructure
NFV-I-PoP	NFV-I points of presence
NFV-O	NFV orchestrator
NG-RAN	new generation RAN
NGMN	next generation mobile networks
NGV	next-generation V2X
NLOS	non-line-of-sight
NLOS-V	vehicular-NLOS
NR	new radio
NRF	network function repository function
NSSF	network slice selection function
NTP	network time protocol
OBU	on-board unit
OCC	orthogonal cover code

OEB	orientation error bound
OEM	original equipment manufacturer
OFDM	orthogonal frequency division multiplexing
OFDMA	orthogonal frequency division multiple access
OLOS	obstructed LOS
OPEX	operation expenditures
ORAN	open RAN
OS	OFDM symbols
OSI	open systems interconnection
OSS	operations support systems
OTA	over-the-air
OTDOA	observed TDOA
OTT	over-the-top
P2PCD	peer-to-peer certificate distribution
PBCH	physical broadcast channel
PBMCH	physical broadcast multicast channel
PC5	shortrange cellular communication (ProSe direct communication interface 5)
PCA	pseudonym certificate authority
PCF	policy control function
PDB	packet delay budget
PDCCH	physical downlink control channel
PDCP	packet data convergence protocol
pdf	probability distribution function
PDP	power delay profile
PDU	protocol data unit
PER	packet error rate
PHD	probability hypotheses density
PHY	physical
PKI	public key infrastructure
PL	path-loss
PLMN	public land mobile network
PMR	professional mobile radio
PNF	physical network function
ProSe	proximity service
PRS	positioning reference signal
PS	public safety
PSBCH	physical sidelink broadcast channel
PSID	provider service identifier
PSM	personal safety message
PT-RS	phase-tracking reference signal
PTS	Swedish Post and Telecom Authority
QAM	quadrature amplitude modulation
QoS	quality of service
QUIC	quick UDP internet connections

R	revolution
R&D	research and development
RA	registration authority
RAN	radio access network
RAT	radio access technology
RB	resource block
RE	resource element
REL	release
RF	radio frequency
RLAN	radio local area network
RLC	radio link control
RMa	rural macro
RMSE	root mean square error
RRC	radio resource control
RRM	radio resource management
RS	reference signal
RSC	Radio Spectrum Committee
RSPG	Radio Spectrum Policy Group
RSRP	reference signal received power
RSTD	relative signal time difference
RSU	road-side unit
RTK	real time kinematics
RTOA	relative time of arrival
RTS	request to send
RTT	round trip time
RTTT	road transport and traffic telematics
rtx	retransmission
RV	remote vehicle
Rx	receiver
Rx UE	receiving UE
S-PSS	sidelink primary synchronization signal
S-SSS	sidelink secondary synchronization signal
SA	system architecture
SAE	Society of Automotive Engineers
SAE-C	SAE for China
SC-FDMA	single-carrier frequency division multiple access
SCMS	Security Credential Management System
SDA	strategic deployment agenda
SDL	supplemental downlink
SDN	software-defined network
SDO	standards developing organization
SDU	service data unit
SDVN	software-defined vehicular networking
SF	shadow fading
SFN	single-frequency network
SGX	software guards extensions

SI	study item
SIC	self-interference cancelation
SIG	special interest group
SIM	subscriber identity module
SINR	signal-to-interference-plus-noise ratio
SL	sidelink
SL_RNTI	sidelink radio network temporary identifier
SLA	service level agreement
SLAM	simultaneous localization and mapping
SLS	service-level specification
SLSS	sidelink synchronization signal
SM	secure message
SMF	session management function
SNR	signal-to-noise ratio
SP	scattering point
SPAT	signal phase and timing
SPS	semi-persistent scheduling
SR	scheduling request
SRS	sounding reference signal
SSE	sum spectral efficiency
SSF	small-scale fading
SSP	service-specific permission
STBC	space-time block codes
TAU	tracking area update
TC	technical component
TC-ITS	Technical Committee ITS
TCO	total cost of ownership
TCP	transmission control protocol
TCU	telematic control unit
TD	time-division
TDD	time-division duplex
TDM	time-division multiplexing
TDOA	time difference of arrival
TF	task force
TG	task group
TIM	traveler information message
TLM	trust list manager
TOA	time of arrival
TOD	teleoperated driving
TR	technical report
TRP	transmission/reception point
TS	technical specification
TSDSI	telecommunications standards development society, India
TSG	technical specification group
TTA	Telecommunication Technology Association
TTC	Telecommunication Technology Committee

TTI	transmission time interval
Tx	transmitter
Tx UE	transmitting UE
UC	use case
UDP	user datagram protocol
UDM	unified data management
UE	user equipment
UICC	universal integrated circuit card
UL	uplink
UMa	urban macro
UMi	urban micro
UMFUS	upper microwave flexible use service
UMTS	universal mobile telecommunications service
UP	user-plane
UPF	user-plane function
URL	unified resource locator
URLLC	ultra-reliable low latency communication
USB	universal serial bus
USIM	universal subscriber identity module
UTC	coordinated universal time
UTRA	universal terrestrial radio access
Uu	air interface between base station and user equipment
UWB	ultra-wide bandwidth
V2I	vehicle-to-infrastructure
V2N	vehicle-to-network
V2P	vehicle-to-pedestrian
V2V	vehicle-to-vehicle
V2X	vehicle-to-everything
VA	virtual anchor
VANET	vehicular adhoc network
VIM	virtualized infrastructure manager
VNF	virtual network functions
VNFM	virtual network function manager
VR	virtual reality
VRU	vulnerable road user
VRUP	vulnerable road user protection
VUE	vehicle UE
WAS	wireless access system
WAVE	wireless access in vehicular environments
WG	working group
WLAN	wireless local area network
WLL	wireless local loop
WRC	World Radio Conference
WSMP	WAVE short message protocol

1

Introduction

Mikael Fallgren[1], Markus Dillinger[2], Toktam Mahmoodi[3], Tommy Svensson[4], Charalampos Kalalas[5], Erik Ström[4], and Antonio Fernandez Barciela[6]

[1] *Ericsson Research, Sweden*
[2] *Huawei German Research Center, Germany*
[3] *King's College London, UK*
[4] *Chalmers University of Technology, Sweden*
[5] *Centre Tecnològic de Telecomunicacions de Catalunya, Spain*
[6] *PSA Group, Spain*

In recent decades, mobile broadband has become an increasingly important part of our daily connected life. Many industries and businesses also rely on wireless mobile communication for connectivity between their sites, services, devices, and users. This connectivity provides an increased awareness that ultimately leads to smarter decisions based on additional information from the surroundings. Such awareness could provide various degrees of assistance to vehicle drivers, including fully autonomous driving, so-called *connected automated driving* (CAD) [1].

CAD is a paradigm brought forward by the convergence of automated driving using on-board sensors, connectivity and cloud access for increased awareness and driving performance. We envision that the fifth generation (5G) and beyond of mobile networks will play a key role with regard to mission-critical services to support CAD. Mobile network technologies enable vehicles to connect other vehicles, networks, and road infrastructure, which in general term is referred to as *vehicle-to-everything* (V2X). Mobile network support of V2X – cellular V2X (C-V2X) – is the topic of this book, with a focus on CAD services but with attention to other *connected road user* (CRU) services as well.

Section 1.1 presents a brief background and motivation for C-V2X, CAD, and CRU services, including key technical terminology. Section 1.2 describes the ambitions of the telecommunication and automotive industries for a joint roadmap for CAD. Section 1.3 goes into more detail about communication technologies for CAD and the standardization organizations. Finally, Section 1.4 outlines the structure of this book.

Cellular V2X for Connected Automated Driving, First Edition.
Edited by Mikael Fallgren, Markus Dillinger, Toktam Mahmoodi, and Tommy Svensson.

1.1 Background and Motivation for C-V2X

In this section, we give some background and motivation for C-V2X starting with intelligent transport system (ITS). We then proceed to discuss CAD and conclude with CRU services.

1.1.1 Intelligent Transport Systems

ITS is a broad term that covers a multitude of applications aimed at making transport systems safer and more efficient. It is hard to nail down an exact definition of ITS, since it has a long and rich history. However, the following definition comes from European Commission (EU) directive 2010/40/EU:

> *ITS integrate telecommunications, electronics and information technologies with transport engineering in order to plan, design, operate, maintain and manage transport systems. The application of information and communication technologies to the road transport sector and its interfaces with other modes of transport will make a significant contribution to improving environmental performance, efficiency, including energy efficiency, safety and security of road transport, including the transport of dangerous goods, public security and passenger and freight mobility, whilst at the same time ensuring the functioning of the internal market as well as increased levels of competitiveness and employment. However, ITS applications should be without prejudice to matters concerning national security or which are necessary in the interest of defence.*

Applications in which vehicles and road infrastructure cooperate are called cooperative ITS (C-ITS). There are two broad categories: traffic safety and traffic efficiency applications. Traffic safety applications aim to reduce the number and severity of road accidents, while traffic efficiency applications target reducing fuel consumption and exhaust emissions and the more efficient use of road infrastructure (e.g. increasing traffic flow on existing roads). Additional potential benefits of C-ITS include improved driver comfort [2, 3]. An early C-ITS service, already included in European cars since 2018, is the availability of connectivity for emergency call (eCall) services.

In 2014, the European Commission set up a C-ITS Deployment Platform as a cooperative framework including national authorities, C-ITS stakeholders, and the Commission to develop a shared vision for an interoperable deployment of C-ITS in the EU [3]. The goal was to provide policy recommendations for the development of a roadmap and a deployment strategy for C-ITS in the EU and identify potential solutions to key challenges. During 2016–2017, the platform further developed a shared vision of the interoperable deployment of C-ITS toward cooperative, connected, and automated mobility (CCAM) in the EU. The ultimate goal is interoperable digitization of transportation in the EU, with a special focus on road, rail, maritime, and air transportation, as well as inland navigation.

As we have described, ITS, C-ITS, and CCAM are broad terms that cover a multitude of transport systems beyond road users. This means they refer to many different types of vehicles – cars, trams, trains, boats, drones, flying taxis, etc. Traditionally, though, C-ITS focuses on vehicles and fixed road infrastructure communications, and therefore other road users are not addressed – in particular, not vulnerable road users (VRUs) such as pedestrians.

In this book, we focus on C-ITS solutions for road-bound vehicles (such as cars, motorcycles, buses, and trucks) enabled by mobile networks, i.e. C-V2X. With C-V2X comes the opportunity to include also VRUs in the system. From here on, we refer to these types of solutions and systems as CAD [1]. Note that although it is not an integrated part in CAD, via the mobile network vehicles can still communicate and learn from all types of CCAM devices.

1.1.2 Connected Automated Driving

To enable connected automated driving (CAD) services ranging from advanced driver-assistance systems (ADAS) to fully autonomous vehicles (also referred to as to self-driving cars, autonomous vehicles (AV), connected and autonomous vehicles (CAV), driverless cars, robotic drivers, etc.), the telecommunication and automotive industries have been moving in similar directions, and in cooperation, to enable capabilities for C-ITS with a focus on road vehicles, in particular cars. To this end, a wide range of use cases for CAD has been identified by European Telecommunications Standards Institute (ETSI), ITS [4], and 3GPP standard releases [5–7], which can be clustered as follows:

- *Automated driving*: Includes use cases targeting fully autonomous driving where vehicles are able to coordinate their trajectories or maneuvers for collision avoidance and/or automated overtaking. This use case category is associated with stringent requirements in terms of latency and reliability to guarantee extremely low packet-error rates as well as high positioning accuracy.
- *Cooperative sensing*: Includes use cases where the exchange of sensor information captured by different sources enhances vehicles' perception beyond the capabilities of on-board sensors, helping the driver or automated car to perform critical maneuvers and navigate safely. Low latency and high data rates constitute the key requirements of this use case category.
- *Traffic safety*: Includes mission-critical use cases related to protecting vulnerable users, road hazard warnings, collision avoidance, etc. This use case category has extreme performance requirements in terms of latency/reliability and positioning accuracy.
- *Traffic efficiency*: Includes use cases that involve updating routes and dynamic digital maps. Typically, traffic efficiency use cases are not associated with strict latency/reliability requirements but require high data rates for efficient route selection.

In addition to addressing technical challenges, for CAD services to fully take off, a diverse set of stakeholders need to come together: car manufacturers, road infrastructure operators, mobile network operators (MNOs), standards-developing organizations, policymakers, and end users. Therefore, harmonious collaboration and close synergies among them are necessary to provide answers to a plethora of research questions that the new mobility ecosystem introduces as well as to reduce the time to market of new technologies and facilitate their early adoption.

Two key drivers for such cooperation are improving traffic safety and traffic efficiency; but there are other incentives, such as offering additional driver comfort. Therefore, it is important to understand the vision and perspective of these two industry sectors as major players in delivering CAD, and we elaborate on that in this book.

1.1.3 Connected Road User Services

C-V2X supports not only CAD, but also many other connected road user (CRU) services, such as the combination of information and entertainment (*infotainment*), over-the-air (OTA) software updates, original equipment manufacturer (OEM) clouds for proactive maintenance and diagnosis, weather forecasts, pollution monitoring, tolling for road usage, road maintenance, parking garage business support, services related to parked vehicles (e.g. parking efficiency, charging the car, and protecting the car), insurance businesses, advertising, fleet management, car sharing, and logistics of integrated moving networks [8]. Providing such integrated services over a single network has the capability of minimizing network deployment and operating costs for MNOs.

Of these services, infotainment is likely to be the main driver for mobile broadband services to cars. Infotainment includes use cases for enhancing the travel experience of both drivers and passengers by bringing connectivity to the car and providing smart navigation functionalities as well as entertainment content (e.g. movies and games). Thus, there is a need to support high-resolution multimedia streaming to cars. Although reliability and latency are not critical for infotainment, supporting mobile broadband services to cars would benefit from being integrated in C-V2X, in particular for cost reasons. This could be the case not only with regard to technical implementation, but also in relation to business models, since advanced CAD services will enable more freedom for the driver to consume infotainment and other CRU services on the move. For other vehicles like buses and trams, general mobile broadband services for passengers, along with surveillance cameras on-board, might be the dominating communications need for the vehicle, and thus less tight implementation integration with C-V2X could be envisioned in the vehicle for these services.

As highlighted here, there are several specific challenges for CRU services. In the subsequent sections of this chapter, we focus on C-V2X for CAD; but many of the techniques described in the book will be of importance for efficiently supporting CRU services as well.

1.2 Toward a Joint Telecom and Automotive Roadmap for CAD

Although sensor technologies have been and are used for ITS services, performance is severely degraded in non-line-of-sight (NLOS) situations and certain weather conditions. Hence, radio technologies are considered a crucial enabler for ITS, with a special focus on CAD. This section consists of a description of telecom's ambition for connected driving, followed by automotive's ambitions for automated driving, and ending with a joint roadmap for CAD.

1.2.1 Telecom's Ambitions for Connected Driving

The ongoing interplay between the telecommunication and automotive industries aims to address society's connected mobility and road-safety needs by radically transforming existing transportation systems into fully automated, intelligent, interconnected systems.

The widespread deployment of in-vehicle sensors with integrated sensing, computing, and data-storing capabilities, e.g. cameras, radar, lidar, and ranging devices, is progressively turning vehicles into sophisticated computing and networking hubs, capable of achieving a detailed understanding of their surroundings. As a result, an extensive amount of real-time information is acquired, processed, and exchanged between vehicles; between vehicles and other road users, such as pedestrians and cyclists; as well as with increasingly intelligent roadside infrastructure, e.g. traffic lights and electronic road signs. In addition to making vehicles better connected, the roadside infrastructure is continuously upgraded with additional cameras to monitor traffic conditions, sensors to gauge temperature and driving conditions, and temporary roadwork signs.

In general, three major groups of use cases need to be supported by C-V2X. The three groups with examples are:

- *Safety*: Emergency electronic brake lights
- *Traffic efficiency*: Local hard warnings; high-definition (HD) maps
- *Automated driving support*: Tele-operated driving; trajectory alignment

The expected timeline will cover the current decade and beyond. It is also foreseeable that future 3rd Generation Partnership Project (3GPP) releases will introduce improvements for better driving experiences.

Reliance on capable wireless communication technologies for ubiquitous V2X connectivity is thus becoming a key asset pushing ITS to a major leap forward [9]. In turn, the automotive industry is undoubtedly one of the stakeholders of emerging communication systems (i.e. 5G and beyond 5G), with its unique features in terms of heterogeneity of end users, stakeholders, and technologies; its diverse use cases; and the unprecedented connectivity challenges [10, 11]. Approximately every 18 months, a new 3GPP release is defined; these will continue to bring new solutions and features that will improve automated driving performance.

For the past three decades, the automotive industry has shaped a vision toward cooperative and automated driving, with reliance on on-board sensors complemented by connectivity to remote sensors and cloud and network services. However, enabling this vision would not be possible without overcoming the underlying telecommunication challenges in terms of low latency, high reliability, wide coverage, high data rates, high availability and scalability (i.e. massive access), security, and positioning accuracy. In this context, the telecommunication industry has progressively evolved during the last two decades to cover all aspects of vehicular communication and support the ever-demanding requirements imposed by use cases and ever-increasing sensor capabilities (e.g. higher video resolutions).

Ongoing standardization efforts aim to develop the features of the main V2X radio technologies, i.e. Institute of Electrical and Electronics Engineers (IEEE) 802.11p- and cellular-based, while offering future-proof solutions that are provisioned to act as the catalyst for advanced – currently unrealizable – automotive services, e.g. accurate positioning, trajectory sharing, real-time location updates, and coordinated driving [11, 12]. In addition, the radio access and network architecture enhancements of 5G wireless technology for C-V2X, such as flexible frame structure, multi-antenna and diversity techniques, service-based network architecture, sidelink modes, network orchestration, and more, are all expected to provide a strong foundation for the novel mobility paradigm [13, 14]. In addition to enhanced connectivity functionalities, the ongoing rollout of 5G technology is expected to bring a major shift in the way today's networks operate. Unlike previous generations, rather than serving only end users, 5G systems are specifically designed to serve a number of industry sectors, including automotive/transportation, associated with a multiplicity of performance requirements. Further, the type of spectrum used for communication will also have implications. In unlicensed spectrum (e.g. ITS spectrum) one can communicate short-range without a subscriber identity module (SIM) card, i.e. no subscription cost, but there is no control over which and how many users use the spectrum at the same time. In a licensed spectrum, we envision that for the use of short- and long-range communication, the spectrum owner will control who is assigned to use the spectrum resources, while keeping track of these users via SIM cards (see Section 1.3).

The design targets of future CAD services envisage cooperative and fully autonomous driving, so novel and innovative approaches are continuously being developed to overcome performance bottlenecks. Accompanying the increasing level of driving automation, safety-critical services impose stringent requirements in terms of end-to-end latency (less than 3 ms), reliability (higher than 99.999%), and positioning accuracy (down to 5 cm) [6]. These requirements become even more challenging when considering the fast-varying nature of the propagation environment and ever-changing network topologies as a consequence of vehicles' mobility.

1.2.2 Automotive's Ambitions for Automated Driving

In the last decade, the automotive sector has embarked on a revolution, including trends that are modifying the bases of the traditional business model. A majority of OEMs are companies that have been around for a long time in their industries, and some of these companies still belong to the families of the original founders. However, drivers' behavior and profiles are changing, with new generations moving from car ownership toward shared mobility. Digitalization is affecting the entire sector, from the manufacturing process, with the arrival of the fourth industrial revolution, to the sales channel, which is now online and offers the possibility of pre-launch reservations. Beyond all the trends mentioned, the arrival of autonomous driving is a major disruptor.

The Society of Automotive Engineers (SAE) has introduced six different levels of driving automation in [15], defined in an increasing order of automation: a human monitors the driving environment for level 0 to level 2, while the automated system monitors the driving environment for level 3 to level 5. In level 0, there is no automation, and the human performs all driving tasks. In level 1, there is driver assistance, and the human driver can enjoy vehicle assistance features (such as e.g. cruise control). In level 2, there is partial automation: the human monitors all the tasks and can choose to take control at any time, while the vehicle is able to steer and accelerate. In level 3, there is conditional automation, where the vehicle can perform most driving tasks and has environmental detection capabilities, while the human is still able to take over at any time. In level 4, there is high automation: the vehicle performs the driving within specific circumstances, while the human still can choose to override at any time. In level 5, there is complete automation: the vehicle performs all the driving, and no human interaction is required. Hence, from a human driver perspective, level 5 is driverless.

Although the different levels of automation are clearly defined and many tests are ongoing worldwide, there are still numerous legislative, regulatory, and technical challenges yet to come. First, legislation is not ready today to shift the responsibility from the driver to the OEM: this is the key element to move to higher SAE levels (level 3 and above) where the automated system monitors the driving environment. Each OEM has its own confidential timeline for these automation levels, and we foresee that automation levels will first be available on highways and later in urban, suburban, and rural areas. Second, business models are not completely defined. Finally, the technical challenges are immense and diverse in nature. These challenges make it difficult to establish a clearly defined roadmap for the adoption of autonomous driving.

Today, level 2 is enabled through ADAS based on on-board sensors. Three main sensors are used: cameras, radar, and lidar. Each has its own strength – e.g. being better for classification, speed estimation, or positioning – but they share common limitations. First, the line-of-sight (LOS) is limited to a few hundred meters and is dependent on natural conditions such as weather, lighting, terrain, vegetation, and artificial obstacles (other vehicles, buildings, and so on). Second, all of them suffer from field of view (FoV) limitations in terms of the available angle to perceive information. The FoV limitation can pose additional constraints in intersections or curves with complex angles. To enable CAD, LOS and FoV limitations provide room for connectivity to serve as a complementary off-board sensor that can provide additional vision to LOS and FoV.

Other than playing the role of off-board sensors, connectivity is very important in the automotive sector. For example, connectivity provides the capabilities needed for OEM-related use cases, including remote diagnosis, over-the air-updates, and offloading data from on-board sensors.

1.2.3 Joint Roadmap for CAD

The efforts and collaboration of a diverse set of stakeholders are visible in telecom's ambitions for connected driving and in automotive's ambitions for automated driving. Together, a combined roadmap for CAD is taking shape, motivated by appealing services like traffic safety and traffic efficiency, where technical advancements and coordination strive to develop solutions to enable these promising services. In terms of defining use cases, different bodies such as 3GPP, the 5G Automotive Association (5GAA), and SAE have already conceived different V2X requirements and key performance indicators (KPIs). At this stage it is difficult to determine who is right or wrong. The first derived estimates and assessments of requirements that have been made should serve as current guidelines rather than the last word on the topic.

A vehicle's on-board sensors (i.e. sensors inside the vehicle as well as sensors on the vehicle) together with off-board sensors (i.e. sensors not on that vehicle, e.g. on other vehicles or on the road infrastructure) can collaborate in collecting relevant information for the vehicle to support automation. However, a general design challenge is that the latency of all the different sensor information has to come together at the same time. Another dimension to consider is when to switch from raw data to object data. Raw data typically consists of substantially more data that is needed for liability reasons, whereas object data consists of less data but carries the risk of misinterpretation or not detecting objects.

Figure 1.1 shows the joint roadmap for CAD developed within the 5GCAR project [16]. In it, the automation levels (L) and 3GPP releases (REL) are represented at the base of the roadmap; the different releases are represented based on the specification dates, until the expected arrival in the automotive market. At the top are a timeline and relevant spectrum bands for various International Mobile Telecommunications (IMT) solutions. Identified use case classes and services in the roadmap are positioned in white and highlighted boxes under three categories: network coordinated decision and control (e.g. driverless vehicle on private roads), network-coordinated perception (e.g. cooperative adaptive cruise control, CACC), and network-assisted information exchange (e.g. OTA and infotainment services). The ones highlighted were analyzed in 5GCAR: i.e. private areas remote driving, open roads remote driving, cooperative maneuver, vulnerable road user protection (VRUP), cooperative perception, and HD maps.

In the automation levels, driverless (level 5) starts before conditional automation (level 3) and high/full automation (level 4) because driverless-level automation has already been deployed on private roads and campuses, mainly in shuttles. True level 5 on open roads needs a few more years for the technology to mature due to more challenging environments, but also due to regulatory and business factors. Level 3 can be considered close to market for some OEMs even if the legal framework for this level is not yet complete. Level 3 will not be conditioned on connectivity. Level 1 (driving assistance), such as cruise control, and level 2 (partial automation) enabled through ADAS are both available in cars today. It is also expected that level 3 or level 4 will be supported for highways and, later, urban areas as well. The exact automation timeline is up to the various OEMs and usually is not publicly available.

1.3 Communication Technologies for CAD

CAD connectivity requirements are diverse. Some applications have relaxed latency requirements (seconds, minutes, or even hours), while others need very low-latency communication (on the order of milliseconds). Reliability requirements are also varied, as well as required data rates. Thus, there are various connectivity solutions for CAD, which can be classified as follows:

- *Vehicle-to-vehicle* (V2V), to exchange information using short-range and/or direct communications between vehicles located in close proximity to each other.
- *Vehicle-to-infrastructure* (V2I), to exchange data using short-range and/or direct communications between a vehicle and roadside infrastructure (e.g. traffic lights).
- *Vehicle-to-pedestrian* (V2P), to exchange information using short-range and/or direct communications between vehicles and mobile devices carried by a VRU such as pedestrian (or cyclist, or pets, or driver/passenger of another vehicle via short-range communication).
- *Vehicle-to-network* (V2N), to exchange information using long-range communications between a vehicle and mobile network or Internet-based cloud services (e.g. a backend remote server in the network, to obtain additional services such as map updates, fleet-based data collection, and automotive cloud services).

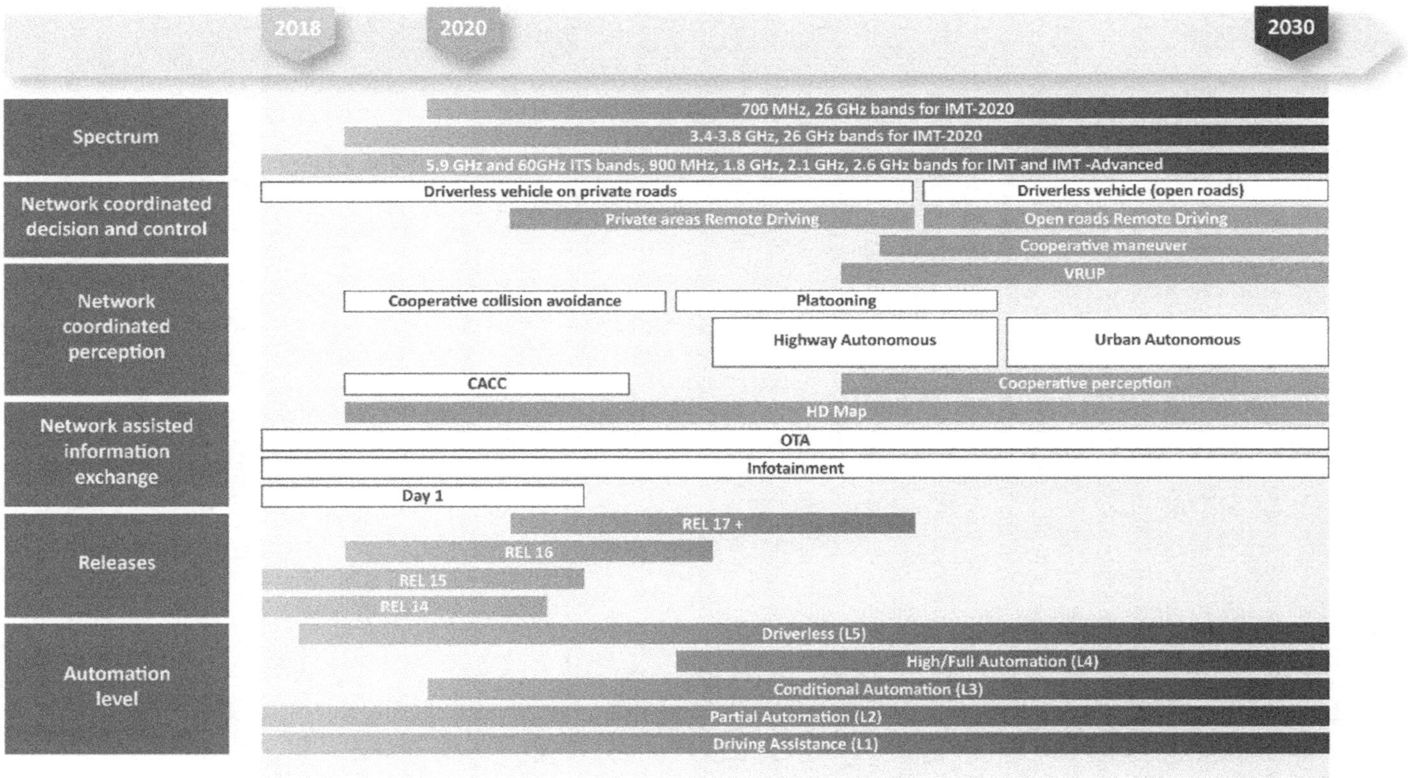

Figure 1.1 Joint roadmap for CAD as developed in the 5GCAR project. Source: [16].

Direct communications can be implemented without any network or infrastructure involved. Short-range communications can be implemented with or without mobile network support, whereas long-range communications rely on mobile networking. In general, all road users would benefit from communicating locally (V2V, V2P), with nearby roadside infrastructure (V2I), and with network and/or cloud services (V2N). In the following, we provide an overview of the primary present-day radio access technologies that are capable of supporting V2X communication.

The first technology considered for vehicles' connectivity was based on the Wi-Fi standard. Hence, in 2004, the IEEE 802.11p task force was formed, shaping the basis for dedicated short-range communications (DSRC) for V2X communication, particularly for applications such as vehicle safety services and commerce transactions via cars. Later, in 2008, the European Commission allocated part of the 5.9 GHz band for priority transport safety applications, to ensure reliable communication. A number of studies have shown that 802.11p can deliver latency on the order of tens of milliseconds [17].

In a parallel effort, the mobile communication industry has started shaping the V2X vision based on cellular networks, and hence in 2014 the first standard for C-V2X was developed based on fourth-generation (4G) long-term evolution (LTE) (also referred to as LTE-V2X). Future generations, such as 5G-V2X, also fall under C-V2X. Since providing guaranteed services has been part of the core design of mobile communication technologies, C-V2X inherits this feature and offers an additional capability of fulfilling the requirements on e.g. latency, reliability, or capacity. However, the primary additional feature offered by C-V2X is long-range communications to the network, V2N. While short-range communications to another vehicle, pedestrian, or road infrastructure are important, the ability to communicate with the network brings the additional advantage of always-on connectivity to vehicles, supported by the mobile networks. C-V2X defined by 3GPP allows a mix of different short- and long-range releases as needed.

IEEE and 3GPP communications technologies have evolved and matured in the past few years. While each comes with advantages and limitations, they can also work in combination to deliver various needs of CAD. In the following, we elaborate on the evolution of these two technologies for CAD and the progress of standardization efforts within IEEE and 3GPP.

1.3.1 Standardization of IEEE V2X

The IEEE standards association has a long history with a number of aspects of ITS and transportation. The well-established 802.11 (wireless local area network) family of standards is undergoing continuous development [18] to support ad hoc V2X communication and networking in rapidly changing mobile environments, as well as the IEEE 1609 family of standards for wireless access in vehicular environments (WAVE). A flavor of IEEE 802.11 known as 802.11p or simply 11p was developed specifically for V2X communication about a decade ago and standardized in the 802.11p-2010 document [19]. The main innovation was to allow for communication outside the context of a basic service set. This enables IEEE 802.11p stations to communicate without first joining an 802.11 network and thereby avoid the time needed for authentication and association. The standard operates in the dedicated 5.9 GHz frequency band[1] reserved for the ITS services.

IEEE 802.11p incorporates a modified version of the IEEE 802.11a orthogonal frequency division multiplexing (OFDM) physical layer (PHY) and uses the 10 MHz channel spacing option [18]. The change from the commonly used 20 MHz to 10 MHz channel spacing was made to cope with longer delay spreads that can occur in outdoor vehicular environments. The medium access control (MAC) layer of 11p is based on enhanced distributed channel access (EDCA), also used in the IEEE 802.11e standard. The EDCA protocol is contention-based and uses carrier sense multiple access (CSMA) with collision avoidance (CSMA/CA). It provides four different access categories for prioritizing data traffic using specific parameters for the contention window size and inter-frame spaces for each access category. Since 802.11 defines only the PHY and MAC layers, IEEE 802.11p needs to be complemented with higher layers and security to form a complete system. There are two main systems based on the IEEE 802.11p PHY and MAC: DSRC in the US [20] and ITS-G5[2] in Europe [21]:

- The DSRC standard suite is primarily developed in the US Department of Transport and a consortium of automotive manufacturers for interoperability tests. At the protocol stack, DSRC utilizes a slightly modified version of IEEE 802.11p for the PHY and MAC layers enabling multichannel operation, while the suite of IEEE 1609.x standards for WAVE is utilized in the higher layers. Above the protocol stack, V2X message sets and related performance requirements are specified by the SAE [15].
- The ITS-G5 standard is mainly developed in Europe by the ETSI and the European Committee for Standardization (CEN) with their relevant technical committees. In recent years, ITS-G5 has undergone a thorough standardization process and extensive field trials to test its performance. The ETSI has initiated pre-standardization studies with the purpose of specifying new ITS services to be applicable in the framework of ETSI ITS Release 2 standard development [22, 23]. In addition to CSMA/CA and EDCA, ITS-G5 introduces MAC layer features for decentralized congestion control (DCC) as specified in ETSI technical specification (TS) 102687 [24], which aims to maintain network stability, throughput efficiency, and fair resource allocation to ITS stations.

These efforts have evolved in the IEEE next-generation V2X (NGV), also referred to as IEEE 802.11bd, by introducing enhancements to both the PHY and MAC layers of 11p, such as an adaptive retransmission scheme as well as adaptive modulation and coding, for enhanced performances, while staying compatible with 11p.

In parallel to the standards developments in Europe and the US, the Japanese research and standardization organization Association of Radio Industries and Businesses (ARIB) STD-T109 [25] has developed a standard for driving safety support systems operating in the 700 MHz band. The standard uses a PHY layer very similar to IEEE 802.11p, but it employs a MAC layer that combines the physical carrier sense function (i.e. CSMA/CA scheme) with time-slotted access to reserve time slots for exclusive use by road-side units (RSUs).

---

1 In Japan, a single 9 MHz frequency channel in the 755.5–764.5 MHz band has also been designated for ITS safety-related applications using V2V and V2I communications.

2 The name *G5* is derived from the frequency band (5.9 GHz).

1.3.2 Standardization and Regulation Aspects of C-V2X

In the following two sections, we introduce existing 3GPP releases and future requirements for Release 17 and beyond. Existing regulations and future demands are also addressed. Note that future aspects of CAD are also discussed in Chapter 10, which describes radio and network requirements for potential future standardization efforts.

1.3.2.1 Available C-V2X Releases and Regulations

In recent years, the 3GPP has raised the need to revisit the design of next-generation mobile networks to efficiently support V2X connectivity. Since Release 14, 3GPP has been working on the development of C-V2X technologies, often referred to as LTE-V2X, as they were initially based on the LTE standard specifications [6, 7, 10, 13, 14, 20, 26–28]. In particular, 3GPP TS 22.185, published in March 2016 (latest updated in 2017), was a Release 14 document that defined key CAD use cases and service requirements for both safety and non-safety (e.g. traffic efficiency) services [7]. Later, 3GPP Release 15 (TS 22.186, published in March 2017) was developed with a specific focus on enhanced use case scenarios, e.g. vehicle platooning and remote driving, including more rigorous functional requirements for advanced features that could not be achieved by earlier standard specifications. In general, the C-V2X radio enhancements include both infrastructure-based solutions in the Uu air interface – i.e. between user equipment (UE) and base stations – and sidelink-based solutions with the PC5 interface (i.e. between vehicles). The main advantage of the Uu interface is that it facilitates a large dissemination range for V2X messages by leveraging the cellular core network. Initially, due to inherent network delay, the Uu interface is expected to be used for less latency-sensitive services such as dynamic high-definition maps, software updates, and infotainment services. However, with architectural enhancements like multi-access edge computing (MEC), delays can be drastically reduced for the Uu interface. On the other hand, the PC5 interface allows for direct communications between network entities (i.e. vehicles, RSUs, and other road users) without requiring messages to pass through the mobile base station. Therefore, the PC5 interface is also suitable for time-critical safety use cases that require low-latency connectivity. Vehicles can use the PC5 interface in the presence or absence of a mobile base station, i.e. with or without cellular coverage. In general, some automotive use cases can already be fulfilled by 4G, but advanced CAD services will require 5G or even beyond. Generally, CAD use cases can make simultaneous use or exclusive use of Uu and/or sidelink links.

The PC5 interface specified in 3GPP Releases 14 and 15 evolved from the device-to-device (D2D) framework – proximity services (ProSe) – standardized in previous releases. 3GPP Release 14 extended ProSe functionality for C-V2X connectivity by adding two new modes (mode 3 and mode 4), which differ in the way transmission resources are assigned to vehicles. In mode 3 (network-controlled) under cellular coverage, selected resources are managed/scheduled by the base station, thus minimizing collisions by realizing a central organization of all communicating entities. On the other hand, in mode 4 (vehicle-controlled) outside cellular coverage, communication resources are autonomously selected by each vehicle using a resource-reservation algorithm that relies on its own channel sensing to ensure that other vehicles reserve orthogonal resources in time and/or frequency to reduce packet collisions.

3GPP has enhanced the protocols governing both communication of vehicles with the network (referred to as Uu) and the direct communication link between vehicles (referred to as PC5 or sidelink) in order to support lower latency and higher reliability. Sidelink was designed to operate either overlaid on the cellular spectrum or in dedicated bands. With the finalization of the 3GPP Release 14 specification at the beginning of 2017, the C-ITS community has a 3GPP-standardized solution supporting both long-range and short-range communication that fulfills the initial requirements of CAD.

In mid-2020, along with the development of 5G protocols for new radio (5G NR), the 5G NR-compliant V2X in Release 16 [26] complements LTE-V2X in Release 14 and offers new features for lower latency, unicast/multicast, a comprehensive quality of service (QoS) framework on sidelink, as well as out-of-coverage solutions for sidelink [29–32]. An important aspect of the evolution from Release 14 (LTE-V2X Phase I) to Release 15 [5] (Phase II) to Release 16 (5G NR or Phase III) is compatibility in terms of coexistence [13, 32].

In Europe, the ITS spectrum is specified in the ETSI standards in 5855–5925 MHz [33]. IEEE standard 802.11p was favored as the communication stack for active safety functions. In general, this frequency band is designated for road safety ITS and other ITS services on a non-exclusive, license-exempt (unlicensed) basis. Since then, authorities, the automotive industry, and road operators have primarily deployed a broader range of communication infrastructure for tests and road trials. In addition to a variety of research prototypes and test deployments, a small number of vehicles and/or RSUs have been commercially equipped with wireless transmitters/receivers.

The European directives define the spectrum usage conditions of this band in a technology-neutral way. In July 2020, the European Commission's Radio Spectrum Committee announced that it would assume C-V2X as another short-range technology in this ITS band [34] being adopted by the EC. The original Delegated Act from the European Commission was rejected in July 2019 because the proposal was not radio technology neutral.

In China, authorities have set aside 20 MHz spectrum in the 5.9 GHz ITS band for coverage expectations across 90% of big cities and major highways in the near future [35].

In the US, the authorities have allocated the frequency band 5850–5925 MHz for non-exclusive, license-based V2X. US regulations make normative references to IEEE 802.11p, with one channel allocated for control and other channels allocated for shared-channel V2X services [36]. However, the FCC has decided to remove DSRC and selected C-V2X as the preferred ITS technology [37].

1.3.2.2 Future Requirements for C-V2X Releases and Regulations

Given that the focus of this book is C-V2X, future developments in this area are discussed here. Note that Chapter 10 also deals with future requirements for radio and networks. Emerging CAD use cases impose requirements that are barely met by existing standard improvements, and C-V2X development is also expected to be one of the major topics in further releases of 3GPP specifications such as upcoming Releases 17 and 18 or beyond: e.g. for high data volumes and upper-bound latency services suitable for sensor-sharing and network-processing tasks. Ongoing 3GPP efforts are aimed at enhancing C-V2X technology in the context of the NR framework [11]. 3GPP is expected to introduce distributed vehicle terminals to improve spectral efficiencies for forecasted increased spectrum demands for V2N services. In particular, Release 16 work items include the design of new V2V broadcast, groupcast, and unicast sidelink communication interfaces to support the increasingly demanding requirements, e.g. in terms of reliability and latency for remote driving and data rate for cooperative perception. Additional areas under consideration include mechanisms to select the best interface (among LTE sidelink, NR sidelink, LTE Uu, and NR Uu) for C-V2X message transmission, configuration/allocation of sidelink resources using the NR Uu interface, and coexistence between LTE-based C-V2X and 5G NR-based C-V2X within a single device. 5G NR considers the use of mmWave bands for C-V2X applications, particularly for applications that require a short-range and high to very high throughputs. A feasibility study on 5G NR-based C-V2X was concluded successfully, and several technical solutions were identified by the 3GPP [14], although 5G NR-based C-V2X is still in its infancy in terms of product availability. Spectrum aspects, e.g. which frequency bands C-V2X sidelink should use, and positioning (absolute and relative) techniques constitute additional topics under discussion for future standardization activities in Release 17. As usual, 3GPP ensures compatibility between older and new releases, i.e. older devices can communicate with new devices (vehicles) and vice versa.

We advocate the definition of additional spectrum requirements and the identification of spectrum bands for full CAD service support. 5GAA has published a study on the spectrum needs of ITS [38], where the amount of bandwidth required for the introduction of day 1 and advanced ITS use cases like CAD is predicted to be 50 MHz < 1 GHz and 500 MHz from 1–7 GHz for long-range communications and 75 MHz in the 5.9 GHz band for short-range communications.

1.4 Structure of this Book

This book consists of 10 chapters describing the ecosystem around CAD, including business aspects and standardization as well as technical solutions enabling successful delivery of such systems from a communication perspective. The structure of this book is as follows, with a summary of each chapter.

Chapter 2, "Business Models," discusses key technology enablers for 5G V2X required rollout conditions, roles of different stakeholders, investments, business models, and expected profit from CAD services. These advanced CAD services, including HD map support, highway chauffeurs, teleoperated driving, and highly and (ultimately) fully autonomous driving, are expected to be enabled through next-generation 5G vehicular networks, starting with 3GPP Release 16. We describe the 5G NR-based V2X ecosystem and stakeholder relationships, different sharing models for network infrastructure, business setup, and, finally, a techno-economic assessment of the investment. The chapter provides insights and triggers discussions on business models for CAD services, 5G NR-based V2X deployment costs, and potential revenues. Available research studies and ongoing discussion within the 5G Public Private Partnership (5G PPP), standardization bodies, and other alliances are used as references to build the arguments in this chapter. The starting point is that, due to the technical requirements of CAD services, the deployment of a so-called 5G digitalized highway is a main enabler. The exemplary highway environment considered through this work includes 5G radio base station sites and fiber backhaul connections. It is further assumed that this investment could be used to a certain extent to provide mobile broadband services in parallel with CAD services. The assumptions made in the context of this chapter rely on working hypotheses and estimates that aim to identify the most suitable investments model(s), but they do not prejudge the effective level of the economic viability of CAD business cases that will result from market developments in coming years. Chapter 2 outlines the 5G CAD landscape of primary stakeholders and relationships. Moreover, an investment and business model is proposed to describe the value flow between the involved actors: it shows under which conditions a return on investment for the 5G-digitalized highway can be expected, depending on investment costs, user fees, and number of users. Moreover, a positive business case can be expected, especially when network infrastructure is shared among different operators. Particularly in the early phase of 5G network deployment, synergies between the private and the public sector could speed up deployment, allowing more users to get access to CAD services with lower charging rates and ultimately leading to much safer roads and more efficient transportation.

Chapter 3, “Standardization and Regulation,” covers standardization aspects and status of the V2X landscape. General aspects of the standardization process are first discussed, and a brief description of the main bodies involved in V2X communications standardization is provided. A deeper view within 3GPP standards organizations is then taken to serve as an example of the various procedures required for communications standardization. Next, current regulatory arrangements for ITS in different regions of the world are discussed, as well as aspects and considerations regarding the radio spectrum dedicated to ITS. Then we give a brief history and overview of the available communication solutions and an overview of higher-layer ITS standardization on a regional level, with special focus given to the details of 3GPP C-V2X standardization. These details include a brief history of 4G solutions, current 5G progress (i.e. Releases 15 and 16) on the LTE and NR technology front, and plans for future releases. Finally, we outline the IEEE and ETSI standards for securing V2X messages and ensuring privacy. More specifically, we cover the protocols used by V2X end entities (e.g. vehicles and road stations) to communicate with public key infrastructure (PKI) authorities, as well as certificates and the secured message format.

Chapter 4, “Spectrum and Channel Modeling,” discusses two important topics related to V2X communication. First, the chapter provides an overview of spectrum aspects of V2X, including frequency bands, their allocation for LTE-based V2X and NR-based V2X, lessons learned from 4G, and spectrum harmonization. Later, the chapter addresses state-of-the-art channel models for V2X communications, including their most relevant components: LoS blockage analysis, path loss and shadow-fading modeling, and fast-fading modeling. Based on existing work, the chapter describes relevant propagation environments for V2X, along with gap analysis in terms of the key missing components required for a complete solution for V2X channel modeling. We cover channel models for link-level and system-level simulations: both single-link and multi-link models for below 6 GHz and above 6 GHz bands, up to 100 GHz.

Chapter 5, “V2X Radio Interface,” provides insights into key aspects of the radio interface design for V2X communications. The major challenge for this design is the provision of stable, highly reliable communication links that need to be maintained even when users are extremely mobile. The first section focuses on the application of beamforming techniques with narrow beams in the mmWave spectrum, where fast beam tracking and adjustment are required. We propose three different beamforming schemes, tailored for unicast, multicast, and broadcast transmissions, respectively. Next, we elaborate on extensions of the lower layers of the protocol stack to address challenges of communication between a large numbers of distributed vehicles in highly mobile environments: we present a theoretical analysis of the trade-off between pilot overhead and spectral efficiency under high mobility along with a corresponding robust receiver design, followed by latest advances for V2V sidelink design, covering reference signal design, synchronization, scheduling, and power control. Finally, we present novel communication concepts facilitated by short-range communication via the V2V sidelink. These concepts cover the cooperation of users in close proximity for uplink and downlink transmissions to improve the reliability of those transmissions at a reasonable system cost, as well as full and flexible duplex, allowing users to transmit and receive at the same time.

Chapter 6, "Network Enhancements," discusses the major network breakthroughs envisioned in 5G to support automotive applications and communications, with an end-to-end perspective. First, we present the challenges introduced on the network by vehicular communications, motivating the need for network enhancements described in the remainder of the chapter. Network slicing is introduced as a way for mobile service providers to offer QoS in the most stable and reliable way possible. The defining impact on future networks of software-defined networking and network function virtualization is discussed: these concepts are of paramount importance for dynamic allocation of network services and for their automated management and orchestration. The "softwarization" of network functions implies deployment on a cloud infrastructure: in this chapter, we present the implications of such an architectural paradigm from both from the core network and the radio access network perspective, each introducing specific sets of challenges. Vehicular communications are characterized by the contemporary coexistence of multiple types of traffic patterns, each meant to convey a different type of information. Some messages are meant to be sent to remote servers, whereas other messages, whose content is functional to support safety-critical automated driving applications, need to be delivered to other road users in proximity with the shortest delay possible. Hence, we discuss local end-to-end paths supporting these specific scenarios. Furthermore, for local communications to be effective, it is essential for road users registered to different mobile network operators to be able to establish very low-latency local communication links between each other. Toward this end, we discuss multi-operator support, including the challenges it poses and propositions to overcome them.

Chapter 7, "Enhancements to Support V2X Application Adaptations," focuses on the QoS framework. QoS handling for V2X poses several unprecedented challenges. First, the huge umbrella of V2X services includes a large set of use cases with heterogeneous QoS requirements, varying significantly among different levels of bandwidth, reliability, and latency. For instance, services such as a lane merge of vehicles coordinated by a road traffic orchestrator, vulnerable pedestrian detection, and remote driving for automated parking would require mostly high reliability and low latency, whereas services such as massive sensors reporting from moving vehicles like the "see-through" service and HD local map acquisition also need higher bandwidth. In addition, the delivery of V2X services introduces new aspects to be considered when handling QoS, such as vehicle mobility patterns, location-based services, use-case-dependent QoS, and joint utilization of different technologies such as long- and short-range communications (e.g. Uu and sidelink). The aim of this chapter is to discuss aspects related to the uniqueness of QoS managements for V2X services, including a series of possible enhancements to the 5G QoS framework to facilitate the fulfillment of QoS in CAD use cases. We provide an overview of the 3GPP QoS framework, introducing the key components in access and core networks already specified in 5G systems for QoS management. The chapter discusses the importance of enhancing the interaction between application and network in terms of mutual sharing of information to facilitate QoS management as well as to increase application awareness about network capabilities in a certain area or time window. Toward this end, we discuss how enhanced application-network interaction could be exploited to implement use-case-aware QoS management to help fulfill QoS until the completion of a particular V2X service (e.g. the whole-lane merging of a vehicle) and how to use information about vehicle trajectory to optimize scheduling mechanisms. The chapter also focuses on several options for joint utilization of long- and short-range transmissions, from solutions implemented at the application layer to solutions implemented in 3GPP's protocol stack.

Chapter 8, "Radio-Based Positioning and Video-Based Positioning," addresses positioning in connected driving through radio signals and camera images. In the past decade, non-satellite positioning has attracted the interest of researchers and industries as one of the key enablers for location-based services. In this regard, a number of solutions as well as commercial products have been developed for a multitude of application scenarios in home, office, warehouse, industrial, and hospital environments. Ultimately, the 3GPP standardization group also initiated a work item on this topic with the aim of empowering 5G technology for a reliable, accurate positioning system that can be used anywhere, anytime. We provide an overview of state-of-art radio and image-based positioning, tailored to the localization of connected autonomous vehicles. More specifically, we address the technology and performance requirements of legacy solutions along with the details of time-based, angular-based, and video-based positioning. Some of the proposed methods closely follow the LTE/NR standard; others are more innovative solutions that can achieve centimeter-level accuracy. Finally, the chapter concludes with a comprehensive set of simulations highlighting pros and cons of each solution.

Chapter 9, "Security and Privacy," discusses these aspects of connected driving. After a brief introduction recalling how security and privacy are critical aspects of C-ITS, the chapter identifies the security and privacy threats and vulnerabilities of C-ITS. We present an overview of state-of-the-art technologies and protocols and European Commission Delegated Act certificate and security policies. Next, we discuss the General Data Protection Regulation (GDPR) and its impacts on C-ITS, as well as some recommendations to fulfill C-ITS promises. We review the security challenges of the three tiers of V2N 5G connections: in-vehicle, network and edge, and cloud security, focusing on isolation challenges between vehicular network slices. Finally, this chapter addresses operational issues with some of the currently proposed security and privacy policies and investigates an alternative approach.

Chapter 10, "Status, Recommendations, and Outlook," analyzes and forecasts further research and development and standardization needs in C-V2X, broader CAD aspects, and

CRU services. Additionally, important CAD stakeholder recommendations are given to accelerate the development of the CAD ecosystem in terms of regulation, manufacturing, and operation. Remaining challenges for CAD are addressed and further elaborated, as well as CRU services. The key areas of these remaining challenges are classified as connectivity solutions; device design; system at large, i.e. beyond technical systems such as network operators and OEMs; and regulation and legislation. Further statements on challenges and opportunities ahead in C-V2X for CAD will stress the need for future C-V2X research covering beyond 5G up to 6G in the coming years. Finally, encouraging perspectives given for the post-pandemic period confirm that the demand for CAD and CCAM may become even stronger: for instance, to improve hygiene standards or measures to combat global warming.

References

1 CAD. (2020). Connected automated driving (CAD). https://connectedautomateddriving.eu/about-cad.
2 European Traffic Safety Council (ETSC). (2017). Cooperative Intelligent Transport Systems (C-ITS).
3 European Commission. (2020). Intelligent transport systems: Cooperative, connected and automated mobility (CCAM). https://ec.europa.eu/transport/themes/its/c-its_en.
4 ETSI. (2009). Intelligent transport systems (ITS); vehicular communications; basic set of applications; definitions. TR 102 638, V1.1.1.
5 3GPP. (2015). Study on LTE support for V2X services (Release 14). TR 22.885, V1.0.0.
6 3GPP. (2017). Service requirements for enhanced V2X scenarios (Release 15). TR 22.186, V15.0.0.
7 3GPP. (2017). Service requirements for V2X services; Stage 1 (Release 14). TS 22.185, V14.3.0.
8 Fallgren, M., Dillinger, M., Alonso-Zarate, J. et al. (2018). Fifth-generation technologies for the connected car: Capable systems for vehicle-to-anything communications. *IEEE Vehicular Technology Magazine* 13 (3): 28–38.
9 ETSI. (2020). Automotive intelligent transport systems (ITS).
10 3GPP. (2018). Service requirements for the 5G system; Stage 1 (Release 16). TS 22.261, V16.4.0.
11 3GPP. (2019). Study on vehicle-to-everything (Release 16). TR 38.885 V1.0.1.
12 IEEE. (2020). Next generation V2X study group project authorization request. P802.11.
13 3GPP. (2018). Study on enhancement of 3GPP support for 5G V2X services (Release 16). TR 22.886, v16.2.0.
14 3GPP. (2018). Study on NR V2X. RAN#80.
15 SAE. (2014). Taxonomy and definitions for terms related to on-road motor vehicle automated driving systems.
16 5GCAR. (2019). Executive summary. Version 1.0. https://5gcar.eu/wp-content/uploads/2019/12/5GCAR-Executive-Summary-White-Paper.pdf.

17 Xu, Z., Li, X., Zhao, X. et al. (2017). DSRC versus 4G-LTE for connected vehicle applications: A study on field experiments of vehicular communication performance. *Journal of Advanced Transportation.*
18 IEEE. (2016). Standard for information technology – Telecommunications and information exchange between systems local and metropolitan area networks – specific requirements – Part 11: Wireless LAN medium access control (MAC) and physical layer (PHY) specifications. IEEE Std. 802.11-2016 (revision of IEEE Std 802.11-2012), 1–3534.
19 IEEE. (2010). Standard for information technology – Local and metropolitan area networks – Specific requirements – Part 11: Wireless LAN medium access control (MAC) and physical layer (PHY) specifications amendment 6: Wireless access in vehicular environments. IEEE Std. 802.11p-2010.
20 Kenney, J.B. (2011). Dedicated short-range communications (DSRC) standards in the United States. *Proceedings of the IEEE* 99 (7): 1162–1182.
21 E. G. Ström. (2011). On medium access and physical layer standards for cooperative intelligent transport systems in Europe. In: *Proceedings of the IEEE* 99 (7): 1183–1188.
22 ETSI. (2016). Intelligent transport systems (ITS); platooning; pre-standardization study. TR 103 298.
23 ETSI. (2016). Intelligent transport systems (ITS); cooperative adaptive cruise control (C-ACC); pre-standardization study. TR 103 299.
24 ETSI. (2011). Intelligent transport systems (ITS); decentralized congestion control mechanisms for intelligent transport systems operating in the 5 GHz range; access layer part. TS 102 687, V1.1.1.
25 ARIB. (2013). 700 MHz band intelligent transport systems. STD T109-v1.2.
26 3GPP. (2016). Initial cellular V2X standard completed (Release 14). Initial C-V2X specification.
27 3GPP. (2016). Technical specification group services and system aspects; study on enhancement of 3GPP support for 5G V2X services (Release 15), V1.1.0.
28 Intelligent Transport Systems (ITS). (2014). Radiocommunications equipment operating in the 5 855 MHz to 5 925 MHz frequency band; harmonised standard covering the essential requirements of article 3.2 of Directive 2014/53/EU.
29 3GPP. (2020). Evolved Universal Terrestrial Radio Access (E-UTRA) and Evolved Universal Terrestrial Radio Access Network (E-UTRAN); Overall description; Stage 2 (Release 16). TS 36.300.
30 3GPP. (2018). Proximity-based services (ProSe); Stage 2 (Release 15). TS 23.303.
31 3GPP. (2014). Study on LTE device to device proximity services; radio aspects (Release 12). TS 36.843.
32 3GPP. (2016). Study on NR vehicle-to-everything (V2X) (Release 16). TR 38.885.
33 ETSI. (2017). Intelligent Transport Systems (ITS); Radiocommunications equipment operating in the 5 855 MHz to 5 925 MHz frequency band; Harmonised Standard covering the essential requirements of article 3.2 of Directive 2014/53/EU.
34 EU Radio Spectrum Committee. (2020). Draft commission implementing decision on the harmonised use of radio spectrum in the 5 875-5 935 MHz frequency band for safety-related applications of intelligent transport systems (ITS) and repealing decision 2008/671/EC.

35 Lawson, S. (2018). C-V2X's momentum in China may drive connected-car development. TU Automotive.

36 Fletcher, B. (2019). FCC looks to dedicate 5.9 GHz for Wi-Fi, C-V2X use. Fierce Wireless.

37 W. Wiquist. (2020). FCC Modernizes 5.9 GHz Band for Wi-Fi and Auto Safety. *FCC News.*

38 5GAA. (2020). Working group standards and spectrum study of spectrum needs for safety related intelligent transportation systems – day 1 and advanced use cases. TR S-200137, version 1.0.

2

Business Models

Antonio Fernandez Barciela[1], Markus Dillinger[2], Jesus Alonso-Zarate[3], Andres Laya[4], Per Lindberg[5], and Konstantinos Manolakis[2]

[1] *Group PSA, Spain*
[2] *Huawei German Research Center, Germany*
[3] *Centre Tecnològic de Telecomunicacions de Catalunya, Spain*
[4] *Ericsson Research, Sweden*
[5] *Volvo Cars, Sweden*

Automotive is a very competitive sector, where margins are narrow and technological adoption always has to be justified by a purpose – primarily profitability and regulatory compliance. Connected vehicles are already a reality in the market based on second, third, and fourth generation (2G, 3G, 4G) technologies. To boost the adoption of fifth generation (5G) in the vertical automotive sector for enabling connected automated driving (CAD) and connected road user (CRU) services (cf. Section 1.1), an effort must be made to emphasize its benefits: not only by original equipment manufacturers (OEMs), but also in the entire value chain – for example, by mobile network operators (MNOs), road operators, mobility service providers, cloud operators, telecom vendors, and infrastructure vendors. Primarily, various possibilities for OEMs and MNOs are shown in this chapter.

To create solutions in the connected car ecosystem, actors need to define their positions, establish business relationships, and coordinate needed activities. Cooperation is required between automotive and telecom actors. The traditional value chains for these industries are transformed into value networks where actors share knowledge and resources to accomplish new services. This transition and some examples of possible ecosystems that can emerge are illustrated in Figure 2.1 [1]. Four different possibilities are shown, depending on the primary actor orchestrating the business relationships (highlighted circle). The focus of this chapter is on cellular vehicle-to-everything (C-V2X), while some of the business and technical aspects have a broader applicability.

In this chapter, the current market analysis is presented in Section 2.1, whereas Section 2.2 contains service definitions from existing services, automated driving features, and convenience services. In Section 2.3, key technology components are highlighted. Section 2.4 outlines practicalities, whereas Section 2.5 contains description of business market opportunities for vehicle-to-everything (V2X). A business model analysis of 5G V2X technical components is presented in Section 2.6, and Section 2.7 ends the chapter with conclusions.

Cellular V2X for Connected Automated Driving, First Edition.
Edited by Mikael Fallgren, Markus Dillinger, Toktam Mahmoodi, and Tommy Svensson.

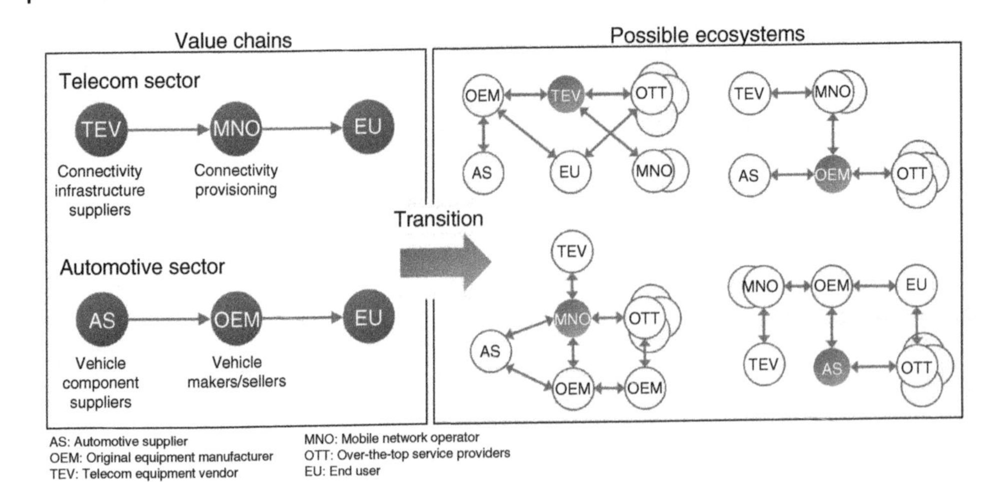

Figure 2.1 Different perspectives on how connectivity and new services can change the automotive and telecom value chains. Source: Adapted from [1].

2.1 Current Market Analysis

To begin, we elaborate on the current V2X market situation and highlight what market players have tested or implemented so far.

Institute of Electrical and Electronics Engineers (IEEE) 802.11p technology has not been significantly adopted [2]. Few automotive OEMs have conducted trials or announced plans to deploy the IEEE 802.11p standard in vehicles: e.g. in May 2017, Cadillac held a demonstration to show 802.11p-based vehicle-to-infrastructure (V2I) communication between vehicles and traffic signals; and as of 2019, VW started to provide its Golf 8 model car with 802.11p capabilities as a standard feature and plans to provide embedded mobile only for connectivity in Europe, because China and US have decided for C-V2X already.

Mobile technology has also been deployed in vehicles for some time [2]. Some automotive OEMs have implemented pre-release 14 long term evolution (LTE) with vehicle-to-vehicle (V2V) capability, such as BMW. The BMW connected drive (launched in 2001) enables radio access information for in-car use. In 2008, BMW was the first automotive OEM to allow unlimited in-car internet usage; since 2015, all new BMWs in certain markets have been equipped with a BMW connected drive. Standard features include access to vehicle information and remote-control feature via a smartphone application, and additional premium features including a real-time information concierge, remote personal assistant, and infotainment. Since 2014, Mercedes Benz has provided most of its new car models with mobile connectivity, offering a suite of services (navigation, infotainment, and remote customer support).

Qualcomm was the first chipset manufacturer to announce an LTE direct-mode/proximity service direct communication interface 5 (PC5) chipset for short range cellular communication; others, like HiSilicon, Samsung, and Datang have also announced their products. PSA performed obstacle signaling tests by a first vehicle to a second one, and alert when an emergency vehicle approaches another car. These tests illustrate PSA's work on intelligent transportation system (ITS), which are supposed to improve safety in its cars. C-V2X is ready for commercial deployment in vehicles since beginning of 2020. 5GAA (5G

Automotive Association) has helped to define/test the protocols/specifications of C-V2X technology by bringing together diverse players from different industries (automotive OEMs, vendors, operators, and regulatory).

The first commercial deployment (Release 14-based C-V2X in vehicles and infrastructure) started in China during the second-half of 2020. In Europe, C-V2X-enabled vehicles and infrastructure are expected in the 2020-2025 timeframe.

Without any regulation, the evolution of C-V2X penetration will rely only on the interest that automotive OEMs have in pushing this technology according to their time scale to implement it in their new vehicles. Automotive OEMs have not yet announced adoption time frames, but in 5GAA, a market roadmap for C-V2X adoption has been agreed. Regarding the status for C-V2X, the chipset vendors that are 5GAA members (CATT, Huawei, Intel, Qualcomm, and Samsung) have expressed their commitment to providing C-V2X chipsets [2]. By the end of 2017, 3rd Generation Partnership Project (3GPP) Release 14 chipsets were available for tests, which coincides with planned validations and testing activities by automotive suppliers (ASs) and OEMs (Audi, Continental, Ford, Nissan, PSA, SAIC etc.). In addition, interoperability tests among suppliers and OEMs were conducted in 2018.

C-V2X technology (5G, including 3GPP Release 16 and above) is expected to enable cooperative ITS (C-ITS) services-including see-through, platooning, teleoperated driving (TOD), highly automated driving (HAD), and fully automated driving (FAD)-but the required roll-out investments, business models and revenues, on which 5G V2X deployment is strongly dependent, are still unclear. Without clear benefit, MNOs (including road infrastructure operators) will not be encouraged yet to start 5G deployments dedicated to CAD and CRU services. The 5G infrastructure Public Private Partnership (5G PPP) Automotive working group (WG) has provided a white paper providing first insights concerning the deployment costs for 5G V2X and a revenue analysis for financially and socially beneficial commercialization [3].

To highlight the potential impacts of 5G's arrival into the automotive sector from a business model perspective, two representative applications have been selected: over-the-air (OTA) updates and automated driving. In each application, we will trace the evolution from a traditional business model toward new possibilities enabled by the technological components to accompany 5G.

2.2 Services Definition for CAD and CRU

We will consider a high-level description of a number of services that can be used when examining business models. The services identified here expose all the elements that constitute a complete business model and specifically highlight new features in 5G that were not available in previous generations of mobile technology. 5G enables a new way to consume existing services and new services. Note that this is by no means an exhaustive list of services; the services included in this section have been selected as examples. They are divided into three categories:

- Existing CAD and CRU services
- Emerging CAD services
- Emerging CRU services

Existing CAD and CRU services are those already available on the connected car market. There are opportunities to enhance quality, lower costs, or in other ways modify the delivery and value chain using 5G technologies for this category of services.

Emerging CAD services are a set of services intended to enable the use of automated driving vehicles. Some automated driving features may be mandatory for higher levels of autonomy; other services may be used to enhance the user experience. This part examines how 5G can enable new business model elements in supplying these type of services.

An important subset of emerging CRU services is convenience services that are intended to enhance the in-vehicle user experience and are not directly related to the task of transportation. These services, in some cases, may only be supplied by using 5G; other services can be supported by earlier generations and enhanced by using 5G.

2.2.1 Existing CAD and CRU Services

In the following subsections, we elaborate on established CAD and CRU services related to safety, convenience for drivers, and remote maintenance. Depending on the business model, some of them are included in packages that can be purchased from car dealers or are included in vehicle prices. Others are based on after-market offerings to drivers.

2.2.1.1 Emergency Call

An *emergency call* (eCall) is part of the safety services within CAD (cf. Section 1.1) that automatically reports an accident, triggered by the supplemental restraint system in the vehicle. Information about the accident event is sent to an actor that can take suitable measures to assist the vehicle passenger(s). The actor could be a private or a public call center. 5G can enrich this service by, for example, providing:

- A highly reliable connection to the call center or emergency services
- High-definition video communication support to improve the deployment and operation of emergency services
- Data communication on top of voice calls to provide information related to the car and environment prior to the accident

2.2.1.2 Remote Diagnostics

Remote diagnostics is the ability to remotely analyze vehicle information and potentially remedy certain failures. It combines uplink data transmission from the vehicle to the mobile network (referred to here as a service center), intelligent data processing at this service center, and downlink communication to the vehicle: e.g. to modify internal parameters of the car or provide information to the user. Note that the difference between remote diagnostics and predictive maintenance is that remote diagnostics are initiated on request by the end user. 5G can enrich this service by, for example, providing:

- A low-latency, high data-rate, reliable connection between the vehicle and the service center
- A local distributed service center through mobile edge computing

2.2.1.3 Car Sharing

Car sharing means a car is shared among several users. When using the car, a given user's preferences, background information, and specific configuration can be loaded according to their profile. 5G can help to scale up this kind of service by providing massive connectivity to cars, as well as secure, low-latency, high-bandwidth communication to the car, enabling instantaneous transfer of the user's preferences to the car. Conversely, while the car is being operated, the user's related data can be collected to, for example, simplify billing (based on e.g. the actual distance traveled and speed).

2.2.1.4 OTA Software Updates

An OTA software upgrade consists of upgrading or modifying a piece of software in the car without taking the car to a garage. With previous communication systems (2G and 3G), an OTA upgrade required the user to download software on a universal serial bus (USB) stick when at home and then plug the stick into the car to perform the software upgrade. With 5G, OTA software updates will become really OTA, without active user participation (although the user may still need to approve the update). Using wireless 5G high-speed, reliable data transmission, OTA upgrades can be performed efficiently and reliably, including also vital embedded SW.

2.2.1.5 Predictive Maintenance

Predictive maintenance means analyzing the behavior of the vehicle and vehicle component(s) in order to anticipate any fault, failure, or misbehavior of the vehicle. Based on sensors, vehicle data surveillance, machine learning, big data analysis, and multisource information processing (history of the vehicle, warranty, etc.), almost real-time warning information can be provided, or a simpler message can be sent to the user regarding maintenance. Examples of such 5G-enabled-technologies are:

- Massive sensing and data collection, in real time or as background (low rate, low power)
- Mobile edge computing to reduce transmission latencies between vehicle and network for offloading car analysis tasks to mobile edge computing by putting analysis close to the vehicle
- Data fusion and network slicing

2.2.1.6 Real-Time Road Traffic Management and Vehicle Guidance

This service provides the vehicle with all the information needed for automated driving. It ranges from basic information to support smart decisions for traffic management (selecting a proper itinerary based on local traffic information or road status) to almost real-time information about nearby events. 5G technology will enhance existing information exchange by providing low-latency and a high volume of reliable and location-aware information to the vehicles.

2.2.2 Emerging CAD Services

In the following subsections, we elaborate on fundamental services for automated driving that become more important for higher levels of automation (cf. Section 1.2.2). Related business models are still to be developed and introduced to the market when such future vehicles are available.

2.2.2.1 Perception by Wireless Connectivity and Sensor Sharing

Even if a vehicle has many on-board sensors, these sensors can only perceive the local environment. As a result, to better support automated driving, it is useful to add a communication service as a kind of new extended sensor to get information beyond the local environment.

This wireless communication service can provide information from the network, road infrastructure, or neighboring vehicles. As introduced in Section 1.3, we can use the connectivity links in vehicle-to-network (V2N) communication, vehicle-to-infrastructure (V2I) communication, vehicle-to-vehicle (V2V) communication, and vehicle-to-pedestrian (V2P) communication for the environmental perception task.

5G will provide advanced V2X technology thanks to low-latency communication between the car and the network or infrastructure, or a direct communication link between devices, with or without network support.

2.2.2.2 High-Definition Maps

High definition (HD) local map acquisition provides the automated driving control with a real-time, up-to-date map of the vehicle's environment. The local map is reconstructed from various flows of information collected by the vehicle: local information from sensors as well as remote information from communication pipes.

2.2.3 Emerging CRU Services

The mobile network should support also convenience services to the driver and the passengers. In the following, we elaborate on two important emerging convenience services. Such services will be offered by various mobile network operators and require an extra subscription.

2.2.3.1 Video Streaming and Gaming

With the advent of automated driving, driver expectations will increase regarding information and entertainment (infotainment) services, because drivers will be less and less active in the driving task. Moreover, passengers will want to consume services similar to those available at home. As a result, there will be increasing demand in terms of bandwidth from and to the car to enable high-quality user services (such as videos, gaming, etc.). Similar requirements for video streaming will result from V2X use cases such as see-through or remote driving. 5G technology is the natural evolution of 4G to provide such capacity, rates, and quality of service (QoS) with increased bandwidth, slicing, and content that is closer to users through multi-access edge computing (MEC).

2.2.3.2 Parking Reservations and Payment

Drivers will no longer need to bother about parking reservations and payment; they will be included in the vehicle experience. The service will not only help to identify the best parking location and availability, reserve a spot, and pay for it, but also help the driver find on-street/off-street parking places and assist with parking the vehicle. 5G will enable this type of service by exchanging real-time information with parking and city databases as well as by providing remote vehicle control that can drive the vehicle from, for example, a garage entrance to the desired location.

2.3 Technical Components

Holistically, it is widely envisioned that 5G will act as a technical enabler for the creation of innovative business models on top of new services or existing services. To be more specific, we can say that 5G will include technical enablers that allow for the creation of new technological components that can be used to create new services or enhance existing services (as described in the previous section) upon which business models will be created. It is important to separate technical enablers, which are transversal innovations, from technological components, which are built on top of the technical enablers.

On a high level, we can argue that the two key technical *enablers* are:

- An innovative radio access technology (RAT) for V2X communications enabling both infrastructure-based and sidelink-based communications (direct communication between devices), a.k.a. device-to-device (D2D) communications. This enabler, together with high-accuracy positioning techniques, has been under development in recent years and is particularly targeted at future scenarios.
- Virtualization of the communications network, from the radio to the core. The virtualization of all the elements of the end-to-end communication network allows for a highly dynamic, reconfigurable infrastructure. This technical enabler has also been under development in recent years.

These two key technical enablers allow for the definition of innovative *technological components* that can be used to support new services. Key technological components highlighted here are characterized as follows:

- *Network slicing*: It will be possible to have various simultaneous and different instances of a communication network running on top of a common infrastructure. This feature will help to boost new business models by reducing infrastructure costs and improving flexibility to cope with dynamic demands from services.
- *Mobile edge computing*: Computing capabilities are installed and operated closer to the final user of a radio communication technology. Having computing capabilities closer to the user reduces latency, improves reliability, and offloads the core and transport networks.
- *Mobile network radio-based positioning and tracking*: 5G radio technologies can provide better performance than LTE and complement existing solutions based on the global navigation satellite system (GNSS). This feature may become crucial for automated driving.
- *Sidelink*: A complementary short-range link in mobile network coverage without using the infrastructure, at least for user traffic. By offloading the mobile network and supporting the automation levels, one can enable new use cases and business models. More details on the four different sidelink modes can be found in Section 1.3.2.1.
- *Moving networks*: Enabling moving vehicles to act as temporary nomadic infrastructure for the communication network can provide an effective extension of mobile network coverage and improve dynamic network deployment.

2.4 Practicalities

This section provides a brief look at primary practical aspects that will affect the definition of the business model for the automotive sector. Elements such as the provision of connectivity, continuity of service when roaming (especially now that a single European digital market is available), and coverage availability are crucial and may jeopardize new businesses.

2.4.1 Profile and SIM Card Provisioning

Cellular communication systems rely on a unique identity provided by a universal integrated circuit card (UICC), i.e. a subscriber identity module (SIM) card. It ensures integrity and security and can embed several services. It usually must be provisioned and associated with a given operator. However, with enhanced SIM (eSIM) [4], the SIM is not a separate piece but rather is directly included in the device as a chip and supports remote provisioning of credentials. Integrated UICC (iUICC) provides a further integration step where the SIM functionality is by design integrated into the telematic control unit (TCU) hardware, but the concept is still under development and not fully endorsed yet – for example, by the Global System for Mobile Communications Association (GSMA).

We assume that an enhanced UICC (eUICC) will be used, as this is the current best practice in the automotive industry. We do not foresee that 5G will change this. The physical component will be sourced by the car OEM directly from an eSIM vendor and will be under the control of the car OEM, but it will allow profile management by fleet managers or service providers. The eSIM device will be integrated in the on-board unit (OBU) by the OBU vendor on behalf of the car OEM. The eUICC will be provisioned with an initial profile (*bootstrap profile*) when supplied from the eSIM vendor. This initial profile will be supplied by an MNO contracted by the car OEM. But the eUICC can store multiple profiles. There may be cars/OBUs equipped with multiple SIM cards and eSIMs under the control of the car OEM and/or end customer. Furthermore, the end customer should be able to use a private profile on the eUICC to connect to an MNO of their choice (i.e. a private eSIM).

2.4.2 Routing Strategy

A connected vehicle has several interfaces available. In terms of connectivity, a global strategy has to be defined to decide which interface to use, depending on the service and connectivity technologies available.

The core of the 5G architecture design is the integration of different radio access technologies. This should be understood in a broad sense. The purpose is to provide the best performance available depending on the frequency bands used (licensed, unlicensed), the needs of the service (e.g. throughput and bandwidth), and the environment (such as high/low-density traffic area, or rural compared to city areas).

The network control can coordinate several interfaces belonging to different MNOs' networks. The network performance and continuity of service of a given MNO will be improved in order to fulfill specific service level agreements (SLAs) cooperatively as defined by the car OEM with other MNOs.

From an automotive perspective, connected vehicles began with a 2G interface; but right now in the market, vehicles can connect to at least the 3G–4G mobile network and Wi-Fi. With the arrival of 3GPP Release 14 compatible components, a new interface providing direct communication between vehicles will be available.

Then there will be at least three interfaces (cellular, Wi-Fi, and sidelink) to take into account. Each OEM will define rules for using the different technologies available. These criteria will depend on elements such as economic prioritization, SLAs, and bandwidth requirements. The impacts of these interface-selection policies will vary for the service domains:

- Safety
- Infotainment
- Automated vehicles

2.4.3 Roaming and Inter-operator Cooperation

To provide a seamless experience for automotive services, global roaming is highly desirable, particularly in the case of the European Union. There are several considerations to keep in mind. For instance, on the one hand, the impact cost of handling all possible bands can be restrictive in a car. From an automotive perspective, the ability to roam is a must for any pan-European service imagined.

In addition to the evident concern for global or pan-European roaming, there is a concern regarding the interoperation of vehicles connected to different MNOs, especially when considering the possible direct link between vehicles (without going through the network infrastructure) and when using a licensed spectrum. There should be a clear and homogenous solution to enable such interoperation, which can be provided on the communication plane when possible or with the service plane providing the required connectivity performance.

The topic of coordination should also consider any possible alignment between unlicensed 5.9 GHz and licensed spectrum operation. 5.9 GHz can be used for direct V2V and V2I communication in certain cases while operating in the licensed bands for V2N communication. However, it is not clear whether an actor should coordinate communication between the license and unlicensed spectrums.

Finally, we underline that the licensed spectrum may be used for V2I as well as V2V communications since licensed bands have the benefit of being able to offer a (higher) SLA. The primary challenge is determining viable options to find the right business models to justify using licensed bands for V2V.

2.4.4 Possible Business Model Evolution

Today, various business models exist for well-known vehicle and passenger services. In the following subsections, we present potential new business models for advanced future vehicles related to partial or full automation. It remains to be proven which are the most sustainable and profitable business models for future mobility solutions.

2.4.4.1 OTA Software Updates

This is an example of a business model with significant potential benefits for OEMs. The need was initially based on car flaws found by OEMs that had to be corrected, along with the desire to change or add new functionality. Such problems were solved by the vehicle owner visiting a garage, which replaced a specific module of the car. The organization of such module-replacing recall campaigns is extremely expensive and negative for OEMs: expensive due to the mobilization of resources in garages near the customer, and negative in terms of brand image because the customer has to go to a garage and leave the car for the intervention. With a recall, it is also not possible to reach 100% of affected vehicles. Some customers will choose not to participate or cannot go to a shop to get updated software.

Thanks to the progressive softwarization of vehicles, new possibilities have arisen in this area. On the one hand, significant flaws can be corrected by a software update. On the other hand, the software itself lets the OEM create new services or features that may profit the customer during the lifetime of the vehicle. Car modules that can be updated by software have a rollback feature, which can return to the previous software version in case of a faulty upgrade.

In both situations and without connectivity, three primary stakeholders are involved: the automotive supplier (many times known as Tier 1) providing new software, the OEM responsible for its validation and distribution, and the garage as the actor that performs the update in the customer's vehicle.

For this business model analysis, the focus is only on the software distribution cost (not costs related to software development), which is where connectivity and, more precisely, 5G will have added value. In this exercise, the software update will be delivered from the automotive supplier to the OEM and from there to the vehicle. This is the normal situation so far due to OEM validation requirements and liability. Other situations may be foreseen, but they are out of the scope of connectivity applied to OTA updates.

With the arrival of connectivity, two new applications are added to the first value chain described. The connected vehicle is a car equipped with a SIM card and an OBU, something that has been a must in Europe since April 2018 (eCall in emergency situations). Connectivity is provided by an MNO. The OEM pays the connectivity costs, and the update is made seamless to the customer, other than granting permission and being notified at the end.

However, this MNO connectivity has limitations: the software distribution cost business model is based on the volume of data transmitted, roaming costs are high, and coverage and bandwidth availability are limited or unknown depending on the geographic location of the vehicle (more details can be found in Figure 2.5). Also, the business relationships between actors in the value chain are very rigid: e.g. MNO migration is difficult when using a physical SIM card (a SIM card with the nominated MNO's profile is "locked" to the vehicle for its lifetime). Recent models come with eSIM: a programmable SIM card that can be remotely programmed to migrate from one operator to another. This will be a factor in dynamizing the connectivity market for OEMs. The current costs of OTA solutions, c.f. Section 2.3, have led OEMs to propose a different software update possibility where the customer is involved. This can be done by the customer downloading the software from the OEM cloud at home and then using a USB stick to perform the update or a private Wi-Fi network to get free connectivity. This application does not provide the best customer experience, but it is still more comfortable than the traditional visit to the garage and much more cost-effective for the OEM than connectivity provided by the network operator. It is time to analyze the arrival of 5G and its technological components and how they may affect the value chain defined so far.

In 5G, three important technological components modify the business model definition for automotive connectivity:

- *Mobile edge computing*: This feature will bring higher performance in terms of reactivity, and there are software upgrades that are only interesting with this type of technology available, such as HD map distribution. The OEM will bring the software to be updated near cars to download the new patch. In terms of the value chain, the network load and end-to-end connectivity can be reduced due to the proximity of the download by increasing the MEC capacity side. This function can be provided by the telecom vendor but will initially take place through the MNO. Other situations can be imagined.
- *Network slicing*: This is another promising feature that will modulate the business model for connectivity. There are many questions to be solved about the way it will be deployed, but it may help to reserve resources according to specific needs; this will arrive at optimal usage, making connectivity more competitive for some applications. Slices create virtual networks that are used for specific services. Today, it is not common for OEMs to go beyond a customer/supplier relationship with MNOs, but this is foreseen to change in the future with the arrival of network slicing.
- *Sidelink*: Starting with 3GPP Release 14, the possibility of a direct dialog between vehicles is possible. This type of communication can be deployed on a licensed or unlicensed spectrum, with or without network control, only for ITS or beyond this use. All of these options will strongly modify the connectivity value chain as it is known today. Taking this into account, new actors will emerge, providing connectivity without the need to be network operators.

These three disruptor factors enable new possibilities where the OEM counts on a cloud provider and then on a road operator to provide connectivity. Some of this is already happening, and the OEMs no longer talk about their own cloud but about a provided service. Depending on the three components, the road operator can be a network operator in a restricted area or any other stakeholder not related to the standard connectivity business.

A natural evolution of this model is the arrival of a primary stakeholder, an over-the-top (OTT) communication provider for connectivity, responsible for providing the full OTA update service to the OEM with an abstraction layer completely agnostic of the connectivity technology or the cloud provider. This role can be played by a network operator, a cloud provider, or any other company interested in this market.

2.4.4.2 CAD Services and Related Automation Levels

We detail the CAD services and related automation levels defined by Society of Automotive Engineers (SAE) in [5] (cf. Section 1.2.2) in the following subsections. SAE levels 3, 4, and 5 are still to come to the market. It is expected that some highway automation will be available in the 2020–2025 timeframe and that advanced urban automation will mature after 2025, depending on the vehicle manufacturer.

CAD for SAE Levels 0 and 1

SAE level 0 means no functions of the driving task are automated. The driver is in full control of all functionality. SAE level 1 means a task is automated: for instance, with adaptive cruise control, the vehicle can control its speed in traffic, and the driver has full control over steering.

In these applications, connectivity is not required. The automated functions are provided by sensors, and data processing performed in the vehicle itself. Connectivity can still enhance the driving experience by, for example, providing real-time traffic information, but that is not considered in this analysis. The driver/user has a relationship with the car OEM (typically through a dealer) for car ownership. The driver also has relationships with an insurance company as well as independent mobility service providers (MSPs) unrelated to the vehicle, such as taxis.

CAD for SAE Level 2

SAE level 2 means both steering and acceleration/deceleration are automated under certain circumstances. However, the driver is always responsible for control of the vehicle and must be able to take control at all times. A typical example of SAE level 2 is a vehicle with adaptive cruise control and lane-centering functionality.

In this application, connectivity is not by definition mandatory. Connectivity is included to provide e.g. real-time traffic information, emergency calls, and other functionality not related to the driving task. Connectivity is typically provided by the OEM, which signs contracts with MNOs to provide the required services. The connectivity provider is typically unknown to the end customer, who is only interested in the services provided on top of the connectivity.

As discussed in Section 2.5.2, the business relationship between OEM and MNO has previously been inflexible and existed for the lifetime of the vehicle, which has sometimes hindered the development of new services (due to high connectivity costs) as well as roll-out to new markets. We anticipate the introduction of eSIM, as described in Section 2.2.3, to facilitate a more flexible relationship.

CAD for SAE Level 3

With SAE level 3, vehicles are able to take control of all safety-critical functions required for driving under certain conditions. The driver must still monitor the driving and be ready to take control after a specific time if the vehicle indicates that it cannot handle the situation. This is the first step where driving responsibility during automated driving is shifted from the driver to the OEM.

In SAE level 3, connectivity becomes a mandatory component for providing the required functionality. Connectivity is required to monitor and control the geographic area where the vehicle is allowed to operate automatically in certain situations. Here we see a number of new actors appear in the value chain: cloud providers providing the computational infrastructure for service production and delivery, map providers, and road operators to provide geocoded information to facilitate automation functionality. In this application, the OEM has direct relationships with cloud providers and map providers. The relationship to the road operator is more uncertain and can depend on the roll-out of standardized

V2X services and the role MNOs take. The OEM can also have a direct relationship with insurance companies to provide insurance while the vehicle is in automated mode.

5G can affect this value chain in several ways: e.g. through MEC in MNOs taking an active part in providing a V2X infrastructure. Network slicing can enhance the QoS and thereby the user's trust in the information received from off-board, potentially allowing the vehicle to operate in automated mode more frequently. SAE level 3 is, however, possible to reach using earlier generations of mobile technology, such as LTE.

CAD for SAE Level 4 and 5, Private Vehicle

SAE level 4 means the vehicle is completely automated under all normal driving circumstances, and the driver is not expected to intervene. However, in certain conditions, the vehicle may not be able to operate in automated mode, such as during severe weather, on dirty or snowy roads, etc., and the driver will have to take control.

In SAE level 5, the vehicle is truly automated in all driving conditions and situations. The driver is no longer required to operate the vehicle. In this scenario, it is assumed that the overall business model does not change from previous applications: the driver (end customer) still buys (or leases) a vehicle from an OEM, directly or through a re-seller, and operates the vehicle as their own.

We do not see the connectivity value chain changing dramatically from SAE level 3. In SAE levels 4 and 5, the importance of getting information from the environment surrounding the vehicle is increased, and therefore it becomes vital that the OEM, which is assumed to have responsibility for the vehicle in automation mode, has access to information from the road operator. With SAE level 5, the customer may no longer need a relationship with an insurance company, as the responsibility of driving the vehicle will reside with the OEM that provides the automation functionality.

The impact of 5G will likely be significant. Network slicing will make it possible to dedicate slices for different services: e.g. a high-availability, low-latency slice for V2X functionality, a high-bandwidth slice for entertainment services, etc. Costs for connectivity can be distributed to different actors in the value chain. MEC and low-latency access will be vital for building efficient V2X applications, potentially creating new business models, with actors taking new positions in the value chain.

CAD for SAE Level 5, MSP

With SAE level 5, where vehicles are fully automated (autonomous), disruptive new business models around mobility are a possibility. One such model being discussed is the situation where customers buy "mobility as a service" rather than a vehicle.

In this application, the mobility service provider (MSP) takes the central role. The MSP has the primary business relationship with the end customer, as well as relationships with the service providers necessary to provide mobility. Here we need to evaluate who will provide the connectivity to the vehicle, or whether there may even be multiple connectivity links to the vehicle: one for OEM services like OTA software updates and other services like emergency calls, closely related to the vehicle itself; and another between vehicle and MSP, to provide both the services required for autonomous driving as well as services for managing the mobility service, like booking, payments, etc. 5G can enable this more open ecosystem of stakeholders wanting and needing access to connectivity to the vehicle, with various requirements for QoS, coverage, availability, bandwidth, etc. At the same time security and integrity requirements must be met.

2.5 Business Market Opportunities for V2X

Once we have analyzed the business model impacts, a few economic scenarios can be used to illustrate the potential business behind connectivity applied to the automotive sector in the V2X domain. Three market opportunities are analyzed: 5G deployment scenarios for connected and automated mobility services based on the 5G PPP Automotive WG white paper [3], the security cost of the current certificate strategy proposed for V2X communications, and an exercise about the size of the OTA update market.

2.5.1 CAD Business Model Enabled by 5G

The second version of the white paper from the 5G PPP Automotive WG [3] includes further work targeting the description of the 5G V2X ecosystem for enabling CAD services and stakeholder relationships, different sharing models for network infrastructure, as well as a business setup and a techno-economic assessment of the investment. The scope of the paper is in line with the strategic deployment agenda (SDA) initiated by the European Commission; therefore, for the sake of simplicity, it builds the discussion around connected and automated mobility services, 5G V2X deployment costs, and potential revenues.

The exemplary highway environment considered through this work is taken from real conditions, and the deployment under consideration includes 5G radio base station sites, civil work, and fiber backhaul connections. It is further assumed that this investment can be used to provide enhanced mobile broadband (eMBB) and other relevant services, since 5G effectively enables a multi-service environment.

Particularly in the early phase of 5G network deployment, synergies between the private and public sectors can speed up deployment. Different sharing models are discussed, based on [6]:

- *Passive infrastructure sharing*: Each network operator deploys its own network in the service area. Only passive infrastructure elements are shared between operators: e.g. space, masts, power generators, and air conditioning equipment.
- *Active infrastructure sharing, excluding spectrum sharing*: Active elements of the cellular network, such as base stations, are shared. Each operator is still transmitting on its own spectrum.
- *Active infrastructure sharing, including spectrum sharing*: Active elements, such as base stations, are shared. A single operator operates the dedicated spectrum.
- *Core network sharing*: Elements of the core network are shared by more than one network operator. The savings corresponding to only sharing elements of the core network are reported to be very low.

The methodology can be considered to include other services for revenues, as shown in the adapted Figure 2.2.

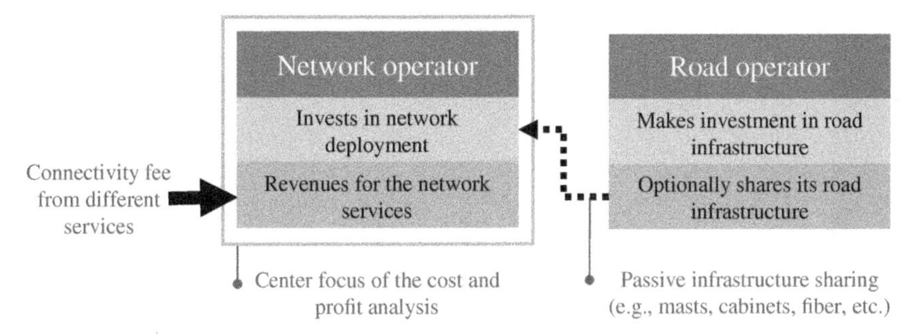

Figure 2.2 Main business setup for the network deployment analysis. The role of Network operator can be covered by different actors in the stakeholder ecosystem [3].

The total cost of ownership (TCO) includes capital expenditures (CAPEX) and operational expenditures (OPEX):

- *CAPEX*: Site infrastructure, 5G base stations, network equipment, cabinets, civil works (physical cabinets, fences, antenna masts, etc.), and fiber backhaul provision along the highway
- *OPEX*: Network operation, maintenance, and replacement, and site lease

In addition, the source of income from the perspective of the network operator is a percentage of the connected and automated mobility service fee, the part associated with communication aspects.

Three deployment alternatives are considered:

- *Deployment 1*: The CAPEX and OPEX investment for the network and fiber backhaul are carried out by a single actor (no infrastructure or network sharing is considered).
- *Deployment 2*: The active elements of the network are deployed by a single actor. The passive elements of the network infrastructure are shared with the road operator.
- *Deployment 3*: In addition to the passive elements, the active elements in the radio access network are shared by more than one network operator.

Also, two different pay-per-use revenue alternatives are considered with different values for the connected and automated mobility service fee and traffic density:

- *Revenue 1*: Baseline scenario with 50 000 vehicles using the highway segment each day and revenue of €0.5 per vehicle for each 1 km of the highway segment. All vehicles are served by a single network operator.
- *Revenue 2*: It is assumed that several network operators provide connectivity along the highway; a single network operator captures only 35% of the vehicle penetration rate.

The evaluation results for the accumulated profit are depicted in Figure 2.3 for an increased user penetration rate of 10% per year. As observed, payback periods between four and eight years are expected, depending on the deployment and revenue alternatives considered.

For the network deployment cost model, no previous network deployment is assumed in the highway considered. In Europe, however, current 4G deployment levels confirm that this is not the case for any important highways. This situation will naturally evolve

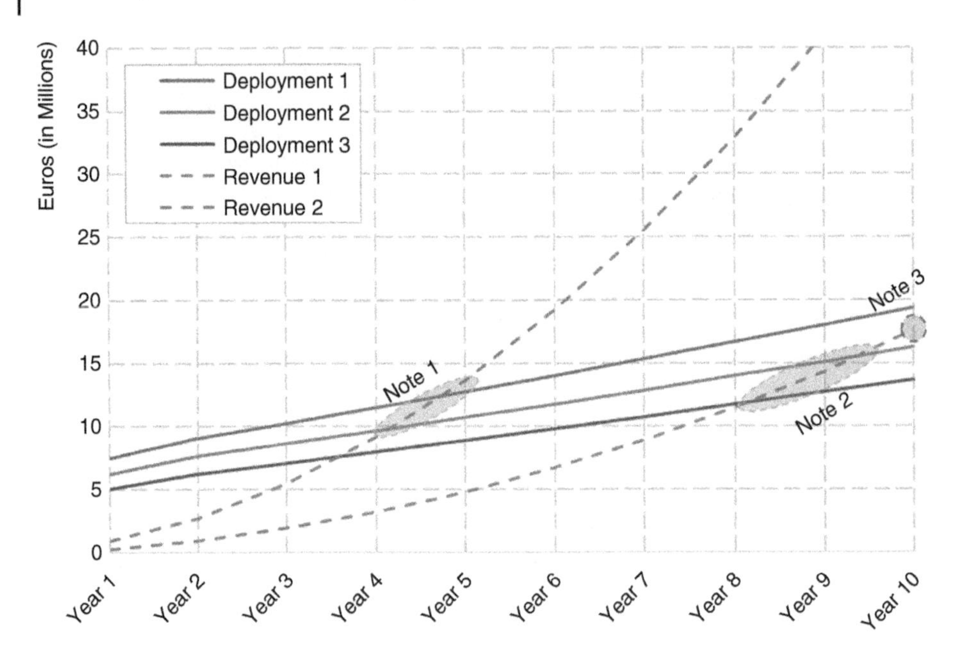

Figure 2.3 Accumulated costs and revenues for different scenarios. Note 1: if a single network operator is able to capitalize on all vehicles in the highway segment (revenue model 1), all deployment options break even within the first five years of service. Note 2: if there is more than one network operator and the vehicle subscriptions are split, the deployment options allowing network sharing break even after 8–10 years of service. Note 3: if no network sharing is allowed and the number of subscribed vehicles is divided among different network operators, no profit is reached within 10 years [3].

for eMBB due to the deployment of 5G, with the new bands already auctioned. Moreover, countries like Germany have included in these auctions specific conditions to reinforce the cellular service level on the main highways. More spectrum details are provided in Chapter 4.

On the other hand, this available cellular network on the main roads may not be able to provide the service levels needed for connected and automated mobility, especially in terms of density, throughput per surface (in some use cases), reliability, and availability. Compared to eMBB, a URLLC slice may be required, where the classical network densification with more sites can be the solution to enhance the service level. Thus an important network deployment cost may be foreseen, as proposed in the white paper. 5G eMBB deployments will likely not bring an important change regarding network deployment, compared to regular 4G deployment. 5G will widely reuse resources already deployed for 4G (passive infrastructure). New deployment approaches will likely occur with ultra-reliable low latency communication (URLLC) deployment, where the business model for URLLC may be different from eMBB. A specific slice or slices can be required, and network resource orchestration using software-defined network (SDN) will allow this flexibility.

As indicated in the Body of European Regulators and Electronic Communications (BEREC) report [6], the primary objective is cost savings. However, BEREC emphasizes the fact that network sharing agreements by slices between operators should not reduce the investment effort of MNOs by much, thereby avoiding less competition for customers.

It is worth analyzing three of the proposed network sharing models to divide the overall cost between different actors.

2.5.1.1 Passive Infrastructure Sharing

Passive sharing is already used by operators to share network costs. In that configuration, the mast has to support several antennas (one for each operator). MNO will have certain regulatory constraints concerning the coverage of highways in 5G, as is the case for 5G for eMBB services. For instance, by 2027, in France, 100% of principal roads have to be covered by every operator in the 800 MHz band [22]. Passive network sharing will not allow reducing the number of antennas; constraints can appear on the available room on the mast. As such, passive network sharing will not allow efficient network densification.

2.5.1.2 Active Infrastructure Sharing, Excluding Spectrum Sharing

In the case of active sharing excluding spectrum sharing, the available spectrum that can be used in a given area is only a sub-part of the global spectrum, and QoS problems can appear. An example of such sharing can be found in the UK between Telefonica (O2) and Vodafone or in Spain between Orange and Vodafone. They have signed a network sharing (active, passive) agreement for medium size towns. This will be extended to 5G for small towns [7].

In the active sharing family, with or without spectrum sharing, we can add geographical sharing, which can be interesting. This may be a solution to enhance reliability. Within a given country, it is called national roaming. Another perspective to consider is the cross-border scenario.

2.5.1.3 Active Infrastructure Sharing, Including Spectrum Sharing

Some examples of network sharing in Europe are available; for instance, in Poland, PTK Centertel (Orange) is sharing network infrastructure with Polska Telefonia Cyfrowa (T-Mobile). Access network and base stations are shared for 15 years, with an estimated savings of €356 million [8].

In addition to safety and improved traffic flow use cases, more advanced services can be envisioned to provide added value for the end customer in terms of increased efficiency, convenience, safety, etc. These services go beyond public benefits and have the potential to create commercial value for the actors in the value chain.

The Automotive WG white paper [3] assumes in the economic model that the end user (car driver, car owner, car passenger) will pay for connected and automated mobility service to a connected and automated MSP. Some other models are possible. Connected and automated mobility service is related to road safety and traffic efficiency and thereby provides value to stakeholders other than the end user: e.g. public authorities, which will get more efficient use of the road infrastructure, lower costs for treating casualties of road traffic accidents, etc. This value can be translated into a business case where, for example, authorities support the network by providing additional infrastructure for different service providers.

The white paper also explores different scenarios of how to share the necessary infrastructure between multiple operators. Depending on what national regulators allow in terms of network sharing, and how willing operators are to use the opportunities provided by network sharing, we will see a major impact on the payback time of the necessary investments. It seems unlikely that all users will use a single MNO on a single stretch of road, through either regulation or competitive pressure, so network sharing will be important to drive down infrastructure costs and make connected and automated mobility a viable business case.

One of the new features of 5G is network slicing, which may be a good solution for sharing costs in developments where there is a network slice operator to provide a dedicated slice to ITS and connected and automated mobility services.

2.5.2 Security Provision

All connected services, including ITS, V2X, and connected and automated mobility services, rely on secure connections between various stakeholders to exchange information. The exchange of information is what creates the value in these services, and it must be protected, along with end users' privacy. Security, however, is not free and can be seen as an additional service layer that must be included in the total service delivery model. This section examines costs related to security and how they may impact the total business model for V2V, V2I, and V2P services.

There are ITS standards (cf. Section 9.2) for providing a dedicated public communication system for cars and road infrastructure. The system provides messages between ITS stations, with different levels of information. Globally, all messages are broadcast using short-range technology. The primary security requirement is authentication (signature) of the base stations and vehicle messages, as well as anonymity of the true vehicle identity document (ID) (pseudonymization). Globally, the ITS communication system will be a peer-to-peer communication system that relies on public key infrastructure (PKI) certificates to provide message authentication and integrity.

As a general rule, all messages should be signed with an authorization ticket certificate for authentication of the ITS sending station. In this system, an ITS station can be a vehicle or a road infrastructure device. Communication can be V2V or V2I.

For these communications, different levels of security services are provided. The primary certificate hierarchy authority is the European Policy Authority, which appoints a trust list manager (TLM). The TLM provides the European Certificate Trust List with a list of trusted root certificate authorities. The trusted root can be European Union-, state-, or private organizations-owned. In such a hierarchy, the car OEM must only provide certificates for its own vehicle fleet, as we can assume the ITS infrastructure to be state-owned or European Union-owned.

For each message, there is a strong need for authentication (signature) and a mid-level need for integrity (hash signature); and for messages that contain personal data, there is a strong need for confidentiality. Therefore, all security services provided by traditional PKI systems will be assumed to be provided, but it should be noticed that the authentication (signature) and integrity service with the anonymity of the vehicle will be the most common in ITS.

2.5.2.1 The PKI Workflow

The public key infrastructure (PKI) workflow is as follows:

- ITS infrastructure road station:
  - Public PKI: RootCA country / European Union delivers authentication certificate

- ITS vehicle:
  - Privately owned PKI RootCA OEM delivers enrollment authority (EA) and authorization authority (AA) certificates
  - Infrastructure EA delivers enrollment certificate for its enrollment domain to an ITS vehicle
  - Infrastructure AA delivers authorization token (certificate with pseudonymization) for an authorization context to an ITS vehicle that provides an enrollment certificate

2.5.2.2 Enrollment of an ITS Station

The vehicle identifies itself with a private key and an intelligent transport system (ITS) ID. This initial authentication can be a specific ITS certificate installed in the factory, or the built-in certificate of the electric control unit (ECU). The ITS station builds an enrollment certificate (EC) request by providing a signature with its private key. The car manufacturer should provide the public keys of its ITS vehicles to the EA. The EA verifies the validity of the request with the corresponding public key. The EA provides the long-term ITS certificate called an EC, which has a lifetime of three years.

As shown in Figure 2.4, the ITS station sends an authorization token certificate request to the AA. The request credentials provided by the ITS station are transmitted securely to the EA to check their validity. The request is signed by the ITS station, and the signature is checked by the AA. When the validity of the request is established, the AA delivers an authorization token. The authorization token can be a single token or a pool of tokens.

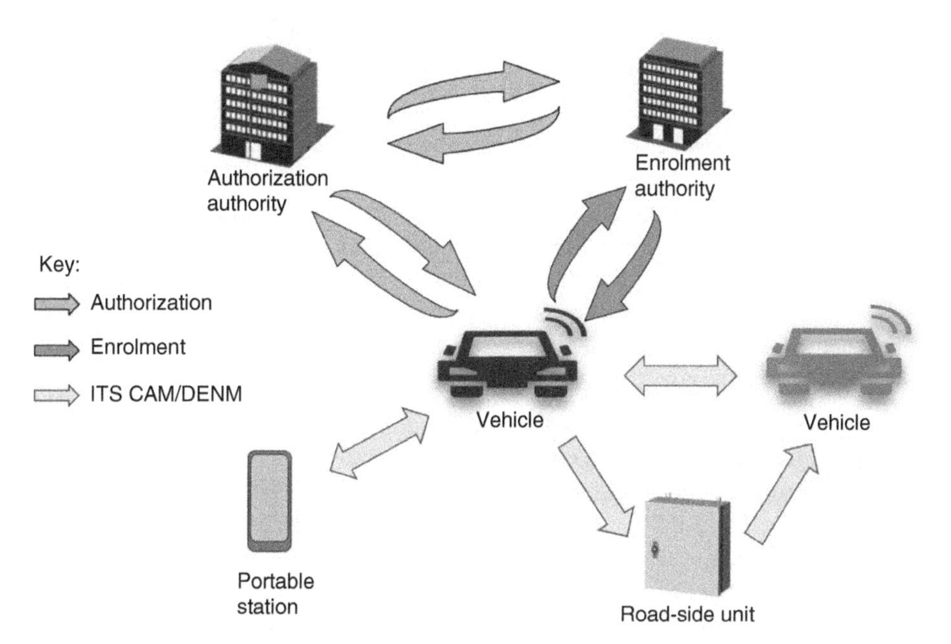

Figure 2.4 Authorization token request procedure. The ITS communication can e.g. consist of cooperative awareness messages (CAMs) or decentralized environmental notification messages (DENMs).

2.5.2.3 Use of Authorizations Tokens

Authorization tickets, are pseudonymized certificates that are used to provide a signature for ITS messages emitted by an ITS station. The signature is in an elliptic curve digital signature algorithm (ECDSA) with the Secure Message (SM) protocol. Authorization tickets have a high rate of usage and, therefore, should be renewed frequently. Different scenarios are taken into account. The certificate volume that needs to be defined is about on-board certificates required by vehicles to be ITS compliant or ITS capable. The primary cybersecurity requirement is to ensure vehicle anonymity, at the same time signing and authenticating ITS messages sent.

For privacy reasons, it has been determined that vehicle anonymity can be achieved by emitting multiple certificates depending on time (x certificate per week), distance (x certificates per km), and reuse condition (true or false). In the Delegated Act [9], released in January 2019 and rejected in July 2019, the recommendation is to use 100 certificates per week.

For this system, the OEM can define its own PKI infrastructure for certificate usage and delivery. The AA and EA are owned by the constructor/OEM PKI in this approach. Multiple certificates volume scenarios can be derived, but in this analysis, only the option of 100 certificates per week will be analyzed.

2.5.2.4 The Cost Hypothesis

The cost hypothesis takes the following elements into account:

- License for certificates.
- Hardware security modules (HSMs) (hardware) for generating keys off-board.
- The certificate is used for pseudonymization: 1 certificate = 1 pseudonym.
- Extra expenses such as PKI integration, deployment, and maintenance are also included.

Taking all of these elements into account and considering with this approach that the service will be provided by the OEM infrastructure, a value of €0.0007 per pseudonym has been estimated. This value is a reference based on the infrastructure needs for an OEM in Europe to provide the certificate generation demanded by ITS. This value is a theoretical estimation with no market reference. Certificate generation can be a service provided by a third party where the service level and optimization of the resources needed can lead to a different value. The purpose of this exercise is to provide a vision of the economic impact of the pseudonym strategy defined so far.

Cost chart assumptions and remarks:

- A linear penetration rate has been adopted, using as a reference the cars sold in 2017 in Europe [10]. Any other vector can be used.
- A final figure is provided for 100% market penetration according to the information available for cars in use in 2016. The purpose of the chart is to provide an overall vision of the market size related to the current ITS certificates strategy.
- If the cost is estimated for one car, there will be 100 certificates per week during the 52 weeks of the year, so 3120 certificates. This provides a cost per year and car of €2.18. From the OEM perspective, to cover a car's average 15 year life cycle, a cost of €33 per car must be taken into account. This will be a major deterrent for ITS adoption unless it becomes regulated. This cost will then probably be passed on to the final customer.

Table 2.1 Cost estimation of the pseudonym certificate policy in Europe.

Year N	New ITS cars/year	ITS cars on roads	Authorization ticket certificates/year	Total certificate cost (in €)
1	17,000,000	17,000,000	88,400,000,000	61,880,000
2	17,000,000	34,000,000	176,800,000,000	123,760,000
3	17,000,000	51,000,000	265,200,000,000	185,640,000
4	17,000,000	68,000,000	353,600,000,000	247,520,000
5	17,000,000	85,000,000	442,000,000,000	309,400,000
6	17,000,000	102,000,000	530,400,000,000	371,280,000
7	17,000,000	119,000,000	618,800,000,000	433,160,000
8	17,000,000	136,000,000	707,200,000,000	495,040,000
9	17,000,000	153,000,000	795,600,000,000	556,920,000
10	17,000,000	170,000,000	884,000,000,000	618,800,000
		298,900,000	1,554,280,000,000	1,087,996,000

Two primary conclusions can be derived from this exercise, as shown in Table 2.1. On one hand, the cost per car (more than €30) will be a deterrent for OEM adoption. This is one reason 5G Communication Automotive Research and innovation (5GCAR) has proposed a technological component with a different and less costly solution [11]. On the other hand, pseudonym generation, as it is defined today, is an opportunity as a service for the telecommunication sector. Current actors or new ones can optimize the cost and service level provided through the synergies of the resources available for providing this service.

2.5.3 OTA Software Updates

OTA is a service promising large value for end customers, in realizing new features in the vehicle after initial purchase, as well as for the OEM in being able to fix quality or other issues in the software without having to bring in vehicles for expensive workshop visits. The software update is related to all kind of services like infotainment and automated driving functions.

The intended use of the mobile network is to enable remote updates and/or configuration of electronic control units in cars via OTA. This is an opportunity for both the automobile industry and the telecommunication industry. The following is a brief reminder of the primary benefits:

- Reduce the resources needed to perform the physical work on cars.
- Avoid the need to have the car physically in a garage to do the update and, if the car has already been purchased, the annoyance to the client when they are contacted about a recall.
- Accelerate the application of updates and correction of bugs, especially critical ones.
- Increase the frequency of updates and new functionalities.
- Allow the remote activation of new services contracted by the client, such as map updates.

With the arrival of 5G, the capacity of the network to handle data traffic is improved using eMBB, but the cost of this volume of data using the mobile carrier network is a key parameter to take into account for OTA deployment over the cellular network. There may be a potential business case, but it has to be profitable for both the automotive industry and the telco industry. If this is not the case, then OTA will be deployed as it is today, based only on unlicensed frequencies or the driver/owner data plan, limiting the possibilities of this use case.

For perspective about the size of the opportunity, an estimate of millions of terabytes has been made as a first step. A constant volume of cars will be assumed for the next 10 years in Western Europe: 17 million per year, all connected to the cellular network and ready for OTA updates. Regarding the upgrades, several assumptions are made:

- Only one upgrade per year is considered.
- No roaming charge is taken into account. All OTA updates are considered in a national environment.
- Cars will be updated during the first five years. For the rest of their lives, only minor updates are foreseen. Two different scenarios for the data volume are studied – the conservative scenario and the non-conservative scenario:
  - Conservative scenario (CS): The size of one car update starts at 5 GB and is increased by 5 GB every three years. The normal product cycle of a car is around six years, and there is normally an embedded electronics update in the middle of this time. This is the reason for the three-year period.
  - Non-conservative scenario (NCS): The size of one car update starts at 10 GB and is increased to 30 GB for cars manufactured the following three years, then reaching 100 GB in the next three-year period.

Figure 2.5 shows the evolution of millions of terabytes during the next 10 years in both scenarios. It is important to note that this volume of millions of terabytes is new consumption in the mobile market that does not exist right now. To avoid any concern on the MNO side in terms of network collapse for a simultaneous update of such a large quantity of information, OEMs confirm that most updates have no stringent deadlines (within a margin of weeks), and the primary focus should be on the price and not on the speed of the update. Obviously, there will also be critical updates (related to safety and cybersecurity) where the OEM will be willing to pay a higher price for faster updates.

To translate this volume into an economic indicator, it is always difficult to get a price per megabyte. These prices are confidential between MNOs and OEM, so to continue this exercise, another assumption is needed. The hypothesis used is to apply the cost of data today in consumer mobile contracts. As we can see in Figure 2.6 (designed using the data from the 2018 research done by Cable in 230 countries [12]), the average cost of 1 GB paid by users in each country is far from being a standard in Europe.

To establish the reference gigabyte price:

- The average cost calculation is performed by reducing the geographical perimeter to six countries – France, Spain, Germany, UK, Italy, and Sweden – that are not at the low or high price borders of the list.
- For each of these six countries the lowest price available is considered, since it is assumed that the volume of the data for car OTA updates is so important that the best price can be obtained in the mobile carrier negotiation.

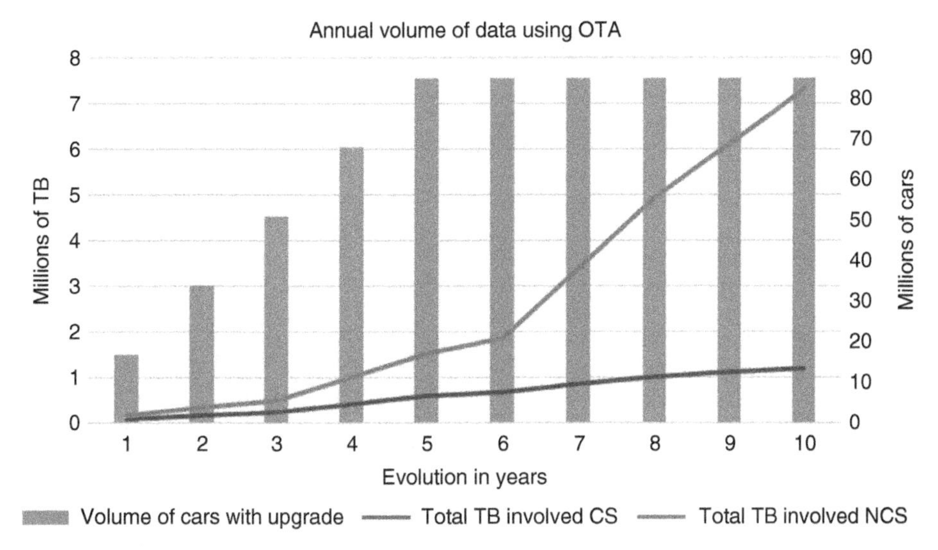

Figure 2.5 Estimation of the annual volume of data in the next 10 years for over-the-air updates.

That gives a reference gigabyte starting price of €0.54. We also consider that the price of data will be reduced in coming years, as mobile telephony cost did in general with an annualized average of a 9% discount during the past decade, as studies by the French Telecommunications Regulatory Authority, Autoritéde Régulation des Communications Electroniques et de la Postes (ARCEP), in [13] and by the Spanish National Commission of Markets and Competition, Comisión Nacional de los Mercados y la Competencia (CNMC), in [14] have shown. Therefore, we assume this discount as a reference for the future, and we apply this discount of 9% per year for the next decade.

In Figure 2.7, we can find the final estimation for the previous volume of data considered for OTA, with the total cost per year, in both conservative and non-conservative scenarios.

The two factors involved, augmentation of data and discount per year, have opposite effects on the cost. This explains why, in the conservative scenario, the cost will decrease toward the end of the period. The annual cost in the non-conservative scenario grows from €91 million to €1,774 million during the next 10 years, showing the significant impact of this new business.

This exercise helps us to understand the business impact of using OTA for updates of the embedded head units. Per car, with the estimation made in this section and without any roaming traffic, the cost starts from €10 for the five-year update in the conservative scenario or more than €100 in the non-conservative scenario. This is not a negligible cost, and it will be the crucial element to evaluate an OTA-based software update solution versus traditional upgrades. The customer is foreseen to not be willing to pay, especially for mandatory updates linked to safety, security, or bug corrections. Only in the case of new features added in an update will the customer viewpoint about the update's value possibly change.

The total amount paid for data traffic 10 years from now could be between €288 million (CS) and €1,774 million (NCS), forcing a huge negotiation with mobile operators in order to obtain lower prices that can make OTA a viable business case for OEMs. The data used for OTA, and the associated negotiation leverage, will also drive down the cost for data for other

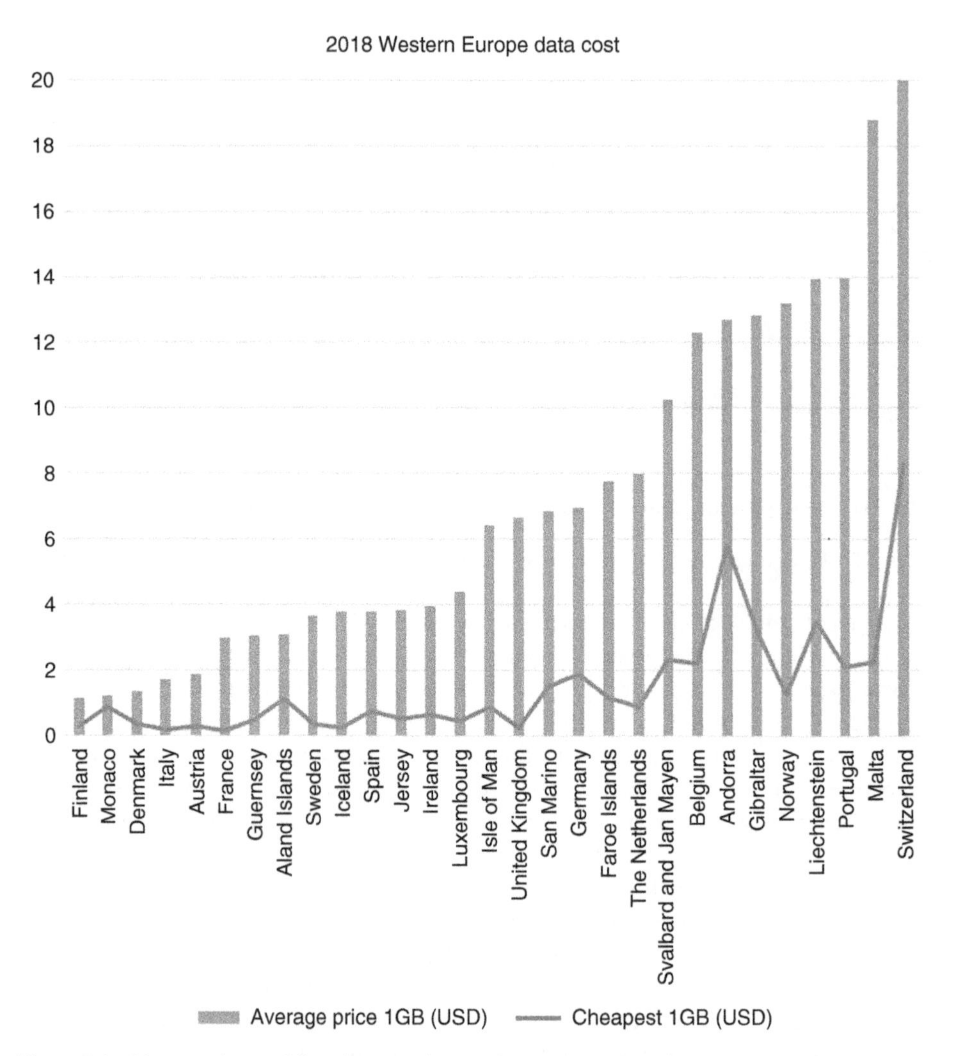

Figure 2.6 Mass market mobile price per gigabyte in the countries of Western Europe.

services, raising the possibility of increasing data offloading from embedded architecture to the cloud.

2.6 Business Model Analysis of 5G V2X Technical Components

The detailed work on defining technical components can be found in the following book chapters. Here, we will analyze technical components from a business model perspective to understand how they can impact the possibilities for creating innovative business models. For each of the technological components, the following items are discussed:

- *Description*: Brief summary of the technology.

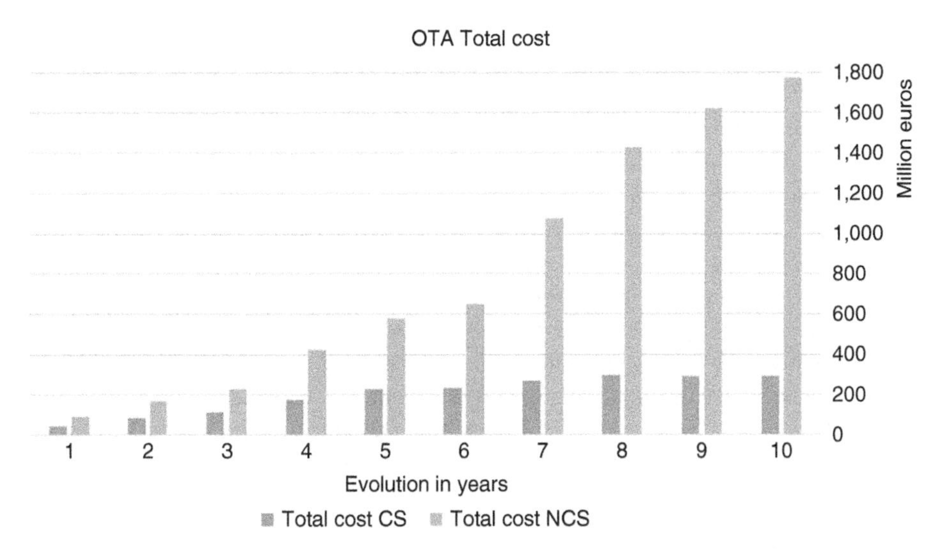

Figure 2.7 Overall cost estimation in the next 10 years for over-the-air updates.

- o *Disruption*: *E* for evolution; *R* for revolution, from existing 3GPP releases (Release 14 and onward).
- o *Value Provided*: Examples of value that can be attained from this technological component. This would be the basis for the creation of new services.
- o *Requirements/Challenges*: Negative factors from a business perspective.
- o *Parties involved in the value chain*: Parties are involved in realizing the value.
- o *Customer*: Who the customers are that benefit from the value created.
- o *Customer benefit (value)*: The perceived value for each customer.
- o *Potential 5G business model element impact*: What element(s) of the business model are impacted by the technical component.

The technical components have been grouped into broader categories where possible, i.e. where technical components do not differentiate the result of the analysis, and where the technical components are related to the same area.

2.6.1 Positioning

As highlighted by the ecosystem (for instance, see [15]), 4G positioning techniques may not be sufficient to serve new uses cases: "In the case where relative positioning information is transposed into absolute 3D positioning information using LTE positioning technologies, the resulting accuracy would not be sufficient for the proposed use case." To satisfy new use cases, vehicle and vulnerable road user (VRU) positions must be provided with higher accuracy to define novel positioning solutions – namely, the following technical components:

- o Trajectory prediction with channel bias compensation and tracking
- o Tracking a vehicle's position and orientation with a single base station in the downlink
- o Beam-based V2X positioning

- Data-aided beam-based V2N positioning
- Enhanced assistance messaging scheme for GNSS and observed time difference of arrival (OTDOA) positioning
- Multi-array 5G V2V relative positioning

All of these technical components aim to provide a more accurate position of the vehicle user equipment (UE) or other types of UE and can contribute in the same way to the business model.

Note that in an automotive context, cars are equipped with onboard global positioning system (GPS) to determine their positions with GPS accuracy. So, the technical components may appear useless, at least from the OEM perspective, because the position is already known from on-board GPS. However, some of the technical components (TCs) not only provide an instantaneous position but also provide trajectory prediction and, more importantly, collision prediction. Such technical components can have value for the OEM.

As a result, we can split the positioning-related TCs into two groups:

- TC providing radio-based positioning
- TC providing trajectory prediction

The characterization and assessment for these TCs are listed in Tables 2.2 and 2.3, and details can be found in Chapter 8.

2.6.2 V2X Radio Design

The TCs proposed in 5GCAR for the radio interface can be grouped in two categories: the infrastructure-based and the sidelink-based solutions. The first group is defined where the network infrastructure is involved. In this group, an MNO will definitely be needed to provide the service. The second group comprises direct communication without network involvement (out-of-coverage) or V2V, V2I, and V2P (in-coverage) over the network. Accordingly, two subgroups have been provided: out-of-coverage sidelink TCs and in-coverage sidelink TCs. In the first subgroup, the OEM is a final customer, and no MNO is involved in the value chain. It should be pointed out that the TCs in the first subgroup can be extended to be applied in the case of MNO involvement as well. However, in the second subgroup, a network service provider is needed to enhance the sidelink characteristics; this may become interesting for a message passing (MP), assuming a competitive advantage compared to another MP using the same OEM vehicles.

Some of the following TCs are discussed in more detail in Chapter 5.

2.6.2.1 Predictor Antenna

A predictor antenna is an extra vehicle antenna that is added to V2X antennas for improving channel estimation to the transmitter side at high speed. Its characterization and assessment are shown in Table 2.4. For recent 5GCAR results on this TC, cf. [16, 17].

2.6.2.2 Beam-Forming

For downloading HD maps or conveying traffic information relevant to a local group of vehicles, x-cast schemes provide efficient use of radio resources. mmWave scenarios are a clear focal point, but x-cast beamforming is also useful below 6 GHz. The characterization and assessment are listed in Table 2.5 and detailed in Section 5.1.

Table 2.2 Positioning.

Technological component	Positioning
Description	Methods to provide accurate positioning of vehicle UE and other UE types
Disruption	E: Radio-based positioning. No need for GPS; the position can be computed by the network or by the UE.
Value provided	Accurate position
Requirements/challenges	Communication-based positioning relies on using a specific communication feature (either specific pilots or messaging) that must be enabled by the equipment. Higher density of antennas and higher carrier frequencies to be used in order to get better accuracy.
Parties involved in the value chain	• Telecom vendor • Tier 1 • Possibly a new actor providing positioning support over the top (database of reference geo-points; computer power, and knowledge to compute triangulation, etc.)
Customer	• Third party (insurance company), service company (e.g. guide, hotel chain) • Public authority may want access to an accurate position (e.g. emergency services)
Customer benefit (value)	Accurate and reliable position opens the door to new services
Potential 5G business model element impact	Possibly disruptive service enabled by accurate and reliable position information

Table 2.3 Positioning and trajectory prediction.

Technological component	Positioning and trajectory prediction
Description	Methods to provide accurate positioning of vehicle UEs or other types of UEs and to predict the trajectory and thus possibility of collision
Disruption	R: Radio-based positioning. Learning techniques to predict the trajectory.
Value Provided	Accurate position and trajectory prediction
Requirements/challenges	Reliability of trajectory prediction; collision detection false alarm or misdetection
Parties involved in the value chain	• Telecom vendor • Third party doing trajectory prediction, or the telecom vendor
Customer	• OEM • Third party (insurance company) • Public authority may want to access accurate position and trajectory history/predictions Mobility provider; this can be an added off-board service independent of the elements provided by on-board sensors
Customer benefit (value)	Trajectory prediction to anticipate collisions or accidents
Potential 5G business model element impact	Possibly disruptive service

Table 2.4 Predictor antenna.

Technological component	Predictor antenna
Description	The predictor antenna concept is used to obtain accurate channel state information at the transmitter side, which is the key for robust and spectrally efficient links.
Disruption	E
Value provided	Adaptive massive multiple input multiple output (M-MIMO) works for very fast-moving connected vehicles.
Requirements/challenges	Antenna cost and the size of the antenna array are always important restrictions for successful deployment in the automotive industry.
Parties involved in the value chain	• Telecom vendor, as the provider of the antenna • Tier 1, as the integrator of the overall communication module to be integrated in the car • OEM
Customer	OEMs, to improve the performance of the connected vehicle while moving
Customer benefit (value)	Communication performance will be better for fast-moving connected vehicles.
Potential 5G business model element impact	None; normal linear business. A new type of antenna will be provided by the telecom vendor, integrated by the Tier 1, and finally included in the vehicle.

Table 2.5 Beam-forming.

Technological components	Beam-forming
Description	Different techniques to improve the management of the beam forming in, for example, mmWave bands from the infrastructure side to optimize the performance of the system
Disruption	E
Value Provided	Improves interference of the system and energy "usage"
Requirements/challenges	Integration of mmWave antennas in the car is under investigation before industry adoption. The spectrum band is not yet harmonized; auction plans may arrive in a few years.
Parties involved in the value chain	• Telco vendors • MNO
Customer	MNO
Customer benefit (value)	The performance of the mmWave communication system will be improved.
Potential 5G business model element impact	Reduced. So far, no clear use cases are identified for mmWave in the V2X domain. Today, mmWave is initially foreseen to cover the MNOs' last mile and increase system capacity.

Table 2.6 Efficiency.

Technological components	Efficiency
Description	Multiplexing and pre-emption mechanisms are used to improve power, latency, and link efficiency.
Disruption	E
Value provided	Improve spectral efficiency and throughput.
Requirements/challenges	No specific challenges
Parties involved in the value chain	• Telecom vendor • MNO
Customer	MNO
Customer benefit (value)	The performance of the system is better in terms of bandwidth available for the final customer.
Potential 5G business model element impact	No important impact. This is the normal linear business model. It may be a competitive advantage for one MNO compared to another.

2.6.2.3 Efficiency

For improving spectral efficiency and supporting fast setups of V2X services (e.g. emergency services), pre-emption and improved pilot channel handling can be introduced to 5G V2X networks. Their characterization and assessment are listed in Table 2.6, and an example is given in Section 5.2.

2.6.2.4 Reliability

For future cars, we need to boost radio reliability, and several mechanisms can be conceived. Their characterization and assessment are listed in Table 2.7 and detailed in Section 5.3.

2.6.2.5 Sidelink Out of Coverage

For out-of-coverage, partial coverage, and the intended non-use of cellular networks for sidelink coordination and management, one must pay attention to improving the sidelink mechanism for improved QoS and reliability at the end for an ad hoc type operation. Several improvement options exist. Their characterization and assessment are listed in Table 2.8, and synchronization aspects are discussed in Section 5.2.

2.6.2.6 Sidelink in Coverage

For in-coverage, partial coverage, and the intended use of cellular networks, one must pay attention to improving the sidelink mechanism for improved QoS and reliability at the end. Several improvement options exist. Their characterization and assessment are listed in Table 2.9, and details can be found in Section 5.3.

2.6.3 Network Procedures

Network procedures are an important aspect of managing a network of connected devices. They describe how various network services will be managed and can have a significant

Table 2.7 Reliability.

Technological components	Reliability
Description	Increase the reliability of the system by using alternative links (V2V as well as V2I) or by changing the code modulation to look for a trade-off between latency and reliability or enhanced control channel reliability.
Disruption	E
Value provided	Maintain reliability for URLLC traffic by exploiting link diversity through UE collaboration. Enable QoS guarantee for URLLC services in V2N communication.
Requirements/challenges	No specific challenges
Parties involved in the value chain	• Telecom vendor • Tier 1 • MNO • OEM
Customer	Mobility provider. The reliability of network communications will benefit the mobility provider and can suppose a competitive advantage between two MPs using the same OEM cars. The added value is off-board, so the customer is not the OEM but the mobility provider.
Customer benefit (value)	Reliable communications for automated driving
Potential 5G business model element impact	An element that may enhance the service of the MP compared to an OEM directly. Change in the linear business as it is defined today.

Table 2.8 Out-of-coverage sidelink.

Technological components	Out-of-coverage sidelink
Description	Enhance sidelink characteristics in out-of-coverage situations. Optimize the synchronization signals, demodulation reference signal (DMRS), distributed resource selection strategy, multiple input multiple output (MIMO) design, and full-duplex operation as well as increase robustness against adverse propagation conditions.
Disruption	E
Value provided	Provide essential sidelink functions and/or better sidelink performance (throughput, resource efficiency, and/or reliability).
Requirements/challenges	No specific challenges
Parties involved in the value chain	• Telco vendor • Tier 1 • OEM
Customer	OEM: Any improvement of sidelink performance without the need for the network will be interesting for the OEM.
Customer benefit (value)	Sidelink may act as an off-board sensor regardless of the network deployment, increasing safety and comfort.
Potential 5G business model element impact	None; linear value chain with a feature provided by a telco vendor and integrated by the OEM through a Tier 1.

Table 2.9 In-coverage sidelink.

Technological components	In-coverage sidelink
Description	Enhance the sidelink characteristics in coverage situations. Optimize sidelink channel interference, discovery signaling overhead, and resource allocation thanks to network-assisted mechanisms; and increase reliability through the combination of sidelink with network links.
Disruption	E
Value provided	Essential sidelink functions and/or better sidelink performance (throughput and reliability) and reliability
Requirements/challenges	Interest from MNOs, due to the complexity of the function and the lack of interest from OEMs to pay for a service available in unlicensed frequency
Parties involved in the value chain	• Telco vendor • Tier 1 • OEM • MNO
Customer	MP: Any improvement of sidelink performance linked to cellular network availability will become a service from the MNO to be sold to the MP.
Customer benefit (value)	Sidelink may act as an off-board sensor regardless of network deployment, increasing safety and comfort.
Potential 5G business model element impact	This service may suppose a difference for an MP compared to another MP even if they use cars from the same OEM.

impact on the effectiveness of service delivery, which in a vehicle environment may be particularly critical, such as in terms of introducing latency. Some of the TCs identified by 5GCAR can have a major impact on the overall service quality experienced by the end customer. In this analysis, some of the TCs have been grouped together, as they are related and target the same customers with similar values. More technical descriptions are available, see e.g. [24], and additional technical background for the following network-related TCs can be found in Chapters 6 and 7.

2.6.3.1 Local Standalone Network Procedures

These are characterized by integrating road-side units (RSUs) in the 5G network architecture for sidelink and Uu (air interface between base station and user equipment) link operations between vehicles, RSUs, and 5G network. Their characterization and assessment are listed in Table 2.10.

2.6.3.2 Network Service Relationship Enhancement

This is characterized by managing local traffic in a decentralized manner with respect to scheduling and V2X service negotiations. Its characterization and assessment are listed in Table 2.11.

Table 2.10 Local standalone network procedure.

Technological components	Local standalone network procedure
Description	• Integrate RSU inside the 5G V2X architecture, acting as UE or as local BS (RSU-based smart zone) • Map sidelink (SL) on 5G RAN using Uu link. Radio protocol is simplified, allowing only local routing in order to support SL like connection. • Assist UE to set up SL (3GPP SL, or PC5) link in unicast (preferred) or multicast mode more efficiently (improving setup time and QoS). This assistance will be performed by a server on the gNB (next generation node B) side.
Disruption	• R: A new element (active RSU) is added to provide road safety, efficient road services, and network infrastructure to be deployed to support a direct connection between vehicles • E: If the TC is not available, V2V link establishment and regular mode should continue to work in a degraded way.
Value provided	Improvement of D2D direct communication for better QoS, availability, and enhancement of network coverage (macrocells), offering a multi-MNO solution; and offloading macro-network, SL, and Uu multi-connectivity for high-reliability and/or high-data-rate V2V communication. Fast application-aware setup of unicast SL.
Requirements/challenges	• Deployment of many RSU(s) with smart zoning capability all along the road • Adaptation to network infrastructure for that specific purpose (local routing, simplification)
Parties involved in the value chain	• RSU provider, possibly embedding Tier 1 technology • Road operator that is deploying a smart zone • MNOs connecting RSU smart zones and deploying features (option 1) • Telecom equipment provider (option 2) • OEM • Tier 1 (implementing specific UE behavior)
Customer	• The customer will depend on the service provided. The customer will be different from the user (this remark is general to road safety/road efficiency). • The customer can be the OEM, which can pay the MNO for this specific TC. • The road operator can deploy simplified BS to support TCs; in that case, the customer is the road operator for a telecom equipment provider (option 2).
Customer benefit (value)	• Road safety enhancement as well as traffic efficiency • Better connectivity for cooperative perception use cases
Potential 5G business model element impact	• The customer of the MNO will probably be a road operator instead of a car driver or car OEM (option 1). • The telecom equipment (option 2) provider can be directly integrated in the value chain as a local road service provider (in a restricted area).

Table 2.11 Network service relationship enhancement.

Technological components	Network service relationship enhancement
Description	• Allow guaranteed transfer of messages and files for a relevant area to a targeted UE. Location-aware scheduling will take into account the network condition and vehicular context to orchestrate messages (files) transfer demands to optimize resource allocation. • Allow transmitting specific service requests with specific requirements (service descriptor) toward network control. Network control can send back information toward the service about the status of the requested service execution, as well as the network condition to adapt the service.
Disruption	E: This TC will run on a specific server (MEC server). This requires an adaptation of the network architecture, an additional cost for this function.
Value provided	• Guarantee of message/file transfer in a relevant area, in the right time • Optimization of network resource allocation • Adapting the service to network conditions
Requirements/challenges	• Deployment of distributed servers. • Being able to retrieve and correlate in real time information from the network (QoS, occupancy) and ITS application service-level specification (SLS)
Parties involved in the value chain	• ITS application provider (car OEM consortium) • MNO that will deploy these TCs
Customer	• The customer will be different from the user (this remark is general to road safety/road efficiency); the car driver probably will not pay for that TC. • The customer can be ITS application provider, which can pay the MNO for this specific TC. • The customer can be nobody, if the TC is seen as a way for the MNO to optimize resource allocation,
Customer benefit (value)	The ITS application provider has a guarantee for service delivery potentially matching the SLA; they can negotiate and monitor service.
Potential 5G business model element impact	• This TC can give additional value for network usage, but it will bring some constraints on the ITS application provider side. • This TC can enable the existence of an independent actor in charge of the service (ITS application provider).

2.6.3.3 Multi-Operator Solutions for V2X Communications

These are characterized by managing national and international roaming conditions due to non-coverage and/or congested areas. Their characterization and assessment are listed in Table 2.12.

2.6.3.4 Network Orchestration and Management

This is characterized by mapping services or network functions dynamically to physical network resources. This is a classical, well-known 5G network feature that can also be useful

Table 2.12 Multi-operator solutions for V2X communications.

Technological components	Multi operator solutions for V2X communications
Description	Address the problem of fast roaming/handover from MNO A to a MNO B: for instance, when the car is leaving the coverage area of MNO A and entering the coverage area of MNO B. In a proposed solution, one UE is able to listen to two radio control channels in the case of shared frequencies, or to two radio frequencies. The UE can register with several operators in a given area.
Disruption	R: High at UE side when the UE has to be populated with two RF modules. High for network management in the case of shared frequencies. High in terms of the deployment scenario if the TC is used as a way to split a geographical area between several operators.
Value provided	• Improvement of network availability and system resilience when coverage is not 100% in a given geographical area. This TC should reduce cost deployment when several MNOs are willing to cover the same area. • Reduction of deployment costs when the TC is used to share a given geographical area among several operators. • Solution for a cross-administrative domain (for instance, cross border).
Requirements/challenges	Deploying this feature on the UE side together with a deployment scenario: for instance, a geographical split will not work properly if a UE doesn't implement the TC.
Parties involved in the value chain	• MNO(s) • OEM
Customer	• OEM • Inter-MNO roaming market (requires roaming agreements between MNOs)
Customer benefit (value)	• Car OEM: better service level (coverage, latency, availability, seamless service, and so on). • MNO will offer a connectivity service without deploying an entire network (depending on the coverage scenario).
Potential 5G business model element impact	• This TC can allow sharing network deployment costs among several MNOs. • A new type of competition between MNOs based on geographical coverage as well as the percentage of coverage in a given area.

to implement V2X services and networking functions. Its characterization and assessment are listed in Table 2.13.

2.6.4 End-to-End Security

Security and privacy are of utmost importance for any connected service, including V2X, ITS, and connected and automated mobility services. Security features add substantial costs to an end-to-end solution, thereby inhibiting the adoption of the related services.

Table 2.13 Network orchestration and management.

Technological components	Network orchestration and management
Description	A set of processes to automate the behavior of the network in order to coordinate hardware and software elements to further support applications and services. The idea is to separate network services from network components, allowing automated network configuration as per the service specifications. Network orchestration requires automation in the operation that can be achieved in different steps: • The introduction of network function virtualization (NFV) reduces costs by reducing the amount of dedicated hardware and bringing agility in the time and effort required to launch new applications. • To handle the mix of physical network functions (PNFs) and virtual network functions (VNF), automation of the network should be taken to the service level, thus increasing efficiency. • Add cloud-native applications to enable continuous integration and delivery. At this point, operators can realize the full potential of the new services and address new revenues.
Disruption	E
Value provided	Optimization of the network resources
Requirements/challenges	It is not an easy migration process since it effectively changes the way a network is managed. To realize the orchestration of 5G networks, operators need to start their automation journey now.
Parties involved in the value chain	Telecom equipment vendors and MNOs
Customer	MNOs
Customer benefit (value)	Improved network management: ability to support varied applications while reducing operational costs
Potential 5G business model element impact	The potential is large, both in savings and in efficiently supporting novel services

5GCAR has proposed a solution to reduce the cost and transmission overhead by using a scheme (for certain suitable applications) where only the initial messages in a communication are signed and permissions are verified, and further messages are only encrypted (Table 2.14).

2.6.5 Edge Computing Enhancements

The availability of edge computing capabilities is one of the really innovative new technologies in 5G, which has a large potential impact on automotive use cases. Edge computing can move services running in the back end (cloud) much closer to the actual UE (vehicle), thereby greatly reducing latency and opening up new ways of distributing functionality

Table 2.14 End-to-end security.

Technological components	End-to-end security
Value provided	Privacy and security in communication are absolute requirements in a V2X system because the entire system is based on trust between various stakeholders that are not previously known to each other. For the end user, it is critical that the V2X service can be trusted; therefore, there is an inherent value in network security, which is realized by the trust and use of the service by the end customer. 5GCAR's proposed innovative scheme of only signing the initial messages in a communication session and then using encryption for further messages has the potential to reduce the cost for end-to-end security.
Requirements/challenges	Maintaining a high level of security typically creates overhead in the communication protocol, leading to longer latency in communications and/or more data to be transmitted. This in turn leads to higher costs for data transmission (if a licensed spectrum is utilized) and increased consumption of shared transmission resources due to the security overhead. Scalability of encryption may be a challenge if the number of keys (driven by the number of UEs) becomes very high. There is a cost for storing keys as well as a cost to generate keys (see Section 3.1.2).
Parties involved in the value chain	• End customer • UE manufacturer, including universal subscriber identity module (USIM) • MNO • V2X service provider (can be the car OEM)
Customer	• End customer • V2X service provider
Customer benefit (value)	• End customer – Access to a trusted V2X service • V2X service provider – Trusted V2X service that is possible to sell • V2X service provider – Protection against liability claims
Potential 5G business model element impact	• Additional cost to extend 5G cryptographic/security functions in the network, to cover the specific V2X case (key management function) • Additional costs due to overhead if using a licensed spectrum (Uu)

between vehicle and back end. It can potentially even allow for functionality that today runs in the in-vehicle systems to be executed at the edge of the back-end infrastructure instead, introducing new ways of launching services and functionality.

TCs described in this section include edge computing in mmWave cellular V2X networks (Table 2.15).

Table 2.15 Edge computing enhancements.

Technological components	Edge computing enhancements
Description	Edge computing is a core new technology concept in 5G, and several benefits for V2X use cases can be seen: • Lane merge: the traffic coordinator can be located close to the intersection. • Cooperative perception based on see-through. • VRU protection.
Disruption	R
Value provided	Edge computing enhancements provide value in enabling certain use cases with very low-latency requirements. They also provide value in being a part of the total off-board computing resource necessary to provide a V2X service. The low latency and high computing capacity offered by edge computing also have the potential to create new possibilities for car OEMs to offload in-vehicle functionality to the edge compute platform, where certain functions can be developed. Typical advantages can be seen in functions requiring data from multiple vehicles or RSUs in the same limited geographic area.
Requirements/ challenges	Handover between base stations connected to different MEC-local data centers. The applications and their internal state need to be transferred to the next data center with the shortest delay possible. This should ideally be performed before the handover is completed. Handover between networks (roaming), and inter-operator roaming (between MNOs from different countries). Both technical and commercial challenges. May look different depending on what type of service provider supplies the MEC service. MNOs are typically restricted by national borders, due to spectrum licenses, whereas other over-the-top providers may be more international in how they can provide a MEC service.
Parties involved in the value chain	• Car OEM • MNO • Network equipment vendor • Cloud service provider • Data center provider
Customer	• Car OEM • MNO
Customer benefit (value)	• Car OEM: Access to additional computing resources (low-latency functions) • Car OEM: Ability to create new functions (low latency, multiple local data sources) • MNO: Frees up resources in the transport link toward the central data center
Potential 5G business model element impact	The business model for MEC is so far very uncertain. MNOs see this as an opportunity to increase the value provided to customers and also operate (sometimes also own) important parts of the required infrastructure. However, MNOs are usually national entities, whereas the required services to be provided by MECs are international and need to function across borders. This can open the market for cloud or data center providers that already operate international service platforms. A MEC service offer can be seen as a logical extension of more centralized computing platforms. Network equipment vendors may also take a role in providing MEC services, although this may be less likely due to their existing relationship as suppliers to MNOs. They would likely not want to disturb that existing business.

2.6.6 Summary

The TCs identified by the 5GCAR project can be divided into the following major categories:

- Positioning
- V2X radio design
- Network procedure
- End-to-end security
- Edge computing enhancements

Of these categories, some contain TCs with the potential to "revolutionize," i.e. strongly affect, existing business models or create new business opportunities with major revenue potential (positioning, edge computing enhancements).

Other categories contain TCs that will enhance existing business models and potentially add new value to existing services by enhancing capabilities or adding new features to an existing base technology (V2X radio design, end-to-end security, positioning, network procedure).

It is not possible to predict in detail how the value of TCs will be realized in actual business cases. The value will likely be created on the service level, where multiple TCs may be included as parts of a single service. The value is determined by negotiations by the actors in the value chain and may be very different depending on the use case, the business model of each actor in the value chain, time, the regulatory landscape, and many other factors. The analysis in this section, however, indicates that most TCs will create value for the vendor or integrator of the technologies. For instance, a TC will increase the cost and, consequently, the value of the car, the module in the car, or the network. These TCs will thus have an impact on the pricing of the car, connectivity, and/or network equipment. Some presented TCs will open the ecosystem for new stakeholders: ITS application providers, localization providers, etc. Finally, some TCs may change the place of stakeholder in the ecosystem (MNO, OEM). The value chain/ecosystem is very immature, and we will most likely see many different attempts at creating new business models from various stakeholders.

2.7 Conclusions

In this chapter, we have introduced important CAD and CRU services and technical concepts. These concepts will be further elaborated in the following chapters of the book. In Chapter 10, we summarize the status of these TCs and give an outlook on future development within CAD and CRU services and C-V2X.

CAD and CRU services provide social benefits, because they will create a positive impact in different domains: traffic efficiency, CO_2 reduction, and safety improvements. Thanks to V2X communications, traffic information shared in urban environments will become more precise and up to date, thus helping improve traffic efficiency and reducing CO_2 emissions. The level of urbanization in the EU is greater than 75% of the population and is expected to be over 80% in 2020 [19]. A reduction of 700,000 tons of CO_2 is expected in cities each year thanks to V2X, valued at over €22 million per year. There are also substantial benefits for rural areas, where the arrival of direct short-range communications will enable V2X adoption for some use cases before the infrastructure deployment. For both environments, the information shared between vehicles will lead to a 5% reduction in road accidents in 2025, with an estimated value of €4.1 billion.

Thanks to the telecommunication ecosystem, C-V2X can be integrated into smartphones, enabling V2P communications and providing protection to VRUs. In particular, potential collisions with pedestrians can be predicted early enough, even in non-line-of-sight (NLOS) conditions, and avoided, hence helping to reduce pedestrian fatalities. In 2016, more than 5000 pedestrians were killed in road accidents (21% of all road fatalities) in the European Union alone [20].

C-V2X, compared to other technologies, has defined a clear roadmap to serve basic safety features and automated services. It can provide better spectrum usage and enable other services beyond safety, i.e. comfort and efficiency. On top of meeting a set of requirements needed for C-V2X, e.g. low latency, availability, and high reliability, there is still work to be done related to the cross-MNOs, cross-OEMs, or cross-border communication aspects. These elements have high importance and will be studied in the coming 3GPP releases.

Roadmaps are impacted in their execution by regional decisions, especially when regulations and the spectrum are under discussion. Some regions have firmly decided to implement a clear V2X strategy, such as China and USA, while other regions are still evaluating various options. The Delegated Act and the program clean air for Europe (CAFÉ) are important legal regulations that will modulate the V2X automotive use case roadmap in Europe over the next decade.

The roadmap will also be conditioned by network deployment. A number of approaches are identified to improve coverage expansion as fast as possible: e.g. by contributing to rolling out the infrastructure deployment, facilitating it with proper regulations, prioritizing the network deployment for highways or main roads where more advanced C-V2X use cases are anticipated in the initial phase and benefits are expected to be higher, and introducing scenario-specific features like those for rural areas where a huge effort is required to provide full coverage; there, we can aim for very basic day-1 type features using unlicensed spectrum and ad hoc mode.

As presented in the 5G PPP Automotive WG white paper, the primary takeaway is that opening up the possibility for network and infrastructure sharing has the evident effect of improving return on the network investment, even if only passive infrastructure elements are shared between operators: e.g. space, masts, power generators, and air conditioning equipment. In spite of the uncertainty in the timeline for advanced C-V2X use cases, other business opportunities are available in the short term. On one hand, the current security strategy proposed in the Delegated Act (rejected in July 2019) may lead to an important new market for certificate management for telecom companies. If this security policy is finally confirmed, market competition could be the only way to reduce prices for pseudonym adoption and mitigate its deterrent effect in the OEM adoption decision.

On the other hand, OTA is a service where a business case may be profitable for both parties – automotive and telecom stakeholders – but it needs a new business model definition to be viable. In its current status, connectivity is unaffordable as an OTA solution. The new on-board electronic and electric architecture of next-generation vehicles will be able to manage a huge quantity of data. As an example, GM [18] has announced processing power capacity up to 4.5 TB h^{-1} in its new electrical architecture. In a connected vehicle paradigm, more data exchanged on-board will naturally mean an increase in data off-loading needs toward the environment. This confirms the business opportunity for the telecom market in the coming years. In the latest report from Ericsson [23], whereas for the smartphone market, consumers are willing to pay up to 20% more for the arrival of 5G, opinions in the telecom market claim that the connectivity business opportunities of connected homes and cars will not happen if the current model of paying for a bucket of gigabytes is maintained.

5GCAR has proposed a list of significant TCs applied in different axes (radio interface, network architecture, security, etc.). They have been analyzed from a business perspective, and the result shows that some of them can create value for both telecom and automotive stakeholders, and others may lead to changes in the ITS ecosystem and value chain.

References

1 Martínez de Aragón, B., Alonso-Zarate, J., and Laya, A. (2018). How connectivity is transforming the automotive ecosystem. *Internet Technology Letters* 1 (1): e14.

2 5GAA. (2017). Timeline for deployment of LTE-V2X.

3 5G PPP Automotive Working Group. (2019). Business feasibility study for 5G V2X deployment.

4 GSMA. (2016). Remote provisioning architecture for embedded UICC technical specification. SGP.02 v3.1.

5 SAE. (2014). Taxonomy and definitions for terms related to on-road motor vehicle automated driving systems. https://www.sae.org/standards/content/j3016_201401.

6 Body of European Regulators for Electronic Communications (BEREC). (2018). BEREC report on infrastructure sharing. Document number: BoR (18) 116.

7 Mobile World Live. (2019). Vodafone, Orange extend Spanish pact to 5G.

8 Capacity. (2019). Poland telecoms market.

9 Council of the European Union. (2019). EN 9207/19 (OR. En) – Outcome of the council meeting – 3688th Council meeting – Foreign Affairs.

10 European Automobile Manufacturers Association (ACEA). (2019). Key figures: Europe leads the way with clean, efficient, high quality motor vehicles that keep Europe moving.

11 5GCAR Deliverable 4.2. (2019). 5GCAR final design and evaluation of the 5G V2X system level architecture and security framework.

12 Cable. (2018). Worldwide mobile data pricing: The cost of 1GB of mobile data in 230 countries.

13 ARCEP. (2017). Évolution des prix des services de communications électroniques.

14 CNMC. (2017). Informe económico sectorial de las telecomunicaciones y el audiovisual.

15 3GPP. (2018). Study on positioning use cases; Stage 1. TR 22.872, v16.1.0.

16 Guo, H., Makki, B., and Svensson, T. Rate adaptation in predictor antenna systems. *IEEE Wireless Communication Letters* 9 (4): 448–451.

17 Phan-Huy, D.-T., Wesemann, S., Björsell, J., and Sternad, M. (2018). Adaptive massive MIMO for fast moving connected vehicles: It will work with predictor antennas! In: *Proceedings of Workshop on Smart Antennas (WSA).*

18 Lienert, P. (2019). GM digital vehicle platform debuts, enables adoption of future technologies: Installed on newly-unveiled Cadillac CT5, with rollout to most GM vehicles globally by 2023. *GM Corporate Newsroom.*

19 European Commission. (2014). Identification and quantification of key socio-economic data to support strategic planning for the introduction of 5G in Europe. Final report.

20 European Commission Road Safety Authority. (2018). Traffic safety basic facts on pedestrians.

21 Arcep. (2019). Le suivi des obligations de déploiements des opérateurs.

22 Ericsson. (2019). 5G consumer potential.

23 Condoluci, M., Gallo, L., Mussot, L. et al. (2019). 5G V2X-system level architecture of 5GCAR project. *MDPI Future Internet* 11 (10): 217.

3

Standardization and Regulation

Efstathios Katranaras[1], Markus Dillinger[2], Taimoor Abbas[3], Remi Theillaud[4], Jose Leon Calvo[5], and Yunpeng Zang[5]

[1] *Sequans Communications, France*
[2] *Huawei German Research Center, Germany*
[3] *Huawei Technologies, Sweden*
[4] *Marben Products, France*
[5] *Ericsson Research, Germany*

Vehicle-to-everything (V2X) technologies are generally classified into short-range or wide-area technologies. Wide-area technologies (i.e. cellular networks over the licensed spectrum) are best suited for vehicle-to-network service requirements and, as the deployment of intelligent transport systems (ITS) is expected to be affected by cost, such cell technologies should provide deployment benefits, given the available 4G cellular network with ongoing evolution to 5G. A crucial advantage of cellular V2X (C-V2X) is also its backward compatibility. These C-V2X standards serve as a baseline for connected automated driving (CAD) and connected road user (CRU) services.

To maximize market uptake of C-V2X, joint standardization efforts between automotive and telecom sectors need to take place. Such an effort has been undertaken, for example, by the 5GAA, which has helped to define and test the protocols/specifications of C-V2X technology by bringing together diverse players from different industries (automotive OEMs, vendors, operators, and regulatory).

Standardization for V2X communication guarantees interoperability of implementations from diverse stakeholders. It supports regulations and legislation and also enlarges markets. This chapter covers the standardization aspects and status of the V2X landscape and outlines the relevant standardization bodies which work on V2X communication. Section 3.1 presents an overview of the standardization process and the primary bodies involved in V2X communications standardization. Then, Section 3.2 discusses the current regulatory arrangements for ITS in the different regions of the world as well as aspects and considerations of the radio spectrum dedicated to ITS. Next, in Section 3.3 we give a brief history and overview of the available communication solutions and an overview of higher-layer ITS standardization on a regional level, with special focus on the standardization status of C-V2X in 3GPP. These details include a brief history of 4G solutions, 5G progress (i.e. Releases 15 and 16) on the long-term evolution (LTE) and new radio (NR) technology front, and plans for future releases. Finally, in Section 3.4, we outline the Institute of Electric

Cellular V2X for Connected Automated Driving, First Edition.
Edited by Mikael Fallgren, Markus Dillinger, Toktam Mahmoodi, and Tommy Svensson.

and Electronics Engineers (IEEE) and European Telecommunications Standards Institute (ETSI) standards for securing V2X messages and ensuring privacy. More specifically, the protocols used by V2X end entities (e.g. vehicles and road stations) to communicate with the public key infrastructure (PKI) authorities are covered, as well as certificates and the secured messages format.

3.1 Standardization Process Overview

This section discusses general aspects of the standardization process and provides a brief description of the primary bodies involved in V2X communications standardization. A deeper view of the 3rd Generation Partnership Project (3GPP) standards organization is then presented to serve as an example of the various procedures required for communications standardization.

3.1.1 General Aspects

Standards, in the telecommunication arena, are essentially "laws" for telecom networks and equipment. They have the form of codified public agreements and specifications models that make wide-area communications possible. In general, standardization procedures within a standards developing organization (SDO) embrace the following values that ensure efficiency and longevity as well as the fair, safe, high-quality development of communication solutions:

- Work is coordinated to avoid multiple standards for the same function.
- Standards are kept maintained as long as there is a minimal level of use.
- Democracy is ensured for any group participating in the SDO.
- The process includes the identification of intellectual property rights (IPRs).
- The impact on society and the environment is always considered.

It is important to mention here that the specifications model is a guideline to be followed by implementation entities; an actual implementation may depart from this guideline (due to implementation constraints) as long as it behaves according to the specifications model.

Since V2X standards consist of a large number of base and test specifications from diverse domains ranging from radio and protocols to security and applications, international standards are paramount for the deployment of V2X communication systems. Such standards deliver specifications to guarantee the interconnection of V2X systems and components as well as implementations' interoperability. They also engender customer trust in services and products and thus lower development costs, and they allow the creation of larger markets with increased and fair competition among vendors instead of only proprietary systems.

3.1.2 Standardization and Regulation Bodies Relevant to ITS Specifications

ITS standardization is primarily a matter of official SDOs, where V2X communication is only one of multiple aspects of their standardization efforts. Multiple topics are related to ITS, covering wide areas of interest ranging from the telecommunications industry to

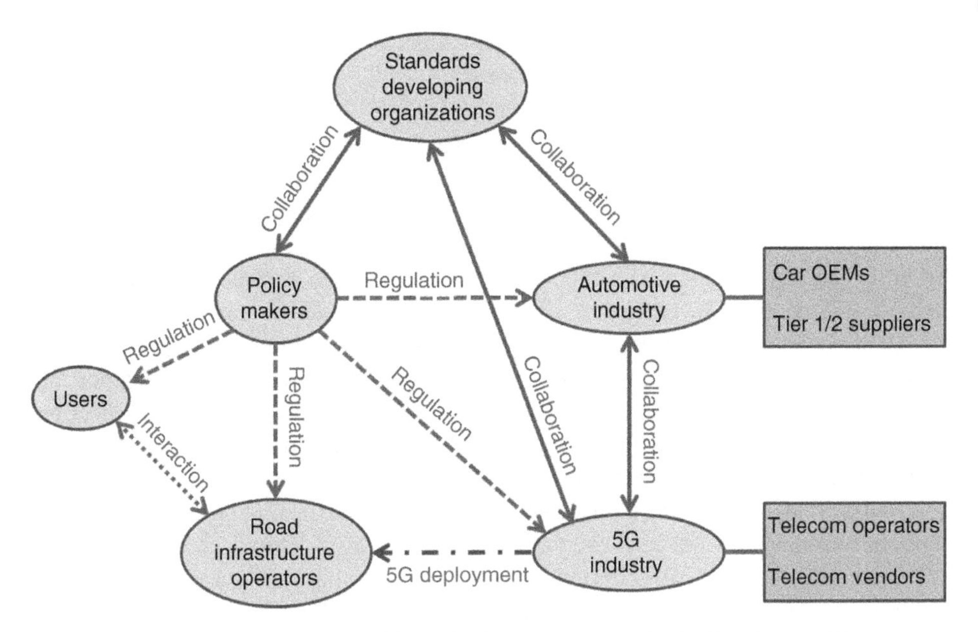

Figure 3.1 Relationships between SDOs and other stakeholders for 5G ITS. Source: [1].

automakers. Thus, these various ITS domains are covered by multiple standardization and regulation bodies, which deal, either directly or indirectly, with V2X communications (often producing overlapping specifications, which require harmonization). [1] provides a 5G ITS landscape, including the main stakeholders and relationships. The automotive industry including original equipment manufacturers (OEMs) and automotive suppliers, 5G industry including mobile network operators (MNOs) and telecom vendors, road infrastructure operators, policymakers, SDOs, and users have been identified as principal stakeholders. The relationships among the various stakeholders are categorized in the form of collaboration, deployment, interaction, and regulation, and are illustrated in Figure 3.1.

The following discussion presents a brief overview of relevant groups responsible for the regulation and/or development of ITS specifications.

3.1.2.1 International Telecommunication Union

The International Telecommunication Union (ITU) is a United Nations specialized agency for information and communication technologies. It is an important regulatory body that identifies and manages additional frequency bands. The ITU Radiocommunication Sector (ITU-R) is responsible for radio communication aspects such as managing the international radio-frequency spectrum, while the ITU Telecommunication Standardization Sector (ITU-T) develops international standards (known as ITU-T recommendations) defining the global infrastructure of information and communication technologies (ICT) and the requirements for every communication generation. ITU-R recommendation M.2083-0 [2] describes the vision of a framework and the overall objectives of the future development of International Mobile Telecommunications (IMT) systems for 2020 and beyond; this constitutes the foundational motivation for the 5G technologies being developed today. The document describes a broad variety of capabilities associated with envisaged usage scenarios, including connected cars and driverless cars.

3.1.2.2 Regional Standards Developing Organizations

Several SDOs around the globe have been officially recognized within their region. Here, the main regional standards bodies that shape research and standardization for ITS are presented.

ETSI is a standardization organization for ICT standards in **Europe**. The ETSI Technical Committee ITS (TC-ITS) aims to achieve global standards for cooperative ITS [56]. Applications include road safety, traffic control, fleet and freight management, and location-based services, providing driver assistance and hazard warnings and supporting emergency services. TC-ITS develops standards related to the overall communication architecture, management (including e.g. decentralized congestion control), security, and related access layer–agnostic protocols: the physical layer (C-V2X or ITS-G5[1]), network layer, transport layer (e.g. with the GeoNetworking protocol), facility layer, (e.g. with the definition of facility services such as cooperative awareness, decentralized environmental notification, and cooperative perception, used by ITS applications). Other topics addressed include platooning, specifications to protect vulnerable road users such as cyclists and motorcycle riders, specifications for cooperative adaptive cruise control, and multichannel operation. For all of those standardization activities, TC-ITS also develops conformance test specifications, which are crucial for the commercial deployment of the technology. TC-ITS is also involved in the related spectrum requirements.

Based in **North America**, the Alliance for Telecommunications Industry Solutions (ATIS) includes several industry committees and fora and develops technical and operational standards and solutions for the ICT industry. One of its strategic initiatives is the Connected Car – Cybersecurity Ad Hoc Group with scope to advance industry-to-industry dialogue between ICT experts and vehicle OEMs focusing on improving cybersecurity for connected vehicles. The Ad Hoc Group is set to analyze the communications paths to the connected car, develop a cybersecurity threat model, and identify potential solutions (e.g. network services) that the ICT industry can offer.

Research and standardization of C-V2X in **China** are based on the general technical specifications of LTE-V2X (air interface, network) and China Communications Standards Association (CCSA). For the application layer and security framework, local activities work closely with the ETSI TC-ITS and the Society of Automotive Engineers for China (SAE-C).

In **Japan**, research and standardization of C-V2X is based on the Association of Radio Industries and Businesses (ARIB) and the Telecommunication Technology Committee (TTC). LTE-V2X activities include the Cellular System Technical Group, Radio System Technology TG, and 5G Connected Vehicle in 5GMF (Fifth Generation Mobile Communication Promotion Forum). These efforts are undertaken in collaboration with the Japanese Ministry of Internal Affairs and Communications.

In the **Republic of Korea**, the Telecommunication Technology Association (TTA) establishes standards in the areas of information technology convergence, telecommunication network, security, software contents, broadcasting, cloud and big data, Internet of

---

1 The name *G5* is derived from the frequency band (5.9 GHz).

Things, and mobile communication technology. Most recent C-V2X activities include the specification of various layers for vehicle communication systems based on 3GPP LTE-V2X wireless access technology as well as a service framework to enable V2X in 3GPP.

The Telecommunications Standards Development Society, **India** (TSDSI) was assembled recently to contribute to next-generation telecom standards and drive the ecosystem of IP creation in India. Concerning V2X, an initial study has started on device-to-device (D2D) communications for V2X: "D2D Communication: Study of Unicast, Relays and V2X."

3.1.2.3 3GPP, IEEE, and SAE

3GPP develops standards for mobile telecommunications systems based on cellular technologies. It unites telecommunication SDOs from around the world and provides its members with a stable environment to produce the reports and specifications that define 3GPP technologies. 3GPP covers cellular telecommunications network technologies, including radio access, the core transport network, and service capabilities, and thus provides complete system specifications. To define technical specifications that will be implemented into network equipment and user terminals, 3GPP takes into account the results of studies produced by the aforementioned organizations. Since 2016, 3GPP has been performing normative work on the development of C-V2X technologies, covering V2X support with both Uu air interface (linking the base station to the UE) and sidelink-based solutions (PC5 interface, directly linking UEs).

IEEE is a professional association for electronic engineering, electrical engineering, and associated disciplines. The IEEE Standards Association (IEEE-SA) is an organization within IEEE that develops global standards in a broad range of industries, including telecommunication. In 2012, IEEE published the specification for wireless local area network (WLAN) WiFi-based V2X (IEEE 802.11p) supporting one of the two key radio access technologies that enable V2X communications: dedicated short-range communications (DSRC). IEEE 802.11p vehicular standard development was focused on vehicular safety, efficient traffic management, and added-value applications (e.g. parking and vehicular diagnostics). In 2019, the IEEE 802.11bd Task Group was created to enhance 802.11p and DSRC-based support of vehicular communications.

The Society of Automotive Engineers (SAE) International is a global association of engineers and related technical experts in the aerospace, automotive, and commercial-vehicle industries with tasks including voluntary consensus standards development. SAE International formed the DSRC Technical Committee in 2014. Since April 2014, the society has been developing DSRC-based V2X application standards in collaboration with ETSI ITS Working Group 1 (WG1) and IEEE1609 WG. The SAE C-V2X Technical Committee was established in 2017 to fill the cellular (and 5G) void in hitherto 802.11p-focused ITS standards.

3.1.2.4 5G PPP and EATA

The 5G infrastructure Public Private Partnership (5G PPP) is a joint initiative between the European Commission and the European ICT industry, which aims to deliver solutions, architectures, technologies, and standards for the ubiquitous next-generation communication infrastructures in the coming decade. With the help of representatives from the automotive and telecom industries, 5G PPP has prepared a detailed vision of how 5G can enable the next generation of connected and automated driving and related critical services that cannot be implemented using today's communication technologies [3]. Significant progress has been made on the characterization of 5G requirements through the identification of several use cases from the automotive vertical that capture the diversity of these requirements.

The European Automotive Telecom Alliance (EATA) promotes the wide deployment of hybrid connectivity for connected and automated driving in Europe. EATA consists of 6 leading associations and more than 38 companies, including telecom vendors, MNOs, and OEMs supplying cars and trucks. EATA seeks to identify and address service and technology roadmaps, safety and security needs, as well as regulatory and business issues.

3.1.2.5 5GAA

5GAA links the mobile communications business with car manufacturers (www.5gaa.org). It was founded in September 2016 as an industry group to develop end-to-end solutions for transportation and transportation services. It includes car manufacturers, operators, Tier 1 suppliers, chipset/communication system suppliers, and operator/infrastructure suppliers. 5GAA is shaping up as a premier pre-standardization association for defining use cases, requirements, technical solutions, and testing for cellular V2X communication. 5GAA is also working together with 3GPP for future 3GPP Releases.

5GAA is working on several parallel activities that will allow the automotive and telecommunications industries to embrace and accelerate the global deployment of intelligent transport and communications solutions. The activities of 5GAA are organized into seven working groups:

- WG1: Use cases and technical requirements
- WG2: System architecture and solutions
- WG3: Performance evaluation and testbeds
- WG4: Standards and spectrum
- WG5: Business models and market entry strategies
- WG6: Policies
- WG7: Security

In addition to several white papers for C-V2X, technical recommendations, spectrum requirements, plugfests, and interoperability tests, 5GAA harmonizes the vehicle and telecommunications industries with respect to agreed market roadmaps for the coming decade. For instance, in December 2017, the roadmap for technology deployment of LTE-V2X was published. Figure 3.2 shows the roadmap for deploying C-V2X technology.

5GAA expects that in the early 2020s, the majority of newly released cars globally will be connected vehicles on the road [53]. Because C-V2X functionality can and will be included as part of the cellular chipsets embedded into vehicles for V2N communications, those vehicles will be able to benefit from the higher level of traffic safety enabled by supporting direct communication between vehicles (V2V) and road infrastructure (V2I), in addition to communication with cloud ITS services (V2N). In addition to the benefit that including C-V2X is evolutionary to the updating of cellular chipsets, there are also cost synergies that will ease the barrier to deployment.

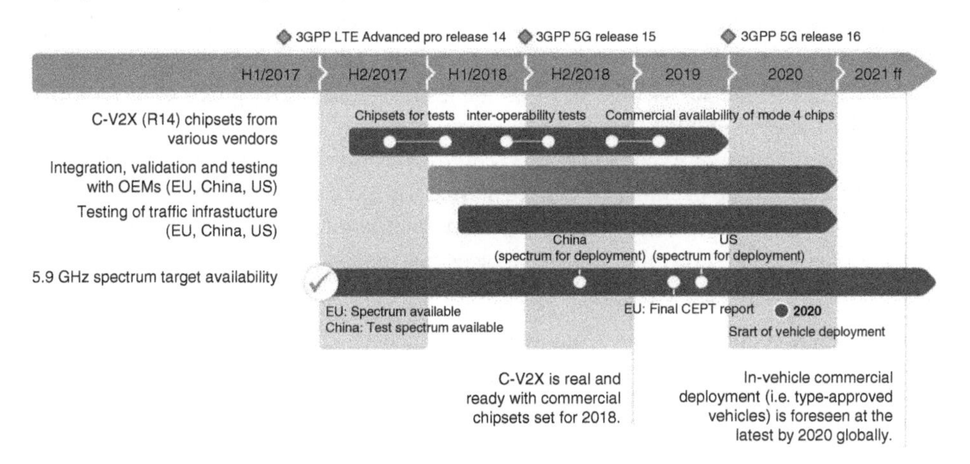

Figure 3.2 5GAA C-V2X roadmap.

3.1.3 3GPP Structure and Standardization Process

3GPP, which has and will have a pivotal role in the specification of V2X communication solutions, is explained here as an example to elaborate on the standardization process in detail [57]. We primarily aim to provide the reader with a clearer glimpse of the procedures and level of structural complexity within standardization bodies.

3GPP is a consortium comprising seven primary organizational partners from regions within North America, Asia, and Europe (such as ETSI), several market representation partners (such as 5GAA), as well as numerous (around 600 in 2020) individual members. Organizational partners control the general procedure and strategy of 3GPP and, in turn, refer to derived specifications for local use. Market representation partners are invited by 3GPP and can offer market advice and requirements. Individual members, i.e. companies, organizations, institutions, etc., participate in 3GPP through their membership as an organizational partner and provide contributions that compose the core of the standardization work.

The specification work is essentially performed at three levels by a corresponding group at each level, including individuals from the various 3GPP members. At the first level, working groups (WGs), via several physical meetings or email discussions during a year, prepare and discuss support for technical aspects or request a change in existing 3GPP specifications. When an aspect or change request is accepted at the WG level, it is noted as an agreement by the WG assigned chairman or vice-chairman. At the second level, technical specification groups (TSGs), via plenary meetings quarterly, approve the aforementioned agreements but can also agree and specify on certain aspects. Finally, the approved agreements are transformed into 3GPP specifications, which subsequently are transposed into deliverables/standards by the organizational partners. There also exists a higher checking level, where the project coordination group approves organizational changes and manages the overall timeframe, work plan, and progress.

Going deeper into the groups' structure, we see that there are three TSGs:

- The radio access network (RAN) TSG specifies the functions, requirements, and interfaces of the Universal Terrestrial Radio Access (UTRA) and Evolved UTRA (E-UTRA) network and beyond. It essentially covers the access stratum functional layer between UE and base station and is composed of five active WGs.
- The Service and System Aspects (SA) TSG coordinates the project and specifies the service requirements and overall architecture of the 3GPP system. It is relevant to the upper signaling layer, non-access stratum (NAS), and is composed of six WGs.
- The Core Network and Terminals (CT) TSG specifies the core network and terminal layer 3 protocols and is composed of five WGs.

The specification documents are organized in series, each covering a theme:

- Technical reports (TRs) are reports based on early exploratory studies on the feasibility of a defined topic or feature and possible technical approaches.
- Technical specifications (TSs) are documents that specify the various functions, procedures, etc. of cellular systems.

3GPP specifications evolve over time and are structured in releases made up of a set of consistent features and specifications. Typically, a new 3GPP release is delivered approximately every 1.5 years. Each release introduces new features and incorporates numerous standards documents, which usually go through several revisions and have the challenge of forward and backward compatibility among releases. Therefore, a market product targets a given release. Timeframes are defined for each release by specifying freezing dates. When a release is denoted as "frozen", only essential corrections are allowed: i.e. new functions or modifications of existing functions are prohibited.

The 3GPP standardization process is also organized in stages (freezing dates are defined for each stage), aligning with the three-stage methodology defined in ITU-T recommendation I.130 [4].

- Stage 1 specifications define the services to offer and their requirements at a high level.
- Stage 2 specifications define the architecture: i.e. high-level design implementation specifications to support stage 1 defined services and requirements.
- Stage 3 specifications define the detailed implementation of the architecture by specifying the necessary protocols in detail.
- Test specifications may also exist, coming after the protocol specifications, which can be viewed as stage 4 of the standardization process.

Apart from the internal 3GPP processes described here, 3GPP also interacts (as input, output, or cross-reference) with various other telecommunications organizations to develop the standards worldwide.

3.2 Regulatory Aspects and Spectrum Allocation

In this section, we present the current regulatory arrangements for ITS in the different regions of the world. In addition, we discuss aspects and considerations regarding the radio spectrum dedicated to ITS.

3.2.1 C-V2X Policy and Regulations in Europe

In August 2008, the European Commission designated 5875–5905 MHz for ITS safety purposes through Decision 2008/671/European Commission. The European Commission adopted the ITS Action Plan (COM (2008) 886) to accelerate the market penetration of ITS services. The European Commission defined Directive 2010/40 as the legal framework for cooperative-ITS (C-ITS) market introduction. In November 2014, the European Commission launched the C-ITS Deployment Platform for Phase 1 from 2014 to 2016 and developed policy recommendations related to technology, law, and commercialization issues.

Phase 2 was published in September 2017 and was based on the output of Phase 1. For upcoming 5G development and commercialization in Europe, in 2016, the European Commission adopted "5G for Europe: An Action Plan" to promote industrial applications. For radio spectrum considerations, the European Commission requested a study in October 2017 from the European Conference of Postal and Telecommunications Administrations (CEPT) to extend the ITS safety-related band at 5.9 GHz for urban rail using communication-based train control (CBTC) in addition to road transport, targeting 5905–5925 MHz. CEPT studies were completed in March 2019. The European Commission also recommends, via Electronic Communications Committee (ECC) Recommendation (08)01, that CEPT administrations should make the frequency band 5855–5875 MHz available for ITS non-safety applications.

These regulatory measures all refer to ETSI Harmonized Standard European Norm (EN) 302571, which defines requirements for operation of ITS equipment in 5855–5925 MHz, covering the essential requirements of article 3.2 of the Radio Equipment Directive (2014/53/EU). According to ECC DEC (08)01 and ECC REC (08)01, ITS equipment complying with EN 302571 is exempt from individual licensing for operating in this band.

It should be emphasized that the principle of technology neutrality in European spectrum regulations implies that any radio technology, which can demonstrate conformance with the essential requirements of the Radio Equipment Directive (e.g. through compliance with EN 302571) can operate in 5855–5925 MHz.

3.2.2 Radio Frequency Spectrum Allocation for V2X Communications

Next, we look at aspects of the radio spectrum for V2X communications. We present the spectrum allocation for IMT systems and 3GPP technologies, the spectrum dedicated for ITS applications, and, finally, considerations for worldwide spectrum harmonization.

3.2.2.1 Spectrum Allocation for IMT Systems and 3GPP Technologies

IMT systems developed by the ITU encompass IMT-2000, IMT-Advanced, and IMT-2020. IMT specifies the sets of requirements for high-speed mobile broadband services and technologies that are marketed as 3G, 4G, and 5G worldwide. The ITU Radio Regulations [5] establish the allocation of frequency bands for each radio service, including the mobile service enabled by IMT technologies. The frequency bands identified by ITU-R for IMT and the related channel arrangements of those bands are documented in ITU-R Recommendation M.1036-5 [6]. In addition to those frequency bands documented in M.1036-5, other frequency bands can also be used and deployed regionally for IMT systems. 3GPP technologies are developed following the requirements from ITU IMT and are designed to operate in the IMT bands identified by ITU-R. As for E-UTRA and UTRA, which are the radio interface technologies of LTE and the Universal Mobile Telecommunications Service (UMTS), respectively, 3GPP has defined the operating bands in specifications [7] and [8]. So far, 3GPP has specified more than 70 frequency bands for LTE within the IMT bands in specific national or regional areas.

The spectrum considered for 5G in Europe includes 703–788 MHz, 3.4–3.8 GHz, and 24.25–27.5 GHz. The 3.4–3.8 GHz band is considered by the Radio Spectrum Policy Group (RSPG), which is a high-level advisory group that assists the European Commission in the development of radio spectrum policy, to be "the primary band suitable for the introduction of 5G-based services in Europe even before 2020 given that it is already harmonized for mobile networks and offers wide channel bandwidths" [9]. Furthermore, according to the RSPG, "5G will need to be deployed also in bands already harmonized below 1 GHz, including particularly the 700 MHz band, in order to enable nation-wide and indoor 5G coverage." Note that the 703–733 MHz uplink (UL) and 758–788 MHz downlink (DL) band has already been auctioned in some countries, e.g. in France and Germany. The RSPG also recognized "the 24.25–27.5 GHz (hereinafter '26 GHz') band as a pioneer band for Europe to be harmonized before 2020."

By the end of 2017, 3GPP approved the first 5G NR radio specifications [10]. In TS 38.101 [11], 3GPP defines two frequency ranges (FRs) for 5G NR operation: FR1: 450–6000 MHz and FR2: 24250–52600 MHz. The lists of NR operating bands in respective FRs contain both newly identified bands for NR, e.g. n77, n78, n79, n257, n258, n260, etc., as well as reused LTE bands, e.g. n1, n2, etc.

3.2.2.2 Dedicated Spectrum for ITS Applications

ITS aim to improve road traffic safety and efficiency. In 1999, the US Federal Communications Commission (FCC) assigned 75 MHz from 5.850 GHz to 5.925 GHz for ITS applications. In Europe, the spectrum 5.855–5.925 GHz has been designated for ITS by the European Commission [12]. In many other countries of the world, e.g. China, Korea, Australia, and Singapore, the 5.9 GHz ITS band has also been designated or is under consideration for ITS applications. Above the 5.9 GHz band, CEPT [13] decided that CEPT administrations should designate the 63–64 GHz for ITS applications, known as the 60 GHz ITS band. In a recent update, CEPT changed the allocation of the 60 GHz ITS band to 63.72–65.88 GHz [14]. The frequency band 5.770–5.850 GHz has been allocated in Japan for ITS applications, e.g. electronic toll collection (ETC). Additionally, a single 9 MHz frequency channel in 755.5–764.5 MHz has also been designated for ITS safety-related applications using vehicle-to-vehicle (V2V) and vehicle-to-infrastructure (V2I) communications in Japan. Unlike the IMT bands discussed earlier, use of these dedicated ITS spectrum bands is license-exempt. Figure 3.3 provides an overview of the dedicated spectrum for ITS applications in different regions of the world.

Spectrum regulations are issued by regional regulators and can vary from region to region. Chapter 4 investigates the spectrum resources that are usable for V2X communication to support ITS and automotive applications in different regions and countries, particularly in Europe. Different systems and technologies operating in the same band or in bands in proximity require special attention to coexistence and compatibility issues, particularly for safety-related applications such as ITS. Chapter 4 also discusses compatibility and technology coexistence issues in the relevant frequency bands.

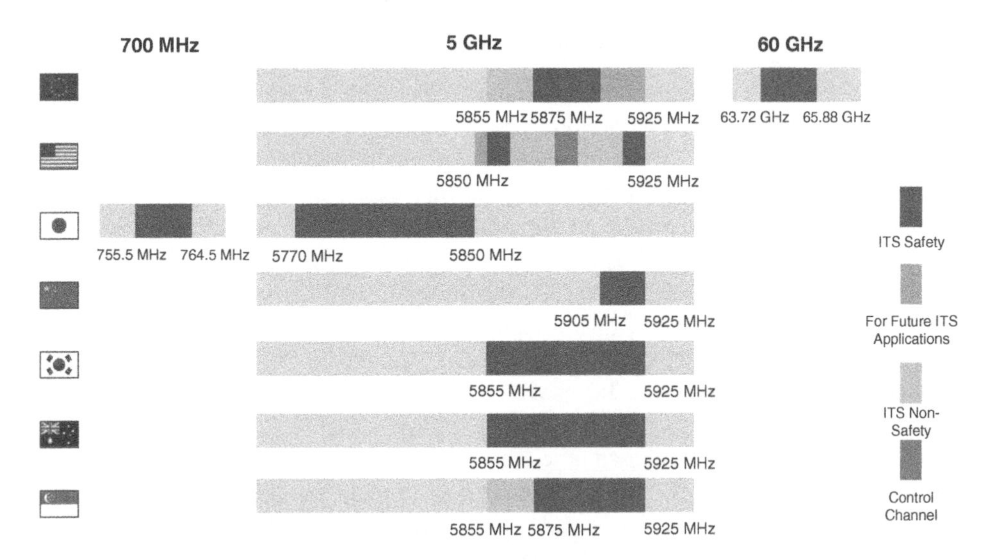

Figure 3.3 Spectrum dedicated to ITS in different world regions.

3.2.2.3 Worldwide Spectrum Harmonization

The IMT spectrum for cellular technologies is harmonized in ITU-R. To consider possible global or regional harmonized frequency bands for the implementation of evolving ITS under the existing mobile service allocation, the World Radio Conference (WRC) 2019 had agenda item AI1.12 for ITS applications [15]. The 5.855–5.925 GHz band has been discussed in different world regions for the harmonized use of ITS. In Europe, CEPT also considers the 60 GHz ITS band under this WRC-19 AI1.12. However, ETC in 5.795–5.815 GHz is not considered by CEPT under AI1.12.

3.3 Standardization of V2X Communication Technology Solutions

Major advances have been made in recent years in the area of wirelessly connected vehicles. IEEE WiFi-based V2X communication reached first a mature stage offering the capability for direct communication among entities with short communication latency and, hence, good support for vehicle safety and traffic efficiency applications. On the other hand, 3GPP has recently embraced vehicular communications and many technical enablers that promise the satisfaction of difficult requirements for latency and reliability for advanced automotive use cases. Such high-end cellular-based V2X communication appears in NR specifications in Release 16 (i.e. in 2020). Here, we follow how communication technologies for V2X have been developing over time and give a brief overview of normative work items on DSRC/C-V2X specifications. Finally, we focus on 3GPP C-V2X standardization, providing a brief history of past releases, current progress (while highlighting the anticipated benefits 3GPP will bring for the 5G support of V2X services), and envisioned future plans.

3.3.1 A Brief History of V2X Communication

The concept of V2X communication (including V2V, vehicle-to-pedestrian (V2P), V2I, and vehicle-to-network (V2N) services) has been maturing for almost two decades. Some state-of-the-art aspects of V2X communications are reported in [16–20]. Generally, V2X technologies can be classified as short-range and wide-area technologies [21]. Short-range technologies are designed to meet high-availability requirements of V2V and V2I services (dedicated to low-power roadside unit [RSU] deployment) and include IEEE 802.11p (WAVE/DSRC in the US and ETSI ITS-G5 in Europe), LTE direct mode (PC5, 3GPP Release 14), and NR direct mode (3GPP Release 16). It is ensured that "older" cars can talk to "newer" cars and vice versa for all cellular V2X modes. Wide-area technologies (e.g. cellular networks over the licensed spectrum) are best suited for V2N service requirements.

The first large-scale services based on connections between vehicles and infrastructure nodes offered ETC and low-rate communication services between vehicle-mounted on-board units and peer vehicles or RSU. These services drove the standardization of IEEE 802.11p, which aimed at providing DSRC capabilities. Initially, the supported data rate was limited to 3 Mb/s (for 10 MHz bandwidth) [22], which was later enhanced to 54 Mb/s (for 20 MHz bandwidth) [23]. In the US, IEEE 802.11p-based DSRC radio and the IEEE 1609-based stack named wireless access for vehicular environments (WAVE) are used, whereas in the EU, the radio is the same, but the stack is called ITS-G5 [24]. Japan uses a modified variant of WAVE for vehicular communication, which is also based on IEEE 802.11p; it is called the ARIB Standard [54]. These early standard sets specified only V2V and V2I communication and enabled applications primarily for driver information and warnings. Moreover, there are several limitations and open issues with current IEEE 802.11p-based V2X systems, which use RSU for extended coverage, have no support for communications requiring ultra-reliability and very low latency, and have no clear evolution path to improve performance in term of latency, reliability, etc.[2] Cellular network-based solutions, on the other hand, have the advantage of almost full radio coverage as well as a guaranteed data rate and/or delay to various applications from coordinated transmissions.

On the C-V2X front, recognizing the increasing demand for vehicular communications, in 2015, 3GPP launched an activity aiming to develop a technology that is from the start integrated into cellular systems and can offer ubiquitous vehicle connections over a wide geographic area. Indeed, Release 14 of LTE includes a full set of technical components for V2X services. These technical components, including radio interface and medium access control layer support, protocols, and management functionalities, together enable V2X communications. 3GPP-based V2X communications can be readily utilized for safety and non-safety (e.g. infotainment) purposes. Thus, the V2X support provided by 3GPP Release 14 makes LTE a suitable technology for meeting the requirements of ETSI for delivering safety messages such as cooperative awareness messages (CAMs) and decentralized environmental notification messages (DENMs) [26, 27].

2 To moderate the performance gap between DSRC and C-V2X as well as support additional modes of operation, an IEEE 802.11 Next Generation V2X study group was formed in March 2018. This resulted in the formation of IEEE Task Group 802.11bd (TGbd) in January 2019 [25].

3.3.2 Overview of DSRC/C-V2X Specifications Around the Globe

Next, we provide an overview of normative work items on DSRC/C-V2X specifications taking place in Europe, the Americas, and Asia.

3.3.2.1 Europe

In Europe, most of the ETSI TC-ITS standardization work is related to C-ITS. The following issues are being addressed: decentralized congestion control, C-ITS security, geonetworking protocol, and C-ITS facility services to be used by ITS applications.

Decentralized Congestion Control (DCC)

Cross-layer DCC provides stability in the ad hoc network by providing resource management when there is a large number of C-ITS messages to avoid and reduce interference and degradation of C-ITS applications.

C-ITS Security

C-ITS and driving safety applications depend upon reliable and trustworthy data transmitted by other vehicles and the infrastructure. In this context, standardized solutions for security and privacy are paramount, and these will be based on the design and implementation of a security management infrastructure for C-ITS. ETSI TC-ITS develops standards defining the security framework for cooperative ITS, including a PKI. This security framework will support PKI trust model requirements from the EU C-ITS deployment platform and bring privacy protection mechanisms for users and drivers, e.g. using pseudonym certificates and regularly changing pseudonyms identifiers in ITS-G5 communications.

GeoNetworking (GN) Protocol

Many ITS applications require the dissemination of information with rapid and direct communication, which can be achieved by ad hoc networking. GN is a network layer protocol for mobile ad hoc communication without the need for a coordinating infrastructure based on wireless technology, such as ITS-G5. It utilizes geographical positions for disseminating information and transporting data packets. It offers communication over multiple wireless hops, where nodes in the network forward data packets on behalf of each other to extend the communication range.

C-ITS Facility Services to Be Used by ITS Applications

ETSI TC-ITS develops and maintains important services to be used by ITS applications. These services include but are not limited to:

- Cooperative awareness, to create and maintain awareness of ITS stations (ITS-S) and support the cooperative performance of vehicles using the road network
- Decentralized environmental notification, to alert road users about detected events using ITS communication technologies
- Cooperative perception, complementing the cooperative awareness service, to specify how an ITS-S can inform another ITS-S about the position, dynamics, and attributes of detected neighboring road users and other objects
- Multimedia content dissemination (MCD), to control the dissemination of information using ITS communication technologies

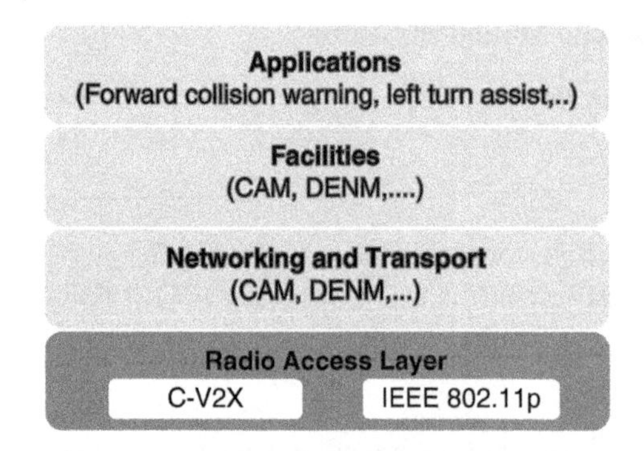

Figure 3.4 ETSI ITS protocol architecture with the two options in the access layer.

Table 3.1 ETSI ITS specifications.

ETSI ITS number	Topic
EN 303 613	LTE-V2X access layer
EN 302 636-4-1	GeoNetworking
EN 302 571	Intelligent transport systems (ITS); radiocommunications equipment operating in the 5855 MHz to 5925 MHz frequency band; harmonized standard covering the essential requirements of article 3.2 of directive 2014/53/EU
TR 103 576-2	Interoperability and backward compatibility
TS 102 941	Security
TR 103 766 and TR 103 667	Co-channel coexistence
TR 103 688	Receiver requirements
TR 102 962 V2	Framework for public mobile networks
ITS WG5 TR 103 630	Study on ITS facility layer security for C-ITS communication using cellular Uu interface
ITS WG5 TS 103 692	Interface between security entity and facilities layer

Figure 3.4 shows the extended ETSI ITS stack with the two options.

Table 3.1 summarizes the most relevant ETSI TC-ITS specifications that are important for C-V2X.

3.3.2.2 The Americas

In the Americas, SAE International formed the DSRC Technical Committee in 2014. Since April 2014, the society has been developing DSRC-based V2X application standards in collaboration with ETSI ITS WG1 and the IEEE1609 WG. Several task forces (TFs) were defined: the Crosscutting TF, V2V Safety Awareness TF, V2V Cooperative Automation TF, and V2I I2V TF. The five task forces of the V2Other TF led to 11 standard specifications, as shown in Table 3.2.

In response to emerging LTE and 5G C-V2X technologies, SAE International announced in June 2018 the founding of the C-V2X Technical Committee. The C-V2X Technical

Table 3.2 SAE DSRC specifications.

SAE number	Topic
J2735	Message set dictionary (BAM, Map, SPaT, PSM, TIM)
J2945/0	Systems engineering process guidance for J2945/x documents and common design concepts
J2945/1	On-board system requirements for V2V safety communications
J2945/2	Performance requirements for V2V safety awareness
J2945/3	Requirements for V2I weather applications
J2945/4	V2I road safety applications
J2945/6	CACC/platooning performance requirements
J2945/9	V2P safety message minimum performance requirements
J2945/10	Map/spat message
J2945/11	Signal preemption
J2945/12	Traffic probe use and operation

Table 3.3 SAE C-V2X specifications.

SAE number	Topic
J3161	Onboard system requirements for LTE-V2X V2V safety communications
J3186	Maneuver sharing and coordinating service

Committee focuses on the application and performance requirements for LTE-V2X and conducts research on C-V2X direct communication, PC5, and 5G-V2X applications. The C-V2X Advanced Application TF focuses on cellular standards related to mobile network operators. Two specifications (see Table 3.3) are defined in cooperation with 3GPP RAN and ITU-R WG5A.

The C-V2X Technical Committee is responsible for supporting media independence to support the complete feature set of IEEE 1609.3 on PC5. In addition to basic safety messages (BSMs), messages for signal phase and timing (SPAT), map data (MAP), and personal safety messages (PSMs) fall under the definition. The complete protocol stack architecture is shown in Figure 3.5: the lower part follows 3GPP specifications considering Uu and PC5 V2X applications, and the upper part reflects the higher V2X protocol layer from IEEE, which is highly reused and adapted for PC5 of C-V2X (LTE-V2X).

3.3.2.3 Asia

In China, the C-V2X WG in the IMT-2020 (5G) promotion group focuses on testing and verifying LTE-V2X, deployment, and networking. It promotes the industrialization of C-V2X through activities such as the definition of business models and service requirements. The WG has tested LTE-V2X in three phases from 2017 onward on allocated 20 MHz bandwidth in the 5905–5925 GHz band. Table 3.4 summarizes the current C-V2X standardization activities in that region.

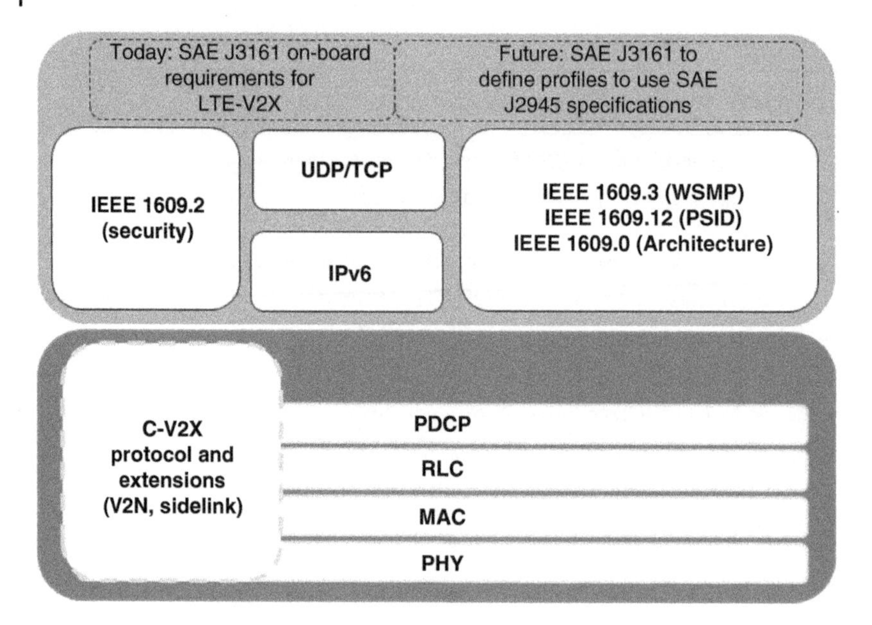

Figure 3.5 SAE C-V2X protocol architecture.

Table 3.4 C-V2X standardization activities in China.

Category	Topic	Type	Involved organizations
Requirements	Overall technical requirements for LTE-based IoV Communications	Industry standard	CCSA
Application layer	Cooperative intelligent transportation system – dedicated short-range communications Part 3: technical requirements for the network layer and application layer	China national standard	TC-ITS, CCSA
	Cooperative Intelligent Transportation System –Application Layer and Application Data Interaction Standards for Vehicle Communications	Common Standards	SAE-C, C-ITS
Network layer	Cooperative intelligent transportation system – dedicated short-range communications Part 3: technical requirements for the network layer and application layer	China national standard	TC-ITS, CCSA
Air interface	Technical air interface requirements for LTE-V2X	Industry standard	CCSA
Security	Technical security requirements for IoV communications based on the public LTE network	Industry standard	CCSA

Table 3.5 TTA V2X specifications.

TTA standard number	Topic
TTAR-06.0146	Data rates for the data and control channels to provide robust V2X communication under vehicle environments
TTAR-06.0204	C-V2X technology and service deployment strategy
TTAK.KO-06.0479	Vehicle communication system stage 3: PHY/MAC (LTE-V2X)
TTAK.KO-06.0482	C-V2X service framework – network architecture and communication procedure

In Japan, ARIB has been investigating 5.9 GHz for ITS service based on C-V2X technology. The coexistence operations of ITS services in this band have been under investigation with tolling systems and satellite and microwave links for broadcasting from March 2018. Furthermore, the 5GMF will be conducting a comprehensive 5G demonstration test that includes real-time communication between vehicles.

In Korea, TTA recently analyzed and compared C-V2X and WAVE technologies. Then, standards were developed to define the specifications of layers – physical (PHY), media access control (MAC), radio link control (RLC), packet data convergence protocol (PDCP), radio resource control (RRC), and NAS – for vehicle communication systems based on 3GPP LTE-V2X wireless access technology. Moreover, TTA completed a specification of the service framework that includes specification of (i) the reference architecture (i.e. protocol stacks, interfaces between network nodes, etc.) and (ii) functional procedures (i.e. V2X message transmission and reception, quality of service [QoS] management, etc.) that are introduced to enable V2X in 3GPP. Table 3.5 summarizes the current C-V2X standardization activities in that region.

Furthermore, at the 52nd China-Japan-Korea IMT-2000 WG Conference, information sharing and technology development of C-V2X among the three countries was agreed on, and a SIG-V2X task force was established to collaborate on standards and standards development. They also contributed to the 5G-V2X report drafted by the Asia-Pacific Telecommunity (APT) Wireless Group.

3.3.3 C-V2X Standardization in 3GPP: Toward and Within 5G

Mobile communication is becoming more and more important in daily life, starting with the 1st generation (1G) and proceeding to the current fifth generation (5G). It should be clarified here that wireless communication generations do not refer directly to technologies but instead to requirements that need to be met by any technology that wants to be characterized as meeting the generation requirements defined by the ITU-T. So, from the 3GPP side, for example, LTE meets the 4G requirements and also the 5G requirements for massive Internet-of-Things (IoT) services, and NR meets the 5G requirements for enhanced mobile broadband (eMBB) and ultra-reliable low-latency (URLLC) services. Each generation has new things to offer at a particular time, and it keeps evolving to offer better services and features for the end customer. Today, the application area of mobile communications is no longer limited to personal communication but has evolved to include many other

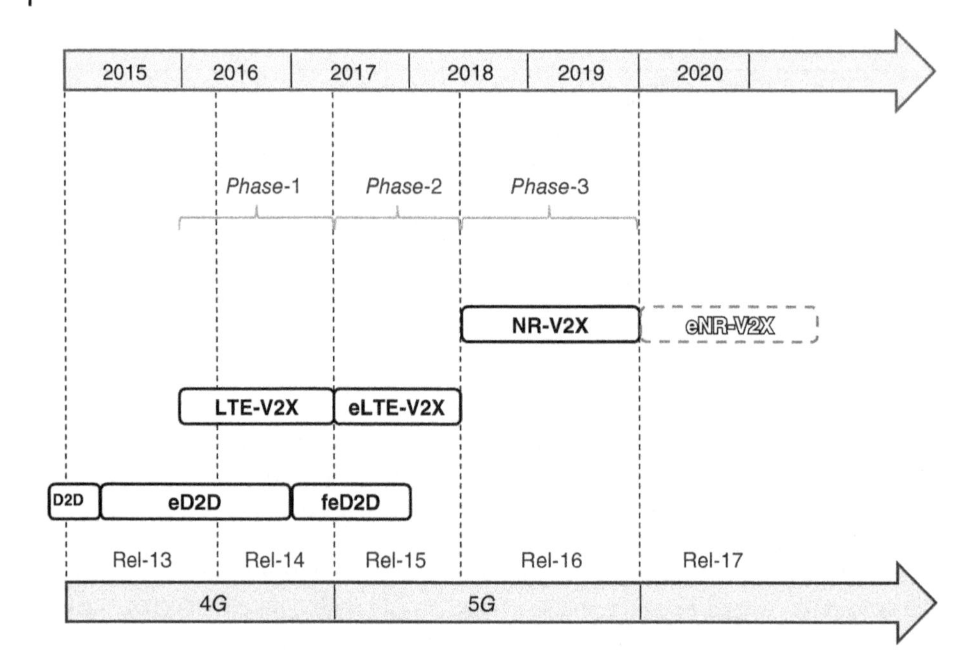

Figure 3.6 C-V2X standardization evolution in 3GPP.

areas, including vehicular communication as one such vertical domain. Before 4G, most vehicular applications were hard to realize due to high mobility and stringent requirements on latency and reliability. However, from Release 14 onward, 3GPP has embraced vehicular communications and many of its technical enablers in its specifications. Figure 3.6 provides an overview of C-V2X standardization evolution within 3GPP releases and generations in recent years.

3.3.3.1 C-V2X in 4G

The first serious 3GPP attempt relevant to vehicular communications appeared in Release 13. A feature called proximity services (ProSe) was introduced to enable D2D communication, thus allowing devices to discover each other's presence and exchange data directly without the intermediate step of network infrastructure [28]. The term *sidelink* (SL) was introduced to complement the conventional uplink and downlink communication directions. In LTE ProSe, SL communication is very similar to uplink (UL) in the sense that the same transmission scheme is used, i.e. Single-Carrier Frequency Division Multiple Access (SC-FDMA), and usable frequency-time resources are a subset of the UL resources. Two modes were also defined for device resource selection/allocation: (i) network-scheduled mode (e.g. when a device is under network coverage) where eNB allocates resources for SL transmission and (ii) autonomous mode (e.g. when the device is out of coverage) where resources are selected from a preconfigured resource pool. Enhancements (i.e. eD2D [29], feD2D [30]) were introduced in that feature stream in following releases. However, ProSe was not designed to perfectly support V2X; scenarios regarding public safety and consumer applications (e.g. wearables) were mainly targeted, including low-mobility

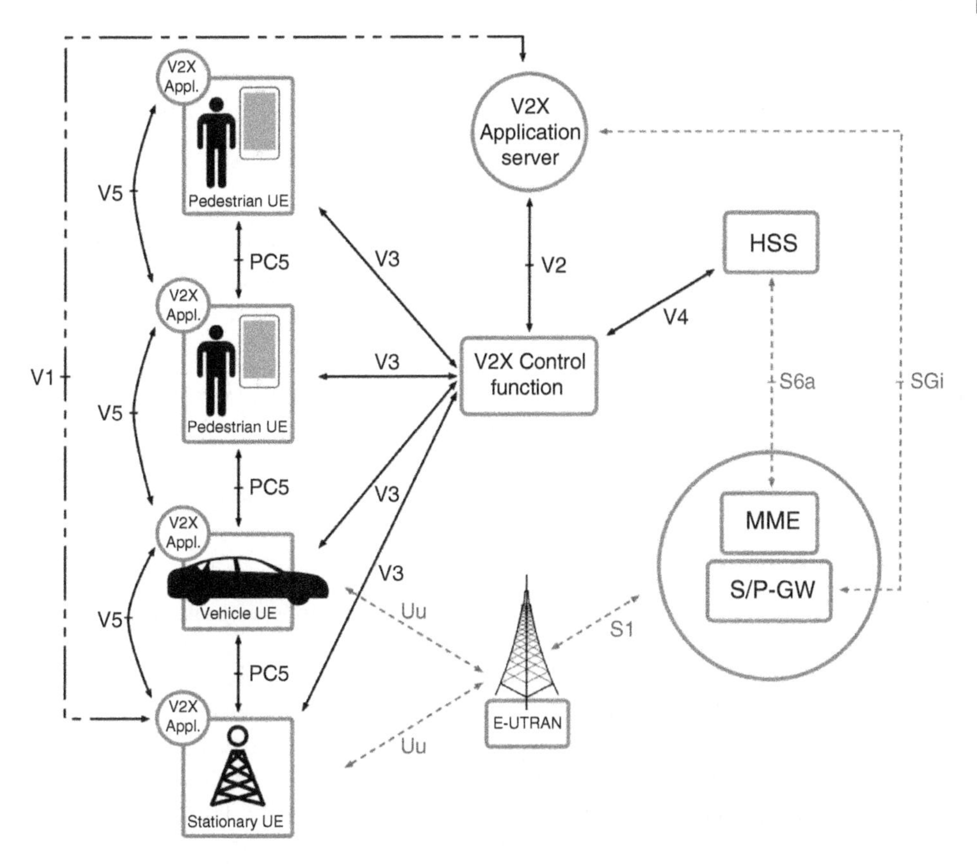

Figure 3.7 Release 14 V2X architecture.

and point-to-multipoint communication, with no strict requirements for latency and reliability.

Nevertheless, ProSe set the foundation for vehicular communication functionality by enabling D2D. Officially, 3GPP has been actively supporting vehicular communications since its Release 14. The 3GPP standard was first expanded to the automotive industry in 2016 to support V2V services, with enhancements to additional V2X operation scenarios in 2017. This standardization work was included as V2X Phase I service in Release 14 LTE, focusing on basic road-safety service. Thus, the first round of specifications revolved around the support of "day 1" safety and non-safety communications, under the name *V2X* [31]. Phase I was completed for LTE in Release 14 [55] and a novel architecture extension was proposed, illustrated in a simplified form in Figure 3.7.

Phase I is specified for supporting limited messaging services such as CAM and DENM. There is no advanced QoS support and no URLLC support. PC5, which is the radio interface specified for ProSe D2D SL in Release 12 and Release 13, is enhanced for high-velocity, high UE density, and shorter-latency V2V support. PC5 enhancement for V2P is mainly from a power efficiency point of view. The Uu interface is enhanced with uplink (UL) semi-persistent scheduling (SPS) and DL multimedia broadcast multicast service (MBMS) for V2X.

3.3.3.2 C-V2X Supported by 5G

With the coming of 5G, advanced use cases for V2X are targeted (with stricter requirements [32] and support of novel, more advanced applications [33]), and enhancement of 3GPP support for V2X services (eV2X services) is envisioned. Both LTE and NR technologies are considered candidate radio access technologies (RATs) to support eV2X.

Initially, in RAN plenary meeting 76 (June 2017), 3GPP agreed to a "study on evaluation methodology of new V2X use cases for LTE and NR" [34]. This study item (SI) was targeted to further improve the evaluation methodology of LTE-V2X and finalize it for NR-based V2X. The SI was completed by RAN plenary meeting 80 (June 2018) and resulted in a technical report [35] and conclusions regarding the following elements:

- Identification of the regulatory requirements and design considerations of potential operation of direct communications between vehicles in the spectrum allocated to ITS beyond 6 GHz
- Evaluation scenarios including performance metrics, vehicle dropping, and traffic models
- Sidelink channel model for spectrum above 6 GHz.

V2X Phase II in Release 15 LTE has introduced some enhanced features in that direction [36].

On the other hand, NR V2X is intended, as V2X phase III, to complement and support interworking with LTE-V2X for eV2X services, while including the possibility to independently support basic safety use cases as well. NR RAT was first introduced in Release 15 and is expected to have higher system capacity and better coverage than LTE. Nevertheless, the advanced V2X use cases beyond those supported by Release 15 LTE-V2X have more stringent requirements. As a consequence, an enhanced NR system and a new NR sidelink needed to be defined. NR V2X is designed to anticipate services with low latency and high reliability requirements in a flexible manner. At the RAN level, a feasibility study on NR V2X [37] was first concluded successfully, and several technical solutions were identified [38]. The study covered aspects including:

- NR sidelink design
- Support of sidelink unicast/groupcast/broadcast
- Physical layer structures and procedures
- Synchronization mechanism
- Resource allocation mechanism
- L2/L3 protocols
- Uu enhancements for advanced V2X use cases
- Uu-based sidelink resource allocation/configuration by LTE or by NR
- RAT and interface selection
- QoS management
- Non-cochannel coexistence between NR and LTE sidelink

Based on the NR V2X feasibility study, Release 16 work item "5G V2X with NR sidelink" addressed the aforementioned aspects and specified radio solutions – including NR sidelink, mainly for V2X – that are necessary to support eV2X services in Release 16 NR [39]. More specifically, this first version of NR sidelink has been developed focusing on support for V2X-related road-safety services, and its design is aimed at providing support for broadcast, groupcast, and unicast communications in both out-of-coverage and in-network coverage scenarios.

It also is worth noting that before NR V2X development, basic support for URLLC services was introduced in Release 15 NR, which included transmission time interval (TTI) structures for low latency, as well as methods for improved reliability. The motivation to support URLLC services came from a broad spectrum of use cases with tight latency and reliability requirements, envisioned to be enabled by NR. Later, in parallel to the V2X study item, a URLLC study was conducted to evaluate Release 15 NR URLLC feasibility for further key use cases, including the transport industry and especially the remote driving use case [40]. This study item concluded that it is beneficial to support a set of enhancements to Release 15 URLLC solutions (eURLLC) [41]; as a result, another Release 16 work item was undertaken to specify these enhancements [42].

3.3.3.3 Future Plans

As we have seen above, the automotive vertical and V2X communications have attracted considerable attention in recent years. V2X sidelink communication has been an important topic since Release 14 (LTE), and the evolution continues in Release 16. It is also expected to be one of the significant topics that drives the work in Release 17 and beyond. In Release 16, for V2X sidelink, the major focus has been on designing basic structure, procedures, and resource allocation for the frequency bands in FR1, i.e. below 6 GHz. For Release 17, the following topics have been discussed and will most likely be essential for future evolution of V2X support:

- Frequency bands above 6 GHz, FR2
- URLLC enhancement over PC5
- Carrier aggregation for sidelink
- Relaying in NR V2X
- Enhanced resource allocation
- Multiple-input multiple-output (MIMO) beamforming, beam management, and multi-panels
- Power consumption optimization
- Positioning support

In December 2019, recently, the Release 17 RAN content was confirmed: it includes the evolution of sidelink technology in the form of the work item "NR Sidelink Enhancement" [43] and the study item "NR Sidelink Relay" [44], with a clear focus on improved support for V2X by NR. While the enhancing normative work targets to provide radio solutions in accordance with the progress in SA works on architecture enhancements [45] to address service requirements and operation scenarios that were not fully supported in Release 16 due to the time limitation, the study item explores sidelink-based relaying functionality for sidelink/network coverage extension and power efficiency improvements, considering a wider range of applications and services from Release 16.

In addition to the continuous evolution of sidelink technology as in previous releases, some new aspects are also under consideration. Sidelink communication that is capable of working in out-of-coverage situations, which has only been used for V2X connectivity so far, could be used as a building block for other application areas, including:

- Public safety (PS), referring to hazard situations such as earthquakes and other natural hazards where parts of public mobile networks stop functioning. Sidelink could be used for UE-to-UE communication to relay messages where there is no coverage.
- Commercialized applications, such as commercial announcements for vehicles and customers that are passing commercial centers or stores.

For these new application areas, the main target is to reuse most of the NR V2X capabilities by maximizing synergies so that changes are minimized.

Table 3.6 captures the various study and work items that are directly associated with C-V2X support and evolution within 3GPP, 3GPP technical reports, and specifications that are important for C-V2X, as well as ongoing and planned relevant standardization activities.

3.4 Application Aspects

V2X applications are applicable to at least one of the communication interfaces, including V2V, V2I, V2P, or V2N. Each application can support one or more use cases in a variety of scenarios and road environments. Use cases linked to V2X applications are often grouped into basic use cases like safety and advanced uses cases like traffic efficiency, infotainment, vehicle operations management, processing off-loading, and society and community groups. All these aforementioned applications are related to CAD and CRU services. The roadmap for introducing V2X applications is often organized by deployment phases called Day 1, Day 2, …, Day N. Such a deployment phase–based approach will be helpful to achieve a smooth transition for V2X communication from supporting only status-awareness messaging to fully autonomous driving.

Each V2X application is designed and developed regardless of the V2X RAT: i.e. to support both cellular and Wi-Fi IEEE 802.11p–based RATs. V2X applications must offer full functionality with the standard compliant V2X software stack for each region, such as US (WAVE), EU (ETSI), and China (SAE-C), including facility, network, and message security layers with the applicable standards: IEEE 1609.x, SAE J2735, ETSI ITS, SAE International, and SAE-C. These standards associations and consortiums have drafted basic requirements for V2X applications to ensure that they are OEM and vendor agnostic: i.e. support interoperability and secure communication. Security and privacy are other key aspects associated with V2X applications for which standards and policies have been published, as introduced here for direct communications and discussed further in Chapter 9.

V2X security aims to ensure message authentication, integrity and resiliency, non-repudiation, and non-replay while preserving privacy when desirable or required for V2X entities. Security for V2X in the US, EU, and China relies on certificates and a PKI. Public/private keys and cryptographic algorithms are used for signing and encrypting V2X exchanges. The PKI is responsible for generating certificates with appropriate validity scopes (dates, geographic, etc.) and permissions. The PKI hierarchical trust chain, topped by root authorities, allows verifying and trusting certificates provided by peer V2X communicating entities. Certificates are assigned to all entities participating in V2X communication, such as vehicles, road stations, application servers, PKI authorities, etc. Certificates specify one or multiple public keys, which can be used to verify the signature

Table 3.6 C-V2X related 3GPP activities and specifications.

SI/WI	3GPP code name	SI/WI Description
Improvement of V2X service Handling	V2XIMP	SP-180247 (SI) SP-181013 (WI)
Application layer support for V2X services	V2XAPP	SP-171071 (SI) SP-180898 (WI) CP-192077 (CT)
Architecture enhancements for 3GPP support of advanced V2X services	eV2XARC	SP-180733 (SI) SP-181121 (WI) CP-192078 (CT) SP-191125 (Security)
Study on NR Vehicle-to-Everything (V2X)	FS_NR_V2X	RP-190224
Study on V2X Media Handling and Interaction	FS_mV2X	SP-170799
Study on Security Aspects of 3GPP support for Advanced V2X Services	FS_eV2X_Sec	SP-190108
5G V2X with NR Sidelink	5G_V2X_NRSL	RP-191723
Study on enhancements to application layer support for V2X services	FS_eV2XAPP (Rel-17)	SP-190477
Study on V2X Services – Phase 2	FS_eV2XARC_Ph2 (Rel-17)	SP-190631
NR Sidelink enhancement	NRSL_enh (Rel-17)	RP-193257
Study on NR Sidelink Relay	FS_NR_SL_Relay (Rel-17)	RP-193253

Specifications and Technical Reports
TR 33.836 "Study on security aspects of 3GPP support for advanced V2X services"
TR 23.795 "Study on application layer support for V2X services"
TS 23.286 "Application layer support for Vehicle-to-Everything (V2X) services; Functional architecture and information flows"
TS 29.486 "Vehicle-to-Everything (V2X) Application Enabler (VAE) service; Stage 3"
TS 24.486 "Vehicle-to-Everything (V2X) Application Enabler (VAE) layer; Protocol aspects; Stage 3"
TR 23.786 "Study on architecture enhancements for the Evolved Packet System (EPS) and the 5G System (5GS) to support advanced V2X services"
TS 33.536 "Security aspects of 3GPP support for advanced Vehicle-to-Everything (V2X) services"
TS 23.287 "Architecture enhancements for 5G System (5GS) to support Vehicle-to-Everything (V2X) services"
TS 24.588 "Vehicle-to-Everything (V2X) services in 5G System (5GS); User Equipment (UE) policies; Stage 3"

(Continued)

Table 3.6 (Continued)

Specifications and Technical Reports
TS 24.587 "Vehicle-to-everything (V2X) services in 5G system (5GS); stage 3"
TS 33.536 " Security aspects of 3GPP support for advanced vehicle-to-everything (V2X) services"
TR 38.885 "Study on NR vehicle-to-everything (V2X)"
TR 26.985 "Vehicle-to-everything (V2X); media handling and interaction"
TR 37.985 "Overall description of radio access network (RAN) aspects for vehicle-to-everything (V2X) based on LTE and NR"
TR 38.886 "V2X services based on NR; user equipment (UE) radio transmission and reception"
TR 23.764 "Study on enhancements to application layer support for V2X services"
TR 23.776 "Study on architecture enhancements for 3GPP support of advanced vehicle-to-everything (V2X) services; phase 2"

of received messages or encrypt messages sent to other entities. Certificates are signed and must be verified. PKI infrastructures also support misbehavior reporting and certificate revocation.

3.4.1 EU Standardization

PKI authorities and protocols that allow such entities to communicate with each other – or that allow V2X end entities (vehicles, road stations, pedestrian devices, etc.) to communicate with PKI authorities – are specified in [46]. The European Commission further published a set of policies for operating such a PKI infrastructure; see [47, 58] (Figure 3.8).

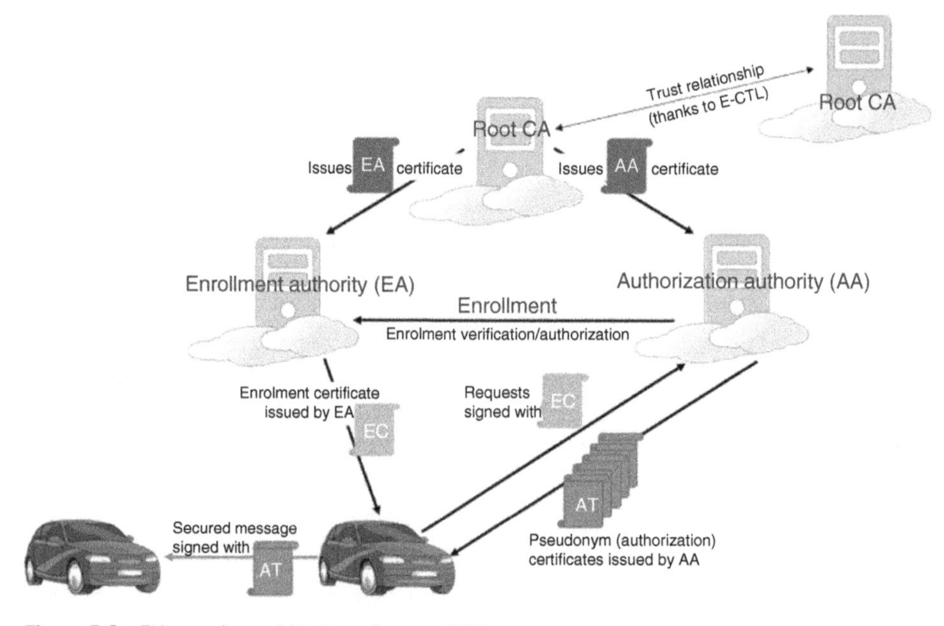

Figure 3.8 EU security architecture. Source: [58].

V2X messages are secured, i.e. signed and/or encrypted, based on the [48] standard, which is a profile of [49] (this amendment was published to address European-specific requirements). PKI messages, exchanged with PKI authorities, are secured based on the same standard. This standard specifies the format of certificates as well. The European Commission further published a set of policies applicable to V2X security in Europe [50].

3.4.2 US Standardization

PKI authorities and the protocols are specified in [51] (Figure 3.9). This specification is currently used by all US Department of Transportation connected vehicle pilots and other V2X projects in the US. A more formal standardization process has started and is being handled by the IEEE 1609 working group; the resulting 1609.2.1 standard is expected in 2020.

V2X messages are secured, i.e. signed and/or encrypted, based on the [52] standard. PKI messages, exchanged with PKI authorities, are secured based on the same standard. This standard specifies the format of certificates as well.

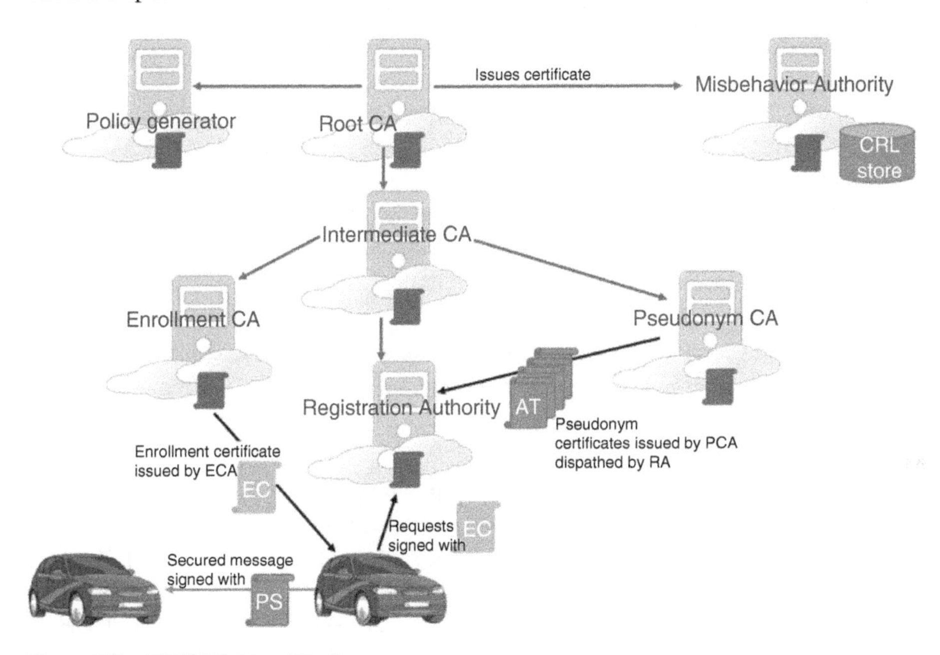

Figure 3.9 US SCMS (simplified).

3.5 Summary

This chapter described the past and present standardization and regulation landscape, which has been essential for defining V2X communications. Coordinated international efforts between several bodies from wireless to the automotive industry have been and continue to be undertaken for the research, development, and implementation of V2X communications systems. These efforts are leading to commonly agreed-on specifications through complex, elaborate procedures. We reviewed the current arrangements in different regions of the world for regulations, considerations for the radio spectrum, communication solutions, and standardization results as well as plans for the future. We took a deeper look at the 3GPP C-V2X front, which promises a technology that will achieve even the most challenging V2X requirements and which has recently gained considerable momentum with support from multiple automakers.

References

1 5G PPP Automotive Working Group. (2019). Business Feasibility Study for 5G V2X Deployment. White paper.
2 ITU-R. (2015). IMT vision - framework and overall objectives of the future development of IMT for 2020 and beyond. ITU-R M.2083–0.
3 5G PPP. (2015). 5G automotive vision. White paper.
4 ITU-T. (1988). Method for the characterization of telecommunication services supported by an ISDN and network capabilities of an ISDN. ITU-T I.130.
5 ITU. (2016). Radio regulations articles.
6 ITU-R. (2017). Frequency arrangements for implementation of the terrestrial component of International Mobile Telecommunications (IMT) in the bands identified for IMT in the Radio Regulations (RR). Working document. ITU-R WP5D.
7 3GPP. (2018). Evolved universal terrestrial radio access (E-UTRA); base station (BS) radio transmission and reception. 3GPP TS 36.104, v14.6.0.
8 3GPP. (2017). Base station (BS) radio transmission and reception (FDD). 3GPP TS 25.104, v14.2.0.
9 European Commission. (2016). Mandate to CEPT to develop harmonised technical conditions for spectrum use in support of the introduction of next-generation (5G) terrestrial wireless systems in the Union.
10 3GPP. (2017). First 5G NR specs Press release.
11 3GPP. (2017). 3GPP technical specification group radio access network; NR; user equipment (UE) radio transmission and reception (Release 15). 3GPP TS 38.101.
12 CEPT ECC. (2015). The harmonised use of the 5875–5925 MHz frequency band for Intelligent Transport Systems (ITS). ECC Decision (08)01.
13 CEPT. (2016). Harmonised use of the 63–64 GHz frequency band for Intelligent Transport Systems (ITS).
14 CEPT. (2019). Harmonised use of the 63.72–65.88 GHz frequency band for Intelligent Transport Systems (ITS).
15 World Radio Conference. (2015). Intelligent transport systems applications. Resolution 237.
16 Hartenstein, H. and Laberteaux, K.P. (2008). A tutorial survey on vehicular ad hoc networks. *IEEE Communications Magazine* 46 (6): 164–171.
17 Harding, J., Powell, G.R., Yoon, R. et al. (2014). Vehicle-to-vehicle communications: Readiness of V2V technology for application. Report DOT HS 812 014. National Highway Traffic Safety Administration, Washington, DC.

18 Chen, W., Delgrossi, L., Kosch, T., and Saito, T. (2014). Automotive networking and applications. *IEEE Communications Magazine* 52 (6): 118.

19 Dressler, F., Hartenstein, H., Altintas, O., and Tonguz, O.K. (2014). Inter-vehicle communication: quo vadis. *IEEE Communications Magazine* 52 (6): 170–177.

20 Festag, A. (2015). Standards for vehicular communication – from IEEE 802.11p to 5G. E & I Elektrotechnik und Informationstechnik.

21 Analysys Mason. (2017). Socio-economic benefits of cellular V2X. Ref 2011027–492.

22 Vinel, A. (2012). 3GPP LTE versus IEEE 802.11p/WAVE: "which technology is able to support cooperative vehicular safety applications?". *IEEE Wireless Communications* 1 (2): 125–128.

23 Lien, S.-Y., Deng, D.-J., Tsai, H.-L. et al. (2017). Vehicular radio access to unlicensed Spectrum. *IEEE Wireless Communications* 24 (6): 46–54.

24 Sjoberg, K., Andres, P., Buburuzan, T., and Brakemeier, A. (2017). Cooperative intelligent transport Systems in Europe: current deployment status and outlook. *IEEE Vehicular Technology Magazine* 12 (2): 89–97.

25 Naik, G., Choudhury, B., and Park, J.N. (2019). IEEE 802.11bd & 5G NR V2X: evolution of radio access technologies for V2X communications. *IEEE Access*.

26 Seo, H., Lee, K.-D., Yasukawa, S. et al. (2016). LTE evolution for vehicular-to-everything services. *IEEE Communications Magazine*: 22–28.

27 Chen, S., Hu, J., Shi, Y. et al. (2017). Vehicle-to-everything (V2X) services supported by LTE-based systems and 5G. *IEEE Communications Standards Magazine*: 70–76.

28 3GPP (2018). Proximity-services (ProSe) user equipment (UE) to ProSe function protocol aspects; Stage 3. 3GPP TS 24.334 v15.2.0.

29 3GPP. (2014). Enhanced LTE device to device proximity services. 3GPP TSG RAN RP-142311.

30 3GPP. (2018). Study on further enhancements to LTE device to device (D2D). UE to network relays for internet of things (IoT) and wearables. 3GPP TR 36.746, v15.1.1.

31 3GPP. (2018). Service requirements for V2X services; Stage 1. 3GPP TS 22.185, v15.0.0.

32 3GPP. (2018). Study on enhancement of 3GPP support for 5G V2X services. 3GPP TR 22.886, v16.2.0.

33 3GPP. (2019). Enhancement of 3GPP support for V2X scenarios. 3GPP TS 22.186, v16.2.0.

34 3GPP. (2017). Study on evaluation methodology of new V2X use cases for LTE and NR. 3GPP TSG RAN RP-171093.

35 3GPP. (2019). Study on evaluation methodology of new vehicle-to-everything V2X use cases for LTE and NR. 3GPP TR 37.885, v15.3.0.

36 3GPP. (2017). V2X phase 2 based on LTE. 3GPP work item description. 3GPP TSG RAN RP-171740.

37 3GPP. (2018). Study on NR V2X. 3GPP study item description. 3GPP TSG RAN RP-181480.

38 3GPP. (2019). Study on NR vehicle-to-everything (V2X). 3GPP TS 38.885, v16.0.0.

39 3GPP. (2019). 5G V2X with NR sidelink. 3GPP work item description. 3GPP TSG RAN RP-190766.

40 3GPP. (2018). Physical layer enhancements for NR URLLC. 3GPP study item description. 3GPP TSG RAN RP-182089.

41 3GPP. (2019). Study on physical layer enhancements for NR ultra-reliable and low latency case (URLLC). 3GPP TR 38.824, v16.0.0.

42 3GPP. (2019). Physical layer enhancements for NR URLLC. 3GPP work item description. 3GPP TSG RAN RP-190726.

43 3GPP. (2019). NR sidelink enhancement. 3GPP work item description. 3GPP TSG RAN RP-193257.

44 3GPP. (2019). Study on NR sidelink relay. 3GPP study item description. 3GPP TSG RAN RP-193253.

45 3GPP. (2019). Study on V2X services – phase 2. 3GPP work item description. 3GPP TSG SA SP-190631.

46 ETSI. (2018). Intelligent transport systems (ITS); security; trust and privacy management ETSI TS 102 941, V1.2.1.

47 European Commission. (2018). Certificate policy for deployment and operation of european cooperative intelligent transport systems (C-ITS). Release 1.1.

48 ETSI. (2017). Intelligent transport systems (its); security; security header and certificate formats. ETSI TS 103 097, V1.3.1.

49 IEEE. (2017). IEEE standard for wireless access in vehicular environments – security services for applications and management messages amendment 1. IEEE 1609.2a-2017.

50 European Commission. (2017). Security policy & governance framework for deployment and operation of european cooperative intelligent transport systems (C-ITS). Release 1.

51 Crash Avoidance Metrics Partnership. (2016). Security credential management system proof–of–concept implementation: EE requirements and specifications supporting SCMS software release 1.2.

52 IEEE. (2016). IEEE standard for wireless access in vehicular environments – security services for applications and management messages. IEEE 1609.2-2016.

53 5GAA. (2017). Timeline for deployment of LTE-V2X. December 2017.

54 Association of Radio Industries and Businesses. (2012). 700 MHz band intelligent transport systems.

55 3GPP. (2019). Architecture enhancements for V2X services. 3GPP TS 23.285, v16.2.0.

56 ETSI. (2020). Technical committee (TC) intelligent transport systems (ITS). https://www.etsi.org/committee/its.

57 3GPP. (2020). About 3GPP. https://www.3gpp.org/about-3gpp.

58 ETSI. (2018). Intelligent Transport Systems (ITS); Security; ITS communications security architecture and security management. ETSI TS 102 940, V1.3.1.

4

Spectrum and Channel Modeling

Taimoor Abbas[1], Mate Boban[2], Jose Leon Calvo[3], Yunpeng Zang[3], and Mikael Nilsson[4]

[1] *Huawei Technologies, Sweden*
[2] *Huawei German Research Center, Germany*
[3] *Ericsson Research, Germany*
[4] *Volvo Cars, Sweden*

This chapter treats two important topics related to vehicle-to-everything (V2X): spectrum and channel modeling. First, the spectrum needs to be associated with V2X communications. From the regulation point of view, spectrum resources are designated for different services, e.g. broadcasting service, satellite service, and mobile service, based on their specific spectrum demands. For V2X, we consider two types of applications: (i) connected automated driving (CAD) and (ii) connected road user (CRU) services. This chapter provides a general overview of the radio frequency spectrum, future needs and availability of the spectrum, as well as the research approach adopted in the fifth generation (5G) Communication Automotive Research and innovation (5GCAR) project for spectrum study. In addition, we have spectrum demand estimates published in June 2020 from 5G Automotive Association (5GAA) [1] that differentiate between basic and advanced use cases.

Next, the state-of-the-art channel models for V2X communications are described, including their most relevant components: line-of-sight (LOS) blockage analysis, path-loss (PL) and shadow-fading modeling, and fast-fading modeling. Based on the existing work, the second part of the chapter describes the gap in terms of the key missing components required for a complete solution for V2X channel modeling. Based on the gap and beyond prior art, the chapter describes (i) new vehicle-to-vehicle (V2V) measurements and characterization of channels above 6 GHz, (ii) multi-link shadowing model based on measurements below 6 GHz, and (iii) channel measurements for massive multiple input multiple output (MIMO) adaptive beamforming.

4.1 Spectrum and Regulations for V2X Communications

This section gives an overview of the spectrum used for V2X communications along with the regulations and recommendations suggested by the different regulatory agencies to achieve spectrum harmonization among the regions and technologies. This chapter focuses mainly

Cellular V2X for Connected Automated Driving, First Edition.
Edited by Mikael Fallgren, Markus Dillinger, Toktam Mahmoodi, and Tommy Svensson.

Table 4.1 European 5.9 GHz ITS band.

Frequency range	Usage	Regulation
5855–5875 MHz	ITS non-safety applications	ECC Recommendation (08)01
5875–5905 MHz	ITS road safety	Commission Decision 2008/671/EC, ECC Decision (08)01
5905–5925 MHz	Future ITS applications	ECC Decision (08)01

on European countries and their regulations, but it also includes detailed information about other relevant areas such as the United States and Asia.

4.1.1 Spectrum Bands in Europe

In the following subsections, we elaborate on spectrum bands in Europe, from the 5.9 GHz intelligent transport systems (ITS) spectrum, through the 5.8 GHz frequency for toll collection and the 60 GHz ITS band, to the International Mobile Telecommunications (IMT) bands.

4.1.1.1 ITS Spectrum at 5.9 GHz

In Europe, the spectrum 5855–5925 MHz, also known as the 5.9 GHz band, has been identified for ITS applications. Additionally, CEPT (European Conference of Postal and Telecommunications Administrations) designates the frequency sub-band 5875–5905 MHz on a non-exclusive basis for ITS traffic safety applications. CEPT administrations will consider in a future review of this decision the designation of the frequency sub-band 5905–5925 MHz for an extension of the ITS spectrum, noting that protection of ITS cannot be ensured in this band. Further, the European Commission decision on 5 August 2008 adopted the designation of the frequency band 5875–5905 MHz for ITS safety-related applications on a non-exclusive basis [2]. The spectrum band 5855–5875 MHz has been recommended to be made available for ITS non-safety applications to support and enhance ITS within CEPT in [3], amended 3 July 2015. According to [3, 4], ITS equipment complying with the European harmonized standard European Norm (EN) 302 571 [5] should be exempted from individual licensing for using the 5855–5925 MHz ITS band. Table 4.1 and Figure 4.1 provide an overview of the regulations and channel layout of the European 5.9 GHz ITS band.

European Telecommunications Standards Institute (ETSI) EN 302 571 is the harmonized standard for the European 5.9 GHz ITS frequency band. It specifies technical requirements and methods of measurement for radio transmitters and receivers of ITS products operating in this band. These technical requirements and methods of measurement are neutral to radio access technologies. Any ITS radio technology can operate in the 5.9 GHz ITS band, as long as it complies with the technical requirements specified in EN 302 571. Currently, both ETSI ITS-G5 and the 3rd Generation Partnership Project (3GPP) cellular V2X (C-V2X) proximity service direct communication interface 5 (PC5), including its first implementation in Release 14 long term evolution (LTE) V2X, are considered radio technologies operating in this band.

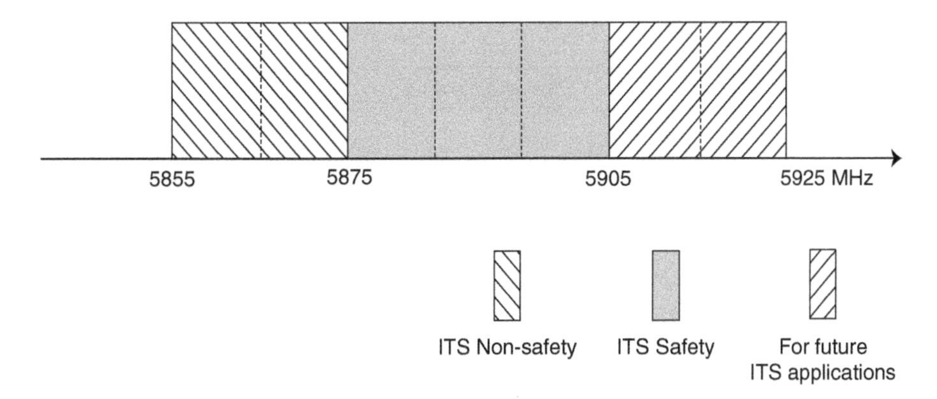

Figure 4.1 Dedicated 5.9 GHz ITS spectrum in Europe.

4.1.1.2 5.8 GHz Frequency for Toll Collection

At 5795–5815 MHz, four 5 MHz channels are designated for road transport and traffic telematics (RTTT) applications, e.g. electronic toll collection (ETC). The system uses passive dedicated short-range communications (DSRC) technology defined by European Committee for Standardization (CEN) standards EN 12253:2004, EN 12795:2003, EN 12834:2003, EN 13372:2004, and ISO 14906:2004. Due to closely allocated frequency bands, the CEN DSRC tolling system may experience interference from ITS-G5 or LTE-V2X stations working in the 5.9 GHz ITS band. Technical solutions ensuring coexistence between the 5.8 GHz tolling system and the 5.9 GHz ITS system are specified in ETSI technical specification (TS) 102 792.

4.1.1.3 60 GHz ITS Band

Above the 5.9 GHz band [6], it has been decided that CEPT administrators should designate the 63.72–65.88 GHz for ITS applications. Directive 98/34/EC of the European Parliament and Council of 22 June 1998 mandated that standards developing organizations (SDOs) like the ETSI will develop technical standards related to corporate communications using radio transmitters and receivers for ITS, which may include V2V and vehicle-to-infrastructure (V2I). These ITS networks operate over a short range with very wideband communications using a variety of directional media and high-gain antennas to enable a high degree of spectrum reuse; and they may use a flexible bandwidth scheme under which they normally operate in a wideband mode and periodically reduce their bandwidth (e.g. for antenna training and other activities).

The technical characteristics of these applications that are meant to operate in the 63.72 – 65.88 GHz band are described in ETSI TR 102 400. ETSI TC ERM TG37 also developed the European harmonized standard EN 302 686 for the 60 GHz ITS band. However, so far, there is no ITS product on the market using this band.

4.1.1.4 IMT Bands in Europe

According to [7], currently, the frequency band for direct V2X communication, known as the PC5 interface, is within Band 47 (5855–5925 MHz). Cellular operation over the Uu

Table 4.2 Frequency bands licensed to MNOs for 3GPP Uu communication in Europe.

3GPP band standardized for Uu	UL operating band F_Uplink_low (MHz)	UL operating band F_Uplink_high (MHz)	DL operating band F_Downlink_low (MHz)	DL operating band F_Downlink_high (MHz)	Mode	In Combination with Band 47 for V2X
Band 1	1920	1980	2110	2170	FDD	No
Band 3	1710	1785	1805	1880	FDD	Yes
Band 7	2500	2570	2620	2690	FDD	Yes
Band 8	880	915	925	960	FDD	Yes
Band 20	832	862	791	821	FDD	No

interface (air interface between base station and user equipment) can take place in various bands. In some of the bands, e.g. Bands 3, 7, 8, 39, and 41, communication over the Uu interface can be combined with communication over the PC5 interface in Band 47.

A list of the licensed bands used by mobile operators in Europe is presented in Table 4.2. The table's far-right column specifies whether the band can be used in combination with the V2X PC5 interface according to 3GPP specifications.

For ITS and automotive applications, public mobile operators' bands may be used for communication through the 3GPP Uu interface based on an agreement between mobile operators and car makers.

4.1.2 Spectrum Bands in Other Regions

In the following subsections, we elaborate on spectrum bands in regions other than Europe. We primarily discuss the US and China, but we also touch on Japan, Singapore, Australia, and Canada.

4.1.2.1 United States

The C-V2X spectrum and IMT bands in the US are outlined in the following subsections.

C-V2X Spectrum in the United States for ITS Applications

In October 1999, the Federal Communication Commission (FCC) allocated the 5.9 GHz band (5850–5925 MHz) for DSRC-based ITS applications and adopted basic technical rules for DSRC operations. Initially, only DSRC was allowed to use the 5.9 GHz band in the US. However, the FCC has decided [41] to operate C-V2X devices and Wi-Fi devices in the allocated 5.9 GHz band. Once the newly proposed FCC rules are applied, ITS operations will consist of C-V2X devices between 5.895 – 5.925 GHz and Wi-Fi devices between 5.850 and 5.895 GHz. The government's radiolocation service, which is used by high-powered military radar systems, and non-government fixed satellite service (FSS) uplink operations are co-primary in the 5.9 GHz band.

The new band plan (high level view) of the 5.9 GHz spectrum in the US is shown in Figure 4.2. It is expected that further details, e.g. required guard band between Wi-Fi and C-V2X or service definitions, will be defined in coming future (after the writing of this book).

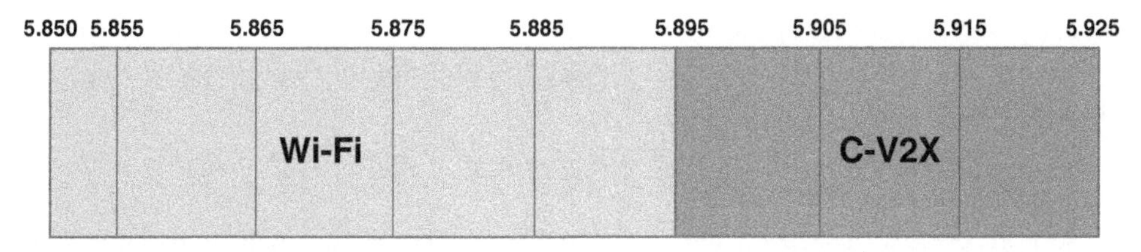

Figure 4.2 New spectrum allocation plan for the 5.850 – 5.925 GHz band in the United States.

IMT Bands in the United States

The use of frequencies in the United States is regulated by the FCC, and spectrum licenses are based on geographical areas. In [8], the suitable spectrum ranges below 6 GHz and above 24 GHz for a variety of 5G applications are specified. Considering the potential spectrum bands for 5G usage and depending on the usage scenarios, various spectrum ranges can be considered. A summary follows:

- Usage scenario eMBB (enhance mobile broadband): For an ultra-high-speed radio link, >24 GHz can be used; and for a high-speed radio link, 3–6 GHz can be used. Basically, the entire spectrum band can be used for ultra-low latency services. Considering reliability, <6 GHz is preferred.
- Usage scenario URLLC (ultra-reliable low-latency communications): >24 GHz can be used for short-range communication, and <6 GHz should be considered for medium-long-range communication.
- Usage scenario mMTC (massive machine type communications): In principle, all frequency bands can be used to support mMTC.

4.1.2.2 China

An overview of regulatory requirements of the 5 GHz band in different regions of the world is presented in [9], including China. In China, the regulation status of the 5 GHz band (5150–5925 MHz) is as follows:

- 5150–5350 MHz: wireless access system (WAS)/radio local area network (RLAN) Indoor only; max effective isotropic radiated power (EIRP): 200 mW
- 5470–5905 MHz: Under investigation

For 5G spectrum development in China, 5905–5925 MHz is assigned for LTE V2X trials, and 2 × 2.3 MHz in the frequency band 800 MHz are planned for narrowband Internet-of-Things (NB-IoT) (for private networks). For 5G eMBB deployment scenarios, the business success of 5G eMBB deployment will be enabled by the harmonization of frequency bands and by finding larger contiguous bandwidth, supporting the aggregation of frequency bands below and above 6 GHz.

On 7 January 2016, the Ministry of Industry and Information Technology (MIIT) launched a 5G compatibility trial using the International Mobile Telecommunications (IMT) 3.4–3.6 GHz band. Preliminary compatibility studies of 5G mm Wave bands have been conducted on certain frequency bands and presented during WRC-19 (Agenda Item 1.13). LTE-V2X trials have been conducted in Shanghai and Chongqing at 5905–5925 MHz using the prototype supported by a National Science and Technology Major Project.

The key points for the 5G spectrum are to support the aggregation of frequency bands; 3.3–3.6 GHz and 4.8–5.0 GHz have been the key 5G bands for initial 5G deployment. V2X will be a key application of 5G URLLC.

4.1.2.3 Other Regions of the World

This section covers the spectrum allocation for ITS and automotive applications using V2X communications in countries and regions other than the European Union, US, and China. Four countries are discussed – Japan, Singapore, Australia, and Canada – and their V2X spectrum allocations are summarized.

In Japan, the frequency band 5770–5850 MHz is allocated for the V2V communication system as well as for ETC using Japanese DSRC technology. Since the spectrum is very congested in that band in several parts of Japan, it is not a preferred band for V2V communication. Therefore, it was decided that 760 MHz (755.5–764.5 MHz band) is allocated for ITS applications. In Japan, a new standard for ITS applications using V2V communication named ARIB STD-T109 [10] was developed for "Driving Safety Support Systems" in the 700 MHz band.

In Singapore, Australia, and Canada, the frequency band 5855–5925 MHz has been allocated for ITS applications. The Australian ITS radiocommunications class license was finalized by the Australian Communications and Media Authority (ACMA) in 2017. According to this ITS class license, to operate in 5855–5925 MHz, ITS stations in Australia must comply with ETSI Standard EN 302 571 [11].

4.1.3 Spectrum Auctions Worldwide

5GAA has published a study on spectrum needs of intelligent transportation systems (ITS) [1], where the amount of bandwidth required for the introduction of day 1 and advanced use case is predicted to be 50 MHz < 1 GHz and 500 MHz from 1–7 GHz. Details for spectrum needs are provided in the reference document, where some are highlighted for selected use cases in Table 4.3.

A primary aspect of achieving spectrum harmonization is internal regulations within the different countries. A common way to organize the spectrum bands in Europe and the United States is by auctioning a specific bandwidth to interested operators that secure a specific spectrum by bidding. On the other hand, in China, the government awards a certain spectrum to operators with regard to their number of costumers and coverage. In Japan, the system is a mixed scheme: there is no auction, but operators are required to fulfill some network requirements once spectrum bands are awarded to them. Finally, in South Korea, spectrum rewarding focuses on the fast deployment of 5G systems, in collaboration with operators.

4.1.3.1 Europe

In this section, we have selected six representative European countries where spectrum either started to be auctioned or rules for mobile airwaves auctions are confirmed at the writing of this book. This region is the main target from this section, and a higher level of detail regarding European auctions is given.

Table 4.3 Spectrum needs for use cases in different geographical settings.

		Spectrum needs (MHz)		
		Rural	**Urban**	**Dense urban**
V2N	Software Update of Reconfigurable Radio System (DL)	12	12	3
	Autonomous Vehicle Disengagement Report (UL)	13	4*	4*
	Patient Transport Monitoring (UL)	28	8*	8*
	Tele-Operated Driving (UL)	100	30*	30*
	Obstructed View Assist (DL)	**	40	90
	Infrastructure Assisted Environment Perception (UL)	**	90–184	
	In-vechicle entertainment (DL)	**	450	900
	High Definition Map Collection and Sharing (UL)	**	360	720
	* The urban and dense urban sectors must both support at least one user.			
	** To meet the use case requirements, an incresed spectrum availability is not sufficient in itself, but needs to be complemented with improved netword and device capabilities, and is under consideration by 5GAA.			
V2V/I/P	70–75 MHz of ITS spectrum in the 5.9 GHz band (allocated in many region and under consideration in other regions) is needed to support the basic safety and advanced use cases under consideration today. This demand is valid for all geographical areas.			

Italy

Italy is one of the first European countries were all the spectrum bands relevant for 5G (700 MHz, 3.4–3.8 GHz, and 26 GHz) have been awarded by 2018. Therefore, the current status of the spectrum bands is final, along with the total cost of the spectrum auctions.

In response to the Commission's study mandate concerning the technical harmonization of the 700 MHz band for broadband wireless land-based electronic communications services, this band is made up of the main portions of coupled spectrum frequency-division duplex (FDD) 703–733 MHz and 758–788 MHz (60 MHz in total) intended for wireless broadband applications of interest for the development of mobile networks. Within these portions, the blocks are assigned according to multiples of 5 MHz. Additional spectral resources contained within the 700 MHz band are represented by the additional portions positioned in the guard bands (694–703 MHz and 788–791 MHz) or in the central range of the band (733–758 MHz).

Regarding the 3.4–3.8 GHz band, the rules for assigning and using the frequencies available in this band for terrestrial electronic communications systems are based, in addition to the applicable technical and regulatory legislation, on the availability of the 3.4–3.8 GHz band in the national territory, deriving from a partial process of freeing the band in question from existing uses. The Italian government had defined, based on the band's state of occupation, a national progressive re-farming action for the existing fixed service utilization, limited to some 30 MHz radio channels positioned in a portion approximately corresponding to the upper part of the band, i.e. sub-band 3700–3800 MHz. In the lower sub-band

3600–3700 MHz, on the other hand, persistence of the fixed service channels made it possible to identify local areas where even making available some portions of the spectrum did not allow generalized use of mobile systems or fixed communications networks.

Finally, the 26 GHz band has been used in Italy, according to the provisions of the CEPT, for applications of the fixed wireless local loop (WLL) service, whose measures for the release of rights to use frequencies available for networks broadband radios have recently been updated by the Italian government. In 2018, the Italian government officially authorized the auction for the 15 years assignment of 5G frequencies [12, 13], which unlocked 1275 MHz of spectrum partitioned as follows:

- 1000 MHz in the 26 GHz band
- 200 MHz in the 3700 MHz band
- 75 MHz in the 700 MHz band (FDD)

During the pre-auction phase, the following companies were prequalified for the auction: Iliad Italia, Fastweb, Telecom Italia (TIM), Vodafone Italia, Wind Tre, Linkem, and Open Fiber. However, at the start of the bidding phase in September, only the first five contenders in the list remained as auction participants. The overall amount after 171 rounds of bidding reached €6,550,422,258 well over the 4 billion minimum base amount established by the government by law. Licenses in the 3.6–3.8 GHz band were the most attractive, achieving combined bids of €4.35 billion. TIM and Vodafone obtained the two largest packets of frequencies in this band, while Wind Tre and Iliad acquired smaller allocations. The 700 MHz auction, which was completed in the first days of the sale process, brought in €2.04 billion, with the spectrum won by TIM, Vodafone, and Iliad. All five were successful in securing 26 GHz frequencies, with the five available blocks raising €163.7 million. Three extra blocks in the 700 MHz band for supplemental downlink (SDL) to enhance FDD 4G network downlink capacity (from Release 10) were also offered, but no operator presented an offer for them. The final assignments indicating block repartition and costs can be checked in [14]. As a summary of the auction, the total repartition by band and contender is shown in Table 4.4.

Sweden

In Sweden, the 700 MHz band was auctioned in December 2018 [15]. PTS (Swedish Post and Telecom Authority) announced the results of the first stage of the auction: a total of SEK 2.825 billion (€274 million) for a total of 40 MHz. After 46 rounds and 6 days, Telia Sweden won 2 × 10 MHz FDD for SEK 1.383 billion (~€132 million), while Net4Mobility won 2 licenses of 2 × 5 MHz FDD for a total price of SEK 1.442 billion (~€138 million). The reserve price was set at 50 MSEK (~€5 million) per 5 MHz. Moreover, coverage obligations are included in Telia's license, with a reserved amount planned for this reason. A summary of the auction is given in Table 4.5.

The 3.4–3.8 GHz frequency band has not been auctioned yet, but some public consultation has been performed, achieving general coordination and recommendations. As of 2020, PTS regulators are planning the frequency allocation in the near future. The allocation of these bands has been postponed by the Swedish PTS due to work being carried out to update the country's Electronic Communications Act (LEK). In this action, up to 15 nationwide licenses will be sold in the 3.5 GHz band (3400 MHz-3720 MHz), incorporating

Table 4.4 Summary of bandwidth and cost by mobile network operator (MNO) in Italy.

Band	Winning bidder	Price (€)	Total spectrum won (MHz)
700 MHz	Vodafone	683,236,396	10
	TIM	680,200,000	10
	Iliad	676,472,792	10
3.7 GHz	TIM	1,694,000,000	80
	Vodafone	1,685,000,000	80
	Wind	483,920,000	20
	Iliad	483,900,000	20
26 GHz	TIM	33,020,000	200
	Iliad	32,900,000	200
	Vodafone	32,586,535	200
	Wind	32,586,535	200
	Fastweb	32,600,000	200

Table 4.5 Summary of bandwidth and cost by MNO in Sweden.

Bidder	State	Price (€)	Amount for coverage (€)
Telia Sverige AB	2 × 10 MHz, FDD 713–723 MHz 768–778 MHz	131,988,500	28,638,000
Net4Mobility HB	2 × 5 MHz, FDD 723–728 MHz 778–783 MHz	68,823,643	-
Net4Mobility HB	2 × 5 MHz, FDD 728–733 MHz 783–788 MHz	68,823,643	-

a maximum of 320 MHz. One license will get 40 MHz of spectrum while the 14 others will have 20 MHz only. The minimum bid is 100 million SEK (9.4 million EUR) per lot, 1.5 billion SEK (140.3 million EUR) in total. Up to 80 MHz will be on offer in the 2.3GHz range, split into eight nationwide concessions, with a reserve price of 20 million SEK (1.9 million EUR) per licence, 160 SEK million (15 million EUR) in total.

In addition to these bands, PTS is also targeting assigning the frequency space 3.7–3.8 GHz for local states. The frequency ranges 3400–3700 MHz and 2300–2380 MHz are available to terrestrial systems capable of providing electronic communications services.

Before awarding the spectrum to companies, PTS is planning to prepare the allocation of local block permits to be assigned without a selection procedure, if possible, and to make 3.7–3.8 GHz band frequency space available to terrestrial electronic communications service providers. 2.3–3.5 GHz will be national licenses, and 2.3 GHz will have certain restrictions protecting other radio applications. For the 2.3 GHz band, there will be eight national licenses of 10 MHz each. A significant aspect of the awarded spectrum bands is that the license duration will be 25 years for both 2.3–3.5 GHz. Noting that there are existing regional/local licenses in the 3.5 GHz band, some spectrum may become available later (2020–2023), with the consequence that the license duration will be shorter in order to have the same expiration date for all licenses.

A spectrum cap will be applied to 3.5 GHz so that at least three licensees/operators can obtain spectrum; the cap will be 120 MHz for this band, while for 2.3 GHz, there will be no such limitation. The current use of 3.5 GHz will be protected until the licenses expire; if not used, PTS intends to enable sharing for secondary use. The investigation of 3.7–3.8 GHz will be coordinated with the pre-study of 3.8–4.2 GHz to determine, among other things, which band is more appropriate for this use and which form of licensing is best. If the answer is 3.8–4.2 GHz, then 3.7–3.8 GHz could become available for national licenses. Responses to the consultation in October 2018 about local licenses, proposed property-based licenses in urban areas and property-based and "regional" licenses outside of those areas. If this is applied, the spectrum would be split between the two uses outside of urban areas. It has been agreed that there will be no requirements on coverage or similar, since the frequency bands are too high to be suited to such requirements. Moreover, it is understood that the offered bands and applications will primarily be used in cities and urban areas where the required coverage is already achieved.

For the 24.25–27.5 GHz band, there were two options: to open 26.5–27.5 GHz for assignment in the future after technical conditions for 5G usage in the band were clarified, or to wait for international harmonization before starting the assignment for the entire band. The Swedish government decided to follow the second option following the conclusion from the World Radiocommunication Conference (WRC) 2019 meeting. The Swedish government decided to assign a part of the 24.25–27.5 GHz band. In this case, PTS plans to assign the 24.25–25.1 GHz band in the near future.

Germany

On 26 November 2018, the President's Chamber determined the award rules and auction rules. Including the already-awarded spectrum for 700 MHz in 2015, the overall situation for the spectrum bands in Germany is given in Table 4.6 [16]. In the middle of 2019, the nationwide use of 2 × 60 MHz (paired) from the 2 GHz band and 300 MHz (unpaired) for the 3.6 GHz band has been auctioned. For industrial purposes, 3.7–3.8 GHz is locally available; i.e. they have not been actioned but can be purchased for selected areas.

In May–June 2015, Europe's first 700 MHz radio spectrum auction was completed in Germany along with the 900, 1800, and 1500 MHz bands. The 700 MHz frequency band has been allocated for 4G, and it can be potentially used for 5G technologies with a validity of 15 years. The current situation for the 700 MHz frequency band is given in [16]. Including the 900–1800 MHz bands, which were also awarded during the same auction in 2015, the total allocated bandwidth was 270 MHz, divided as follows:

Table 4.6 Auctions held and planned in Germany.

Frequency band	Auction details	Notes
700 MHz (paired 703–733 MHz/ 758–788 MHz)	2015	This band was allocated in 2015 for use by 4G LTE and can be re-farmed for 5G.
2.1 GHz (60 MHz paired 1920 MHz–1980 MHz/ 2110 GHz–2170 GHz)	Auction in Spring 2019 (finished June 2019)	
3.6 GHz (300 MHz at 3.4–3.7 GHz)	Auction in Spring 2019 (finished June 2019)	
3.7–3.8 GHz (to be offered for local applications, such as networks for industrial campuses)	Not announced (at the writing of this book)	
24.25–27.5 GHz (to be offered for local applications, such as networks for industrial campuses)	Not announced (at the writing of this book)	

- Telefonica Deutschland obtained 60 MHz divided into three bands (2 × 10 MHz in 700 MHz, 2 × 10 in 900 MHz, and 2 × 10 in 1800 MHz) for a total of ~€1.2 billion.
- Deutsche Telekom obtained 100 MHz divided into four bands (2 × 10 MHz in 700 MHz, 2 × 15 in 900 MHz, 2 × 15 in 1800 MHz, and 20 MHz in 1500 MHz) for a total of ~€1.8 billion.
- Vodafone GmbH obtained 110 MHz divided into four bands (2 × 10 MHz in 700 MHz, 2 × in 900 MHz, 2 × 25 in 1800 MHz, and 20 MHz in 1500 MHz) for a total of ~€2.1 billion.

In March 2019, an auction to award the spectrum in 2 GHz and 3.4–3.7 GHz began in Germany. The operators involved in the auction were Telefonica DE, Vodafone, Deutsche Telekom, and the newcomer 1&1 Drillisch. As part of this spectrum auctions, Federal Network Agency promote the rapid introduction of 5G. Spectrum demands and key performance indicators (KPIs) for the 5G network were prescribed to support CAD for roads, railways and water roads [17]:

- Coverage and delay requirements to be supplied partly by the end of 2022 and partly by the end of 2024:
  - All federal highways with at least 100 Mbit/s
  - All national and state roads with at least 50 Mbit/s
  - The seaports and main waterways with at least 50 Mbit/s as well
  - All other railways with at least 50 Mbit/s
- Roaming and infrastructure sharing:
  - The Federal Network Agency expects cooperation between network operators in areas where expansion by a single network operator is not economically viable. Increased cooperation, such as infrastructure sharing and roaming, can significantly reduce the costs of supplying space.

Table 4.7 Auction results for the 5G auction in Germany (August 2019).

	Drillisch Netz AG (MHz)	Telefónica Germany GmbH & Co. OHG (MHz)	Telekom Deutschland GmbH (MHz)	Vodafone GmbH (MHz)
2 GHz	2 × 10	2 × 10	2 × 20	2 × 20
3.6 GHz	50	70	90	90
Total	70	90	130	130

These requirements are strict conditions to allow any solvent company to bid for frequencies during the auction that started on 19 March 2019. These strict requirements have attracted criticism from parties such as businesses and other actors. While the idea of these requirements given by the Federal Network Agency promotes an equal deployment of 5G technology in all areas, not just picking up business in major urban areas, it could potentially be a showstopper for the commercial deployment of 5G. The outcome of the auction in Germany for 5G is as shown in Table 4.7, with a total combined cost of €6,549,651,000 after 497 bidding rounds.

Spain

In Spain, part of the 3.6 GHz spectrum band (3400–3600 MHz) was awarded for 5G services in 2016, with licenses valid until 2030. This spectrum can already be used to deliver 5G services pursuant to General Telecommunications Law 09/2014. National Plan 5G 2018–2020 [18] contemplated the reordering of the already-occupied part of the 3.6 GHz band to establish larger and more contiguous blocks in line with what is recommended for the development of 5G technology, and the first action was to auction this band. The government has provisions in place to award frequencies in the 700 MHz band, which had been previously used by the digital terrestrial television (DTT) services where new blocks of spectrum will also be auctioned to cover more extensive areas of territory with 5G, with lower user density.

As mentioned before the 700 MHz band is currently used for DTT services, which will be released by the time of the spectrum auction. So that mobile network operators (MNOs) can commercially dispose of the frequencies at the moment of winning the award. That is, the spectrum will be auctioned when it is available to operators, something that did not happen with the auction of 4G (which was auctioned in 2011 but not available to the winning bidders until 2015).

In the 26 GHz band, 400 MHz are available for immediate use on the lower portion of the band and 500 MHz plus another 500 MHz with some limitations on the higher portion of the band. The remaining part of the band is used by radio links in point-to-point fixed services over mobile trunk networks. This band is also included in the National Plan 5G and will be tendered to operators in the medium term, making more contiguous spectrum available.

In 2018, Spain's Ministry of Energy, Tourism, and Digital Agenda launched the auction process for 200 MHz still available in the 3.7 GHz band, defined as the priority band. The auctioned spectrum was divided into 40.5 MHz blocks, each of which will be valid for a period of 20 years (until 2038) and had a starting price of €2.5 million, with a minimum expected value of more than €100 million. No MNO has reached the limit of 120 MHz imposed by the Spanish government. As Vodafone has been the operator that has invested the most in the auction, it will have priority to choose and could have its 90 MHz in a common block, making better use of the spectrum. This is not the case for other MNOs awarded in the 3.7 GHz auction. The auction had 34 rounds, and the total final amount raised was €1410.7 million. The final assignment has the following block repartitions and costs, which can be checked in [19]. In addition to the acquisition costs (437.5 million), the operators will spend €104.6 million due to the interest fee and €868.5 million due to a spectrum reservation fee, making the total cost for the operators €1410.7 million.

United Kingdom

In the UK, free view television and wireless microphones use the 700 MHz band. The government has contributed £500–600 million to clear the spectrum, a process that began in March 2017 with the reconfiguration of a DTT transmitter. The aim is to migrate DTT to the 470-690 MHz spectrum and make alternative spectrum available for wireless microphones.

Ofcom's auction of spectrum in the 2.3 GHz (for 4G improvements) and 3.4 GHz bands concluded in April 2018. The 2.3–3.4 GHz spectrum was previously used by the Ministry of Defense but has been freed up by the government to make it available for civil uses. Part of the 3.4 GHz spectrum is used for 4G wireless broadband, such as by Relish in London, which is now owned by Three following its acquisition of UK Broadband. The spectrum in the 3.6–3.8 GHz bands is partially in use by fixed links and satellite services, but Ofcom also plans to auction 120 MHz of spectrum in this bandwidth. Ofcom published its final rules for the 2.3–3.4 GHz auction in July 2017, designed to reflect recent market developments and safeguard competition now and in the future. The rules imposed two restrictions on bidders to limit the amount of spectrum dominant operators could win:

- No operator would be able to hold more than 255 MHz of immediately usable spectrum, i.e. in the 2.3 GHz band, following the auction.
- No operator would be able to hold more than 340 MHz of the total amount of spectrum following the auction, equivalent to about 37% of all the mobile spectrum useable in 2020. This includes spectrum available in the completed auction and in the 700 MHz band.

Ofcom did not propose any coverage obligations on the winning bidders as it did with the 4G auction in 2013, because the provision of the latest frequencies is more about boosting network capacity than expanding network coverage. The auction of spectrum in the 2.3–3.4 GHz bands concluded in April 2018, with all four of the nation's existing MNOs securing new frequencies. The total amount of the auction was £1,355,744,000 where £1,149,844,000 was for the 5G band, well above previous estimates of £630 million to 1 billion. O2 UK emerged as the biggest spender after offering £205.9 million for all 40 MHz of the 2.3 GHz spectrum on offer, as well as paying £317.7 million for 40 MHz in the 3.4 GHz band. Vodafone UK made the largest offer for 3.4 GHz frequencies, meanwhile, with its £378.2 million bid netting it a 50 MHz block in that band. Everything Everywhere Ltd. bagged a 40 MHz block of 3.4 GHz spectrum with a £302.6 million bid. Three UK walked away with a smaller allocation than any of its rivals, agreeing to pay £151.3 million for a 20 MHz block of 3.4 GHz spectrum, which it accumulated with the 124 MHz it already has from the acquisition of UK Broadband.

France

In January 2019, ARCEP (French Telecommunications Regulatory Authority) announced temporary frequency authorizations to develop 5G pilots in France. The regulator will allow for experimentation in the 3.5 GHz band (3400–3800 MHz) in the metropolitan areas of Lyon, Bordeaux, Nantes, Lille, Le Havre, Saint-Étienne, Douai, Montpellier, and Grenoble.

The 700 MHz band has already been affected by operators during the 4G spectrum auction. It has a limited bandwidth but offers good propagation conditions. France raised €2.79 billion in a 4G spectrum auction in November 2015, according to the country's telecoms regulator. After 11 rounds of bidding, the price of the spectrum reached €466 million for each 5 MHz block. Bouygues Telecom and SFR were awarded a block each, while Free Mobile and Orange were awarded two blocks each. Under the rules of the auction, no bidder could win more than three blocks of 700 MHz spectrum. For the French regulator [20], the 700 MHz band is neutral in terms of technology: if one operator has licensed spectrum in that band, it has the choice to deploy 4G or 5G. Furthermore, technological neutrality will be introduced on 900 and 1800 MHz. After re-farming (the 900 MHz band is used for Global System for Mobile Communications (GSM) and 1800 MHz for LTE), these bands could be used for 5G.

The 3.4–3.8 GHz band will not be completely available before 2026 (it is used by satellites and state and private operators). If a reorganization occurs, 280–340 MHz will be available before 2026; if not, only 220 MHz will be available over 400 for the entire band. The attribution process will be the subject of a second public consultation when after it, the frequencies-attribution process will be launched in the 3.5 GHz band. At the time of writing ARCEP has indicated that the 5G spectrum auction is delayed. ARCEP aspires to run an auction of spectrum in the 3.4 GHz to 3.8 GHz bands.

ARCEP has decided to dedicate 40 MHz to professional mobile radio (PMR) use from 2570 to 2620 MHz. The rest of the band is used for 4G (LTE, LTE-Advanced). The 26 GHz band is already available. The condition of usage of this band is not currently defined in France. The license authorization could be restricted to a geographical area. In January 2019, ARCEP created a request for proposals for innovative projects willing to experiment with 5G mm Waves.

4.1.3.2 United States

The FCC is focusing on making additional low-, mid-, and high-band spectrum available for 5G services. The overview of the 5G spectrum in the US is as follows [21]:

- *High-band*: The FCC has made auctioning high-band, millimeter-wave spectrum a priority. The FCC held its first 5G spectrum auction in 2018 in the 28 GHz band. In 2019, the FCC will hold an auction in the 24 GHz band starting on March 14 and auctions in the upper 37, 39, and 47 GHz bands later in the year. With these auctions, the FCC will release almost 5 GHz of 5G spectrum into the market – more than all other flexible use bands combined. The FCC is working to free up another 2.75 GHz of 5G spectrum in the 26 and 42 GHz bands.
- *Mid-band*: Mid-band spectrum has become a target for 5G buildout, given its balanced coverage and capacity characteristics. With the work on the 2.5, 3.5, and 3.7–4.2 GHz bands, up to 844 MHz could be made available for 5G deployments.
- *Low-band*: The FCC is acting to improve the use of low-band spectrum (useful for wider coverage) for 5G services, with targeted changes to the 600, 800, and 900 MHz bands.

- *Unlicensed*: Recognizing that unlicensed spectrum will be important for 5G, the agency is creating new opportunities for the next generation of Wi-Fi in the 6 GHz and above 95 GHz band.

FCC has completed its first 5G auction, with a sale of 28 GHz spectrum licenses raising $702 million. The 28 GHz auction took 38 days and 176 rounds of bidding to complete. A total of 3072 licenses offered in 425 MHz blocks were up for grabs. Of these, only 107 received no acceptable bids. The FCC auctioned two 425 MHz bands, covering 27.500–27.925 and 27.925–28.350 GHz, by geographic county. The licenses for upper microwave flexible use service authorize both fixed and mobile operations. Winners of the 28 GHz licenses (Auction 101) will not be announced publicly until after the 24 GHz auction concludes, when the names of the winners in both auctions will be released.

The 24 GHz will be also assigned through auction, followed by auctions for three more spectrum bands. Auction 102 is the FCC's second auction of upper microwave flexible use service (UMFUS) licenses [22]. Auction 102 offers 2909 licenses in the 24 GHz band. The lower segment of the 24 GHz band (24.25–24.45 GHz) will be licensed as two 100-MHz blocks, and the upper segment (24.75–25.25 GHz) will be licensed as five 100-MHz blocks. Bidding in Auction 102 began on 14 March 2019.

AT&T led the way with total bids of $982 million in the 24 GHz auction, buying 49% of all the available licenses, while Verizon spent the most in the 28 GHz spectrum auction with total bids of $506 million for the spectrum. Verizon won 72% of all the available 28 GHz licenses. T-Mobile was the second-place bidder in the 24 GHz auction with $803 million in total bids, giving the company 40% of all the available licenses. Meanwhile, T-Mobile spent $39 million on the 28 GHz auction. And US Cellular won the second place in the 28 GHz auction with a total of $129 million in bids, close to its fourth-place finish in the 24 GHz auction.

4.1.3.3 Asia

In the following subsections the Asian spectrum auctions in China, Japan and South Korea are outlined.

China

The MIIT of China has allocated 5G frequencies to the three major operators in China: China Telecom, China Unicom, China Mobile. There is no neutrality in terms of technology used, which means the deployed technology must be 5G. This has been done to hasten 5G deployment in China. Additionally, the attribution process for the spectrum is not clear: the licenses given to the operators are temporary [23]. However, there is no clear vision on the position of the operators for having such a limited time allocation. The government of China believes that 5G will boost the internal economy and industrial development. In contrast to the situation in Europe and the United States, no auction has been organized in China, and therefore, no spectrum cost has been published. The spectrum attribution in 2018 is as follows:

- *China Unicom*: 3.5–3.6 GHz (100 MHz) frequency range for a nationwide 5G trial rollout until June 2020. The telco said that it would gradually cease to use the frequency in the 2.555–2.575 GHz range that it had been using for 5G trials and will progressively return it to the MIIT.

- *China Telecom*: 3.4–3.5 GHz (100 MHz) frequency range for 5G trials in mainland China. Under the agreement with the Chinese government, China Telecom will return its 2.635–2.655 GHz spectrum in 2020.
- *China Mobile*: 2515–2675 MHz and 4.8–4.9 GHz (260 MHz in all) ranges for nationwide 5G trials. The 2575–2635 MHz spectrum frequency is re-farmed from China Mobile's pool of time-division (TD) LTE spectrum.

Currently China Mobile is the biggest operator in the world, encompassing 755 million subscribers, while 200 million 4G users are subscribed to China Unicom. Moreover, China Telecom provides service to 153 million subscribers. Thus, spectrum distribution is consistent with the number of subscribers, since China Telecom has more spectrum allocated than the combination of the other two operators. Nevertheless, it is worth mentioning that China Mobile has no spectrum available in the 3.5 GHz band.

Japan

The Ministry of Internal Affairs and Communications (MIC) announced on 10 April 2019 that all four applicants (NTT Docomo, KDDI, Softbank, and the newcomer Rakuten) were awarded radio frequencies and licenses to roll out 5G services. Each licensee was awarded 400 MHz spectrum on the 28 GHz frequency, while three of them were awarded 200 MHz on 3.7 GHz except Rakuten, which requested 100 MHz. The spectrum distribution process is similar to what happened in China in terms of pricing, where no spectrum auction was conducted. The difference is that the spectrum allocation was organized by the MIC regarding requirements that the operators needed to fulfill. The distribution of spectrum was done in April 2019, and the criteria took into account the commitments of each operator in terms of 5G investments and percentage of population covered by 5G (Table 4.8).

The spectrum band of 27 GHz was also awarded, obtaining the allocation for the four operators given in Table 4.9.

NTT and KDDI committed to population coverage (>90%) and strong investments in 5G. Thus, the results of spectrum attribution seem consistent with the investment commitments made by the four operators. The conditions for the allocation of spectrum included commitments to commence services in every prefecture of the nation within two years and set up 5G base stations in at least half the country within five years. The communications ministry also divided Japan into 4500 blocks, requiring operators to set up base stations in at least half of them within five years. Docomo and KDDI each plan to achieve coverage of more than

Table 4.8 Final spectrum assignments for the 3.5 and 4 GHz bands in Japan [24].

Operator	Spectrum band (GHz)	Bandwidth (MHz)
NTT Docomo	3.6–3.7 and 4.5–4.6	200
KDDI	3.7–3.8 and 4.0–4.1	200
Softbank	3.8–3.9	200
Rakuten	3.9–4.0	100

Table 4.9 Final spectrum assignments for the 27 GHz band in Japan.

Operator	Spectrum band (GHz)	Bandwidth (MHz)
NTT Docomo	27.4–27.8	400
KDDI	27.8–28.2	400
Softbank	29.1–29.5	400
Rakuten	27.0–27.4	400

Table 4.10 Investment engaged for coverage deployment in Japan.

Operator	Planned investments for deployment	Expected coverage
NTT Docomo	7 billion USD	More than 90%
KDDI	4.1 billion USD	More than 90%
Softbank	1.8 billion USD	64%
Rakuten	1.7 billion USD	56%

90% in that time. SoftBank and Rakuten set less ambitious goals, at 64% and 56%, respectively. To achieve the requirements stated by the MIC, the four operators envision investing the amounts in Table 4.10 to enhance and deploy networks linked to the expected coverage.

South Korea

Korea is the first state in the world to commercialize 5G subscriptions for the public. An auction was organized in the summer of 2018. All three operators – SK Telecom, KT, and LG Uplus – agreed to launch their 5G offers the same day (3 April 2019).

280 MHz bandwidth of the 3.5 GHz spectrum and 2400 MHz bandwidth of the 28 GHz spectrum were available in the block auction. Each telco had a 10-block cap per spectrum. SKT emerged as the biggest spender overall, bidding a total of $1.10 billion for 100 MHz in the 3.5 GHz band (3600–3700 MHz) and a further $173 million for 800 MHz of bandwidth in the 28 GHz band (28.1–28.9 GHz). For its part, KT Corp offered a total of $828 million for its 3.5 GHz spectrum (3500 –3600 MHz), although it spent the most of any operator on the 28 GHz frequency band, obtaining the 26.5–27.3 GHz block with a $175 million bid. The final nationwide operator, LG Uplus, spent $680 million on an 80 MHz block in the 3.5 GHz band (3420–3500 MHz) and committed a further $170 million for its 28 GHz allocation (27.3–28.1 GHz). The overall attribution of spectrum in the 3.5 GHz band is given in Table 4.11.

For the 28 GHz band, an overview of the spectrum situation is as shown in Table 4.12.

The spectrum allocation for 5G in Asia has already been completed, showing the involvement of these nations in leading 5G deployment. In China and Japan, as opposed to Europe and the US, there was no public auction, and frequency bands were allocated to operators by a central government (China) or based on the expected coverage and expenditure of the

Table 4.11 Final spectrum assignments for the 3.5 GHz band in South Korea.

Operator	Spectrum band (GHz)	Bandwidth (MHz)	Cost ($)	Period (yr)
SK Telecom	3.6–3.7	100	1.1 billion	10
LG Uplus	3.42–3.5	80	680 million	10
KT Corp	3.5–3.6	100	828 million	10

Table 4.12 Final spectrum assignments for the 28 GHz band in South Korea.

Operator	Spectrum band (GHz)	Bandwidth (MHz)	Cost ($)	Period (yr)
SK Telecom	28.1–28.9	800	173 million	5
LG Uplus	27.3–28.1	800	170 million	5
KT Corp	26.5–27.3	800	175 million	5

5G deployment (Japan). In South Korea, the spectrum attribution for 5G is more balanced than for Japan and China. It is consistent with the network-sharing agreement between mobile operators in South Korea: the priority is deployment speed rather than competition between operators, which is the opposite of the situation in Europe.

4.1.3.4 Summary of Auctions and Cost Comparison Worldwide

After providing a worldwide vision of the 5G auctions that have taken place so far, it is argued that the European Union is leading in 5G band auctions. Significant telecommunication markets such as Italy, Spain, and Germany have already held or are about to finish part of this process, with a significant quantity of spectrum made available (Figure 4.3).

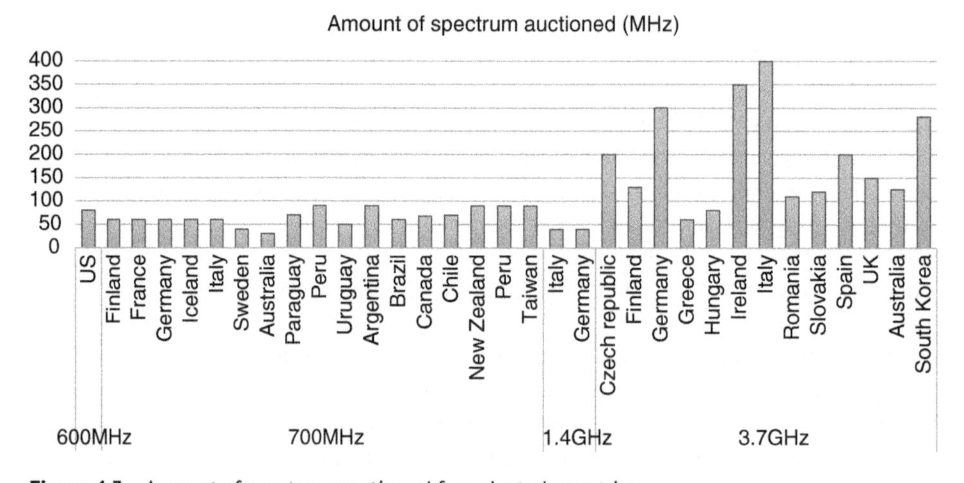

Figure 4.3 Amount of spectrum auctioned for selected countries.

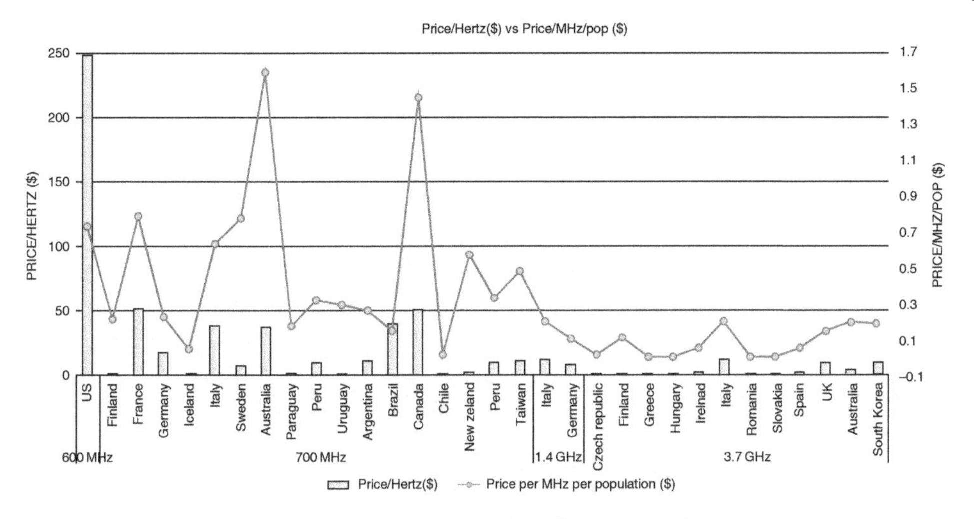

Figure 4.4 Comparison of the price paid per hertz in different countries.

Beyond Europe, the US and South Korea also have a leading role, making the various 5G bands available for MNOs. China is also active, but it is not included in the list because there was no auction but rather frequency assignments for the different MNOs.

Regarding the price per hertz in Figure 4.4, in big countries like the US, Canada, Brazil, and Argentina, the price for the 600 and 700 MHz bands is more expensive than in any other band or country. These bands have very good propagation and are especially important in big countries to fill cellular coverage gaps. If population is introduced as a factor for comparison, then the most developed countries in the world lead the classification, with special remarks for Australia and Canada.

Focusing on the European market, several aspects must be highlighted. 5G is a trending topic with significant public media interest. However, 4G spectrum bands are also being auctioned; those from the last five years must also be considered to understand the complete spectrum landscape. On average, around 200 MHz are made available in the 3.5 GHz band and around 60 MHz in the 700 MHz band. Both 5G and 4G bands are shown in Figure 4.5.

The re-farming of frequencies that are currently being used for second generation (2G), third generation (3G) and fourth generation (4G) is under consideration in many countries. The leading candidates to be used for 5G are the 2.1 and 2.6 GHz 3G bands since the allowed bandwidth in these bands is a good fit for 5G services. Moreover, it is interesting to consider re-farming the 1.8 GHz band once many of the existing systems have migrated to 5G technology. The 1.4 GHz band is also called the L-band. This band is harmonized in Europe and can potentially be re-farmed to be used in 5G for SDL. It is widely used for IMT applications due to its good coverage in outdoor settings and buildings. The spectrum auctions are still ongoing in many countries, but for those already done or about to be finished, a brief comparison is given in Figure 4.6.

As has been explained in this section, Italy has led all the spectrum auctions in all the 5G bands. The first consequence is the high price paid by the operators compared to other important telecom markets like the UK and Spain. In the case of Germany, which is one of the last countries to finish its frequency auction, the cost shows behavior similar to 5G

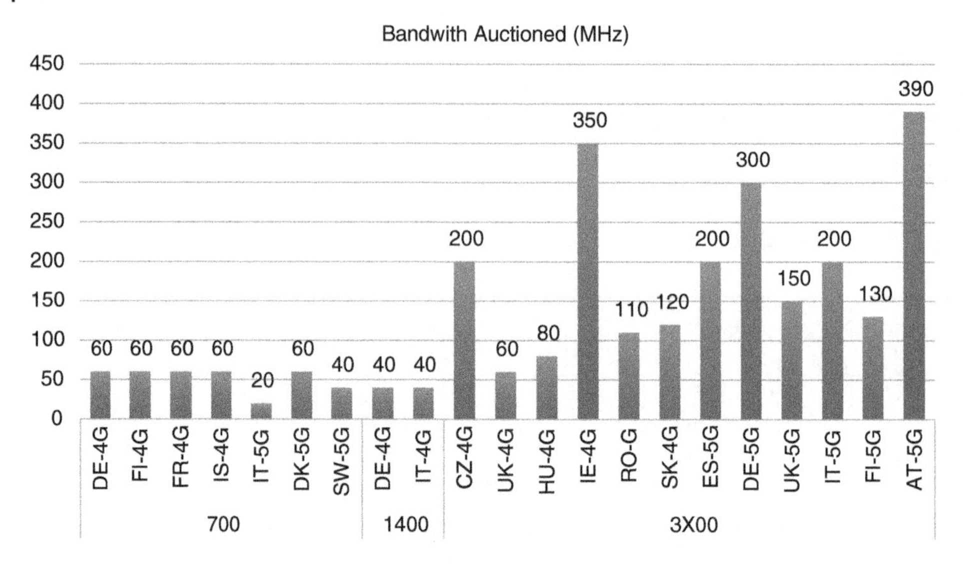

Figure 4.5 Bandwidth auctioned in European countries.

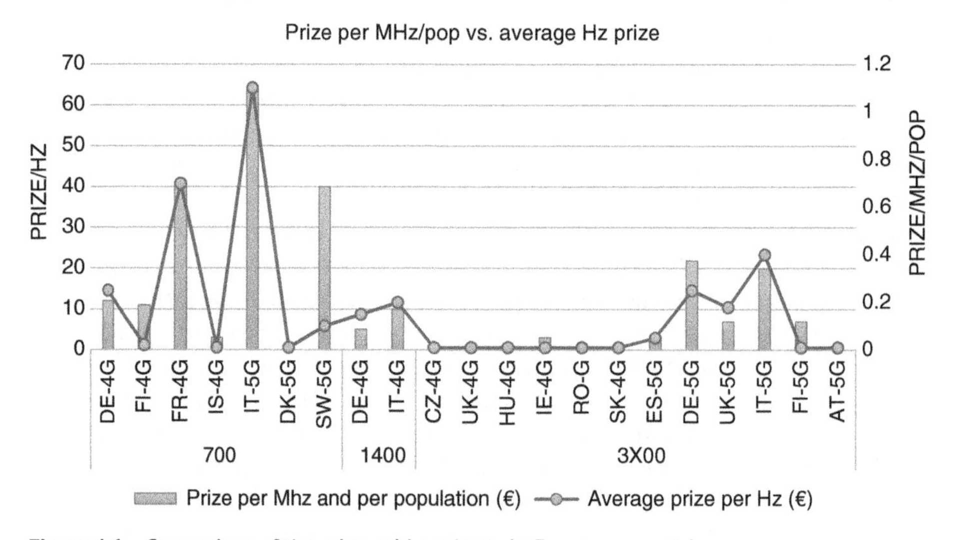

Figure 4.6 Comparison of the price paid per hertz in European countries.

in the UK and is less expensive than in Italy. In our opinion, the cost of this auction could be taken as a significant reference due to the size of the market and the significant technical requirements defined in the auction. In general, it can be observed that the price has increased during the 5G auctions when compared to 4G auctions. Moreover, using as a variable the population of the country, the value per MHz is quite similar in the different European countries: on average, around €0.4/MHz/person.

4.1.4 Spectrum Harmonization Worldwide

In the following subsections the worldwide spectrum harmonization activities digital single market and World Radiocommunication Conference are presented.

4.1.4.1 Europe and Digital Single Market

The digital single market (DSM) strategy aims to open up digital opportunities for people and businesses and enhance Europe's position as a world leader in the digital economy. In the European Union, the Member States coordinate their spectrum management approaches in a common regulatory framework to support the internal market for wireless services and to foster innovation in electronic communications and other sectors. The Member States harmonize spectrum access conditions at European Union level to ensure efficient use of radio spectrum and enable interoperability of underlying equipment and communications services. The European Commission works together with Member States to modernize spectrum management to facilitate spectrum access through more flexibility in usage conditions.

A framework for a radio spectrum policy in the European Union was launched by the 2002 regulatory framework for electronic communications, and particularly by the Radio Spectrum Decision. The Radio Spectrum Decision defines the policy and regulatory tools to ensure the coordination of policy approaches and harmonized conditions for the availability and efficient use of radio spectrum for the internal market. The Radio Spectrum Decision allows the European Commission to implement the decision to harmonize technical conditions about the availability and efficient use of spectrum for the proper functioning of the single market. The European Commission may issue mandates to the CEPT administrators for the preparation of such technical implementing measures. To assist the European Commission, two complementary bodies were set up following the Radio Spectrum Decision to facilitate consultation and to develop and support an European Union Radio Spectrum Policy: the Radio Spectrum Policy Group (RSPG) is a group of high-level national governmental experts to help the European Commission develop general radio spectrum policy at the community level. The second group is the Radio Spectrum Committee (RSC) which assists the Commission in developing technical implementation measures to ensure harmonized conditions across Europe for the availability and efficient use of radio spectrum.

4.1.4.2 World Radiocommunication Conference 2019

At the World Radiocommunication Conference (WRC) that took place in Egypt from 28 October to 22 November 2019, two items about ITS and the RLAN spectrum have been discussed.

Recognizing that harmonized spectrum and international standards would facilitate deployment of ITS radio communications, WRC 19 agenda item 1.12 was approved by WRC-15, and Resolution 237 asked to consider possible global or regional harmonized frequency bands for the implementation of evolving ITS under existing mobile service allocations. The mobile service bands being used by the evolving ITS may also be used by other applications and services, and some of the frequency bands are also being considered under other agenda items. The technical and operational studies performed by ITU Radiocommunication Sector (ITU-R) have indicated that the frequency band

5850–5925 MHz, or parts thereof, have been designated for the implementation of evolving ITS by some administrations. Consequently, the ITU-R has developed Recommendation ITU-R M.2121, "Harmonization of frequency bands for Intelligent Transport Systems in the mobile service," and Report ITU-R.2445, "Intelligent transport systems (ITS) usage."

Agenda item RLAN 1.16 aims to consider issues related to wireless access systems, including radio local area networks (WAS/RLAN), in the frequency bands between 5150 MHz and 5925 MHz, and take the appropriate regulatory actions, including additional spectrum allocations for mobile service, in accordance with Resolution 239 (WRC 15). This resolution states "that the results of ITU-R studies indicate that the minimum spectrum need for WAS/RLAN in the 5 GHz frequency range in the year 2018 is estimated at 880 MHz; this number includes 455–580 MHz already utilized by non-IMT mobile broadband applications operating within the 5 GHz range resulting in 300–425 MHz additional spectrum being required" and called for the ITU-R to:

- Study WAS/RLAN technical characteristics and operational requirements in the 5 GHz frequency range
- Perform sharing and compatibility studies between WAS/RLAN applications and incumbent services in the frequency bands 5150–5350, 5350–5470, 5725–5850, and 5850–5925 MHz while ensuring the protection of incumbent services, including their current and planned use
- Consider enabling outdoor WAS/RLAN operations in the frequency band 5150–5350 MHz
- Consider potential MS allocations to accommodate WAS/RLAN operations in the 5350–5470 MHz and 5725–5850 MHz frequency bands
- Identify potential WAS/RLAN use in the 5850–5925 MHz frequency band

In the CPM, for each of the five bands 5150–5250, 5250–5350, 5350–5470, 5725–5850, and 5850–5925 MHz, evolutions of Resolution 229 (Rev. WRC-12) are proposed, along with the suppression of WRC 2015 Resolution 237. Regarding the 5850–5925 MHz band, applications for mobile service in this frequency band have already been implemented in various countries throughout the world. Therefore, any sharing analysis carried out under this agenda item should not prejudice usages of mobile service and not impose any additional constraints on other services to which the band is allocated. Some concerns were raised about applications operating under the primary mobile service in this band. Some sharing studies carried out so far on a national or regional basis looking at WAS (RLAN) interfering with ITS showed the need for appropriate separation distances, in cases of co-channel operation. As a result, work by some administrations and regional groups on possible mitigation techniques was initiated to help improve compatibility between individual RLAN devices and ITS applications. However, based on the results of these studies so far, conclusions under this agenda item could not be reached.

4.1.5 Summary

As shown in Section 4.1.3, 5G spectrum assignment is ongoing worldwide; in Europe, several countries have completed auctions in various bands. Three main frequency bands are already defined for 5G technology, which can be used for CAD and CRU services: 700 MHz,

3.4–3.8 GHz, and 26 GHz. Clearly, the 3.4–3.8 GHz will be the first band commercially deployed, focusing first on eMBB, which is the current business model in the telecom market. The 700 MHz band has interesting opportunities for improving the cellular network coverage, due to its propagation properties, and could be especially interesting for V2X use cases. This band is made available in Europe from 2020 on. In the spectrum auctions held so far, the case of Italy should be noted, because the price per hertz of this band reached a notably high value. This will be a barrier for new use cases in this band, such as V2X, due to the need for operators to obtain a significant return on investment in these frequencies. The example from Asia should perhaps be taken into account, where deployment has been prioritized over the price of the frequency bands. Spectrum estimates from June 2020 [1] show much higher spectrum demands for C-V2X use cases.

Finally, the millimeter-wave band (26 GHz) has been harmonized by the European Commission, completed in all members by the end of March 2020, with effective usage of 1 GHz by the end of 2020. Countries like Italy have already held auctions for this band. In terms of use cases, the only one clearly identified so far is WLL; while other possible use cases are related to high-definition video, augmented reality (AR) and virtual reality (VR). Other bands that may be relevant for 5G are defined around 1 and 2 GHz. These bands are not harmonized between countries and were initially reserved for LTE. Part of them will be reserved for vertical industries (normally planned for factories). In these bands, re-farming will enable the migration from LTE to 5G technology, increasing spectrum availability and efficiency thanks to New Radio. Some entities like the General Services Administration (GSA) and FCC are looking for new frequency bands beyond those described here; but even if they are finally harmonized and assigned, there will be a significant time to market for any of them.

Despite the expectations of the DSM for a common framework in frequency allocation, the truth is that there is no common European definition for frequency auctions held at a national level. As explained in Section 4.1.3, some countries have defined strong performance and deployment requirements for frequency assignments, while others have no requirements. Even though entities like the Global System for Mobile communications Association (GSMA) have provided some auction best practices [25], there is no harmonization in the process. The case of Germany must be noted, with significant demands related to coverage, data rate, and latency for various types of roads, confirming the interest of the government in enabling critical V2X infrastructure.

4.2 Channel Modeling

This section covers the relevant state-of-the-art channel models for V2X communication, along with describing recent V2V channel measurements and models. First, it describes environments relevant for V2X communication, together with link types subsumed under the term *V2X*. Next, it describes the state-of-the-art channel models, including their most relevant components: LOS blockage analysis, path-loss and shadow-fading modeling, and fast-fading modeling. The contributions beyond the state of the art pertain to (i) new V2V measurements and characterization of channels above 6 GHz, (ii) multi-link shadowing models based on measurements below 6 GHz, and (iii) channel measurements for massive

MIMO adaptive beamforming. Finally, the chapter concludes by analyzing the compliance of recent channel measurements with existing standards and providing possible recommendations for further enhancements.

4.2.1 Propagation Environments

Radio propagation is influenced by the types of objects found in the environment where the communication occurs. In the case of V2X communications, the most significant objects that influence propagation are buildings, vehicles (both static and mobile), and various types of vegetation (e.g. trees, shrubbery, etc.) [26]. Relevant scenarios for V2X channel modeling, link types, vehicle types, and primary propagation states are described in Figure 4.7.

4.2.1.1 Link Types

In terms of the differences between link types, V2V and vehicle-to-pedestrian (V2P) links experience strong clutter on both sides of the links - transmitter (Tx) and receiver (Rx) - whereas V2I, with the infrastructure/base station side most often elevated above the clutter, experiences considerably different channels on the Tx and Rx sides. Figure 4.8 depicts the key characteristics of various link types, along with their most significant differences.

4.2.1.2 Environments

Environments can be divided qualitatively into highway, rural, and (sub)urban, with a word of caution: the division between environments is not exact, and the differences in some instances are hard to define. For example, rural and highway scenarios differ in the number of lanes (typically larger in case of highways) and surroundings (typically more foliage in

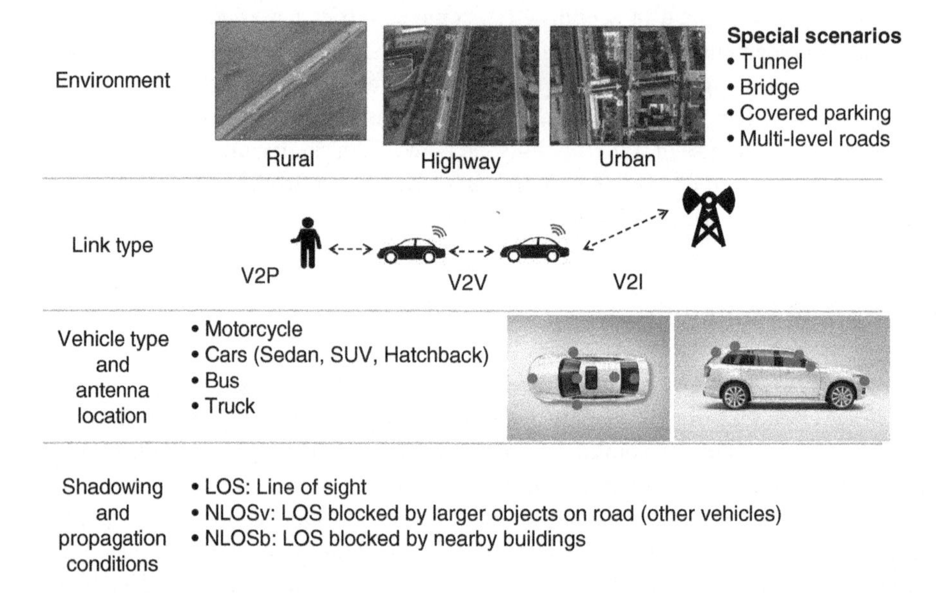

Figure 4.7 Environments, link types, and specific considerations for V2X channel modeling.

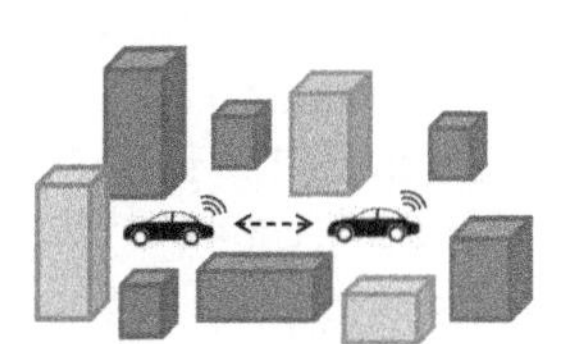

- Antennas close to ground
- Both Tx and Rx surrounded by scatterers
- Dual mobility
- More dynamic

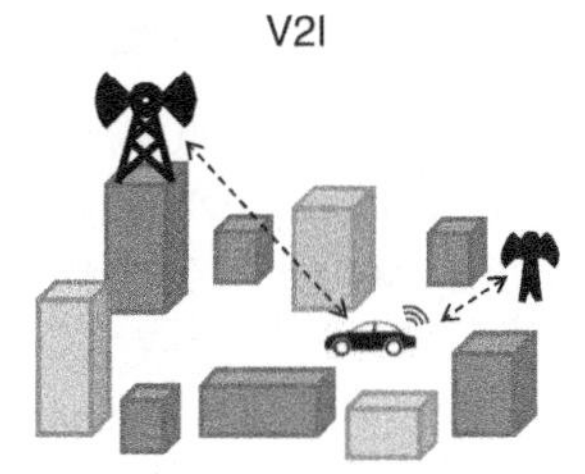

- Base-station at elevated position
- Vehicle surrounded by scatterers
- Infrastructure often scatterer free
- Single mobility
- Less dynamic

Figure 4.8 Main characteristics of V2V/V2P and V2I channels.

rural areas; guardrails more often found on highways), but the two environments can also share many characteristics. Furthermore, the impact of foliage cannot be neglected in any of the environments, as shown in measurement studies [26–28].

Rural

A rural environment is characterized by a road that often has open surroundings, with few or no objects along the roadside. The type of objects near the road can be trees, buildings, railings, etc. Roads are usually a single lane per direction with two-way traffic. Due to the lower density of scattering objects, the experienced number of multi-path components (MPCs) is lower than in an urban environment. The lack of scatterers can cause problems with successful reception when a LOS component is missing. If the Tx and Rx are moving in opposite directions, the Doppler shift can be significant.

Highway

A highway scenario is characterized by a road with two or more lanes per direction. The maximum allowed speeds are higher than in rural environments, making the Doppler shifts and spreads potentially high. Typical scattering objects in the surrounding are metallic guardrails, gantries, and bridges, with any buildings usually a few hundred meters from the roadside. The density of metallic scatterers is higher than in the rural scenario. In extreme cases, the delay spread (DS) can be significant due to the presence of metallic road signs above the road situated further away.

Urban

The urban environment refers to city streets used either for one-way or two-way traffic, along with intersections, roundabouts, etc. In terms of propagation characteristics, there is rich scattering due to many surrounding objects (both static and mobile) affecting wave propagation: buildings, vehicles (mobile and parked), road signs, streetlights, traffic signs, traffic signals, etc. In an urban environment, the probability of blockage of the LOS component is higher than in other environments. This implies that successful reception is often based on strong MPCs. The top right part of Figure 4.7 shows a typical four-way urban intersection.

4.2.2 Channel-Modeling Framework and Gap Analysis

Technical report 3GPP TR 38.901, "Study on channel model for frequencies from 0.5 to 100 GHz," [29] defines a geometry-based stochastic (GBS) framework for channel modeling from 0.5 to 100 GHz. It specifies parameters for LOS probability, path-loss (PL), shadow fading (SF), and fast fading (FF) and covers several scenarios: urban macro (UMa), urban micro (UMi), indoor office, and rural macro (RMa). The report is focused on base station (BS) to user equipment (UE) communication and does not explicitly address V2X channels. Specifically, the following components significant for V2X are not covered:

- No dual mobility (V2V)
- No V2X-specific scenarios (RSU-vehicle, V2V)
- No V2X antenna considerations
- No V2X-specific parameterization (especially for V2V)
- No PL, SF, or FF for V2V
- No LOS probability and blockage evolution for V2V and V2P channels

These components are crucial for realistic modeling of vehicular channels. To enhance the 3GPP framework, this section attempts to fill as many gaps as possible by using state-of-the-art literature on V2X channels and also by providing new channel parameterization based on recent measurements. In the following subsections, we elaborate on specific components of the V2X channel model, predominantly focusing on the V2V case, since [29] covers the V2I channels comprehensively (particularly the vehicle-to-base station case). More specifically, we describe the parameters to be included in [29], as depicted in Figure 4.9, ranging from selecting the correct scenario (first block in Figure 4.9) to LOS propagation condition (block 2), path loss (block 3), etc. In summary, we propose to use the framework in [29] for all channel generation, whereas for specific components of the process, we use the parameters and models listed in the following subsections.

Table 4.13 provides a summary of the availability in the existing literature of models for a specific propagation effect, along with the desirable properties of the models. While it is difficult to provide a comprehensive summary, we attempt to identify which components of V2X channel modeling are available in well-defined, feasible channel models available in the literature.

4.2.3 Path-Loss Models

In the following subsections various path-loss models are described, starting with path-loss for V2V LOS links, followed by shadow-fading models, and fast-fading parameters.

4.2.3.1 Path-Loss for V2V LOS Links

The path loss for V2V LOS links are outlined in the following subsections with free space path loss, two-ray ground-reflection model, path loss for V2V channels obstructed by vehicles, and path loss for V2V channels obstructed by buildings and other objects.

Free Space Path-Loss

The free space path-loss model is the resulting loss in signal strength when the electromagnetic wave traverses from Tx to Rx through free space, without any obstacles nearby

Table 4.13 V2X channel-modeling aspects available in the existing literature.

V2X channel model components and their availability in SOTA		Model exists / effect accounted for in SOTA <6 GHz	>6 GHz
Propagation mechanism	• LOS blockage	+ V2X	+ V2X
	• Path loss	+ V2X	+ V2I; −V2V/P
	• Shadow fading	+ V2X	+ V2I; −V2V/P
	• Small-scale fading	+ V2X	+ V2I: −V2V/P
	• Correlated fading effects for single and multi-links	+ V2I/V; N/A V2P	+ V2I; −V2V/P
Modeling approach	• Non-geometry-based (e.g. TDL)	+ V2I/V; −V2P	+ V2I/V; −V2P
	• Geometry-based deterministic	+ V2X	+ V2I; −V2V/P
	• Geometry-based stochastic	+ V2I/V; −V2P	+ V2I; −V2V/P
Model properties	Must have	Good to have	
	• Spatial-temporal dependencies (esp. forV2V)	• Extensibility	
	• Non-stationarity (esp. for V2V)	• Double-directional antenna configuration dependency	
	• Applicability	• Scalability and complexity	

+: the model exists in the literature; −: the model does not exist in the literature.

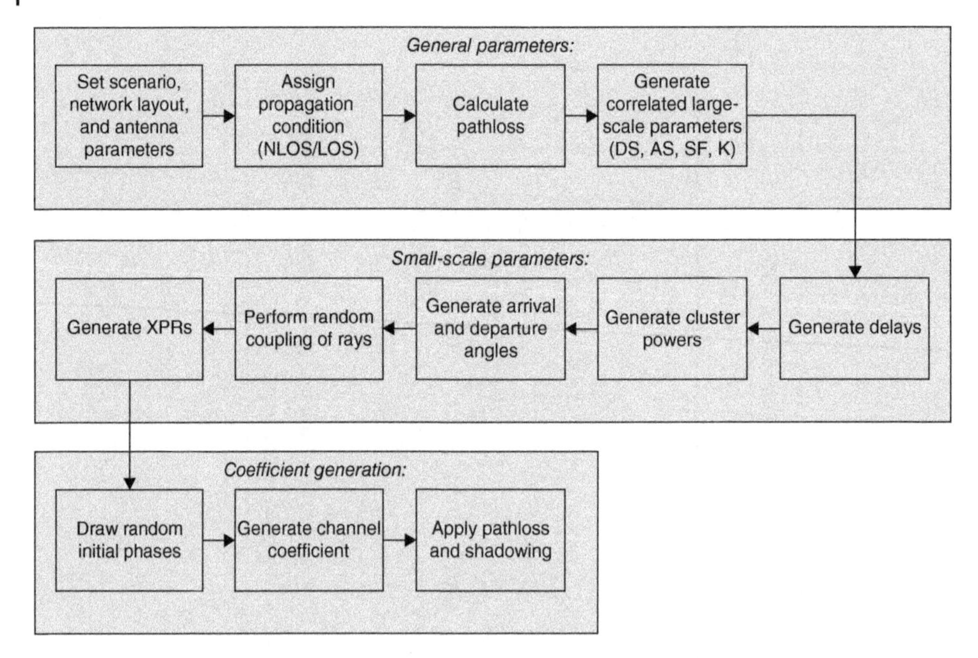

Figure 4.9 The channel coefficient generation process from [29].

that could cause reflections or diffractions. The free space path loss (FSPL) is given by the equation

$$FSPL = 20\ log_{10}(d) + 20\ log_{10}(f) + 20\ log_{10}\left(\frac{4\pi}{c}\right) - G_{Tx} - G_{Rx}$$

where d is the distance between the transmitter and receiver, f is the carrier frequency, and G_{Tx} and G_{Rx} are the gain of the transmitting and receiving antenna, respectively. Free space is a theoretical model that by itself does not model the path loss for V2X channels well since, at the very least, there are perturbations of the free space signal by reflections from the road on which vehicles travel.

Two-Ray Ground-Reflection Model

The free space propagation model assumes the existence of only the LOS ray. However, due to the inherent structure of the environment where V2V communication occurs – over the face of road surface – in the case of LOS communication, the propagation characteristics are most often influenced by at least two dominant rays: the LOS ray and the ground-reflected ray. The two-way ground-reflection model with an appropriately adjusted reflection coefficient was shown as a very good path-loss model for LOS V2V channels [28, 30]. In this scenario, the LOS path interferes with the ground-reflected path. The two rays arrive at the receiver with a different phase and different power.

The different phase leads to constructive and destructive interference, depending on the distance, d, between the receiver and the transmitter, as shown in Figure 4.10. When increasing the distance between the transmitter and the receiver, the alternating pattern of constructive and destructive interference stops at breakpoint d_b. From this distance onward, the length difference between the two rays is smaller than half the wavelength,

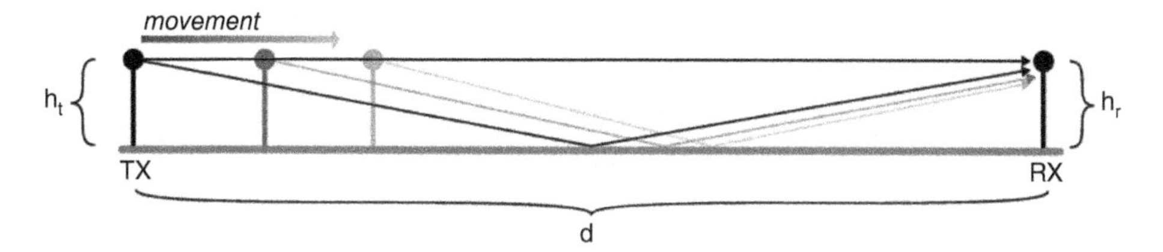

Figure 4.10 The two-ray ground-reflection model.

and the small angle of arrival (AOA) on the ground causes a phase shift of 180° for the reflected wave, leading to destructive interference.

For these two rays and referring to Figure 4.10, the resulting E-field is equal to

$$E_{TOT} = E_{LOS} + E_{Ground} = \frac{E_0 d_0}{d_{LOS}} \cos\left[\omega_c\left(t - \frac{d_{LOS}}{c}\right)\right] + R_{Ground}\frac{E_0 d_0}{d_{ground}} \cos\left[\omega_c\left(t - \frac{d_{ground}}{c}\right)\right] \tag{4.1}$$

where E_{Ground} is the E-field of the ground-reflected ray, R_{Ground} is the ground-reflection coefficient, and d_{ground} is the propagation distance of the ground-reflected ray, where h_t and h_r are the height of the transmitting and receiving antennas, respectively, and d_{LOS} is the ground distance between the antennas (*d* in Figure 4.10). Note that using the exact height of the antennas (h_t and h_r) is important, since a small difference in terms of either h_t or h_r results in significantly different interference relationship between the LOS and ground-reflected rays. When the originating medium is free space, the reflection coefficient R is calculated as follows for vertical and horizontal polarization, respectively:

$$R_{\parallel} = \frac{-\epsilon_r \sin\theta_i + \sqrt{\epsilon_r - \cos^2\theta_i}}{\epsilon_r \sin\theta_i + \sqrt{\epsilon_r - \cos^2\theta_i}} \tag{4.2}$$

and

$$R_{\perp} = \frac{\sin\theta_i + \sqrt{\varepsilon_r - \cos^2\theta_i}}{\sin\theta_i + \sqrt{\varepsilon_r - \cos^2\theta_i}} \tag{4.3}$$

where θ_i is the incident angle and ε_r is the relative permittivity of the material. From the E-fields in Eq. (4.1), the ensuing received power P_r (in watts) is calculated as follows (assuming unit antenna gain at the receiver):

$$P_r = \frac{|E_{TOT}|^2 \lambda^2}{4\pi\eta} \tag{4.4}$$

where λ is the wavelength and η is the intrinsic impedance ($\eta = 120\pi\Omega$ in free space). An appropriate reflection coefficient needs to be used to match the measurements. To that end, in [28], curve fitting of this model to measurement data yielded a ε_r value of 1.003 as the best fit (note that the remaining parameters in the calculation of the reflection coefficient are dependent on geometry only).

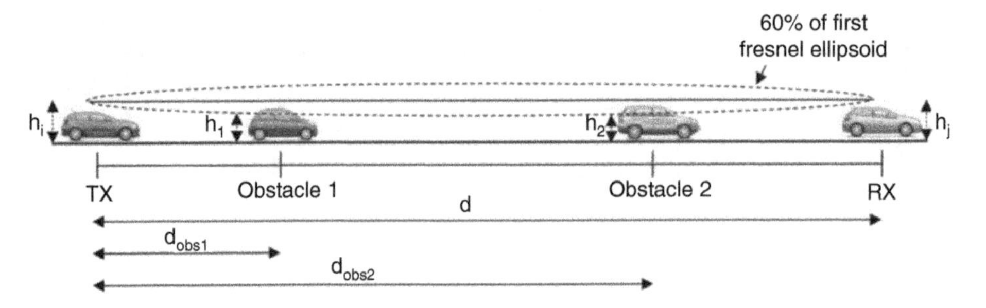

Figure 4.11 Vehicles-as-obstacles path-loss model.

Path-Loss for V2V Channels Obstructed by Vehicles: Vehicles-as-Obstacles

When vehicles are causing blockage to the LOS link, they induce additional attenuation. A model for vehicles-as-obstacles is described in [27], where vehicles are modeled using (multiple) knife-edge diffraction. The model uses a free-space path-loss model as the baseline, with additional attenuation due to each of the vehicles blocking the LOS link. Attenuation (in dB) due to a single knife-edge obstacle A_{sk} is obtained using the following equation:

$$A_{sk}\begin{cases}6.9+20\,\log_{10}\left[\sqrt{(\nu-0.1)^2+1}+\nu-0.1\right] \\ \qquad\qquad for\,\nu > -0.7 \\ \qquad\qquad 0; otherwise,\end{cases} \tag{4.5}$$

where $v = 21/2H/r_f$, H is the difference between the height of the obstacle and the height of the straight line that connects Tx and Rx, and r_f is the Fresnel ellipsoid radius (as shown in Figure 4.11).

To calculate attenuation due to multiple vehicles, the model employs the ITU-R multiple-knife diffraction method [31]. The vehicles-as-obstacles model was validated experimentally and was shown to model well the path loss of V2V channels obstructed by other vehicles [32].

Path-Loss for V2V Channels Obstructed by Buildings and Other Objects: Log-Distance Path-Loss Model

Log-distance path loss is an extension of free-space path loss, where the path-loss exponent does not necessarily equal two (as is the case in free space propagation) but is a function of the environment surrounding Tx and Rx.

The log-distance path-loss model is formally expressed as

$$PL(d) = PL(d_0) + 10\gamma\log(d/d_0) + X_\sigma$$

where PL is the total path-loss measured in decibels (dB), $PL(d_0)$ is the path loss at the reference distance d_0, d is the distance between Tx and Rx, γ is the path-loss exponent, and X_σ describes random shadowing effects.

Finally, the received power P_r is calculated as

$$P_r = P_t + G_t + G_r - PL(d)$$

where P_t is the transmit power, and G_t and G_r are antenna gains in dBi. Log-distance path loss with an appropriate path-loss exponent and shadowing deviation was experimentally

Table 4.14 Shadow-fading parameter σ for V2V communication.

Environment link type	Highway (dB)	Urban (dB)
LOS	3.3	5.2
NLOSv	3.8	5.3
NLOSb	4.1	6.8

shown to model well the path loss for V2V links in NLOS cases [32]. For V2V links at 5.9 GHz, the following values can be used [32]:

- $d_0 = 1$ m
- $PL(d_0) = 47.8649$ dB
- $\gamma = 2.5$ (slight obstruction by building)
- $\gamma = 3$ (strong obstruction by building)

4.2.3.2 Shadow-Fading Models

In the following subsections the single-link shadow fading for V2V and V2I channels, and the multi-link shadow fading for V2V channels are outlined.

Single-Link Shadow Fading for V2V and V2I Channels

In the log-distance path-loss equation, the value of the standard deviation σ of the shadow-fading variable $X_\sigma(0, \sigma)$ can be adjusted so that it better describes a specific environment and link type. For V2V links, [32] contains a detailed measurement-based analysis of σ for both highway and urban environments. The values of σ are shown in Table 4.14.

Due to the differences in terms of antenna height, scatter density, and relative speed, V2I links exhibit different propagation characteristics compared to V2V links. Of particular importance is a subset of V2I links where the infrastructure end of a link is a road-side unit (RSU), since these links are distinguished compared to well-studied cellular V2I links. Aygun et al. [33] used measurement data collected in the urban environment of Bologna to evaluate shadow fading for four propagation conditions of RSUs V2I links: LOS, non-line-of-sight (NLOS) due to vehicles, NLOS due to foliage, and NLOS due to buildings, i.e. NLOS due to static objects – NLOSb (c.f. Figure 4.7) link type – further divided into blockage due to buildings and blockage due to foliage. Assuming the log-normal shadow-fading process, the authors extract the resulting mean, minimum, and maximum of σ, and the standard deviation of the shadow-fading process. The summarized results are shown in Table 4.15, with the following remarks: (i) for a highway environment, the NLOSb state is not applicable; (ii) due to the lack of measurements in a highway environment, LOS and NLOSv (c.f. Figure 4.7) results for the urban environment are reused for highways as well.

Multi-Link Shadow Fading for V2V Channels

Abbas et al. [34] performed V2V measurements and showed that a single vehicle could incur more than 10 dB attenuation, in line with previous results [32]. Based on the measurements,

Table 4.15 Shadow-fading parameter σ for vehicle-to-roadside unit (V2RSU) communication.

Environment link type		Highway (dB)	Urban (dB)
LOS		2.2	2.2
NLOSv		2.6	2.6
NLOSb	Buildings	N/A	3.3
	Foliage	N/A	2.4

the authors designed a GBS propagation model for highway environments that incorporates vehicular obstructions and determines the time duration that the link spends in LOS and NLOS states. By extracting the probability distribution of each state from measurements, it uses a probabilistic model based on Markov chains to transition between various LOS conditions. The model demonstrates the importance of differentiating a LOS link from a NLOS link as well as energy contributed from LOS and NLOS rays. However, this model was developed for single-link communication, and a critical factor in vehicular adhoc network (VANET) simulations, which is often neglected, is to include the cross-correlation of different communication links. This is significant for wireless communication systems using multi-hop techniques to overcome the issue with shadowed vehicles in V2V systems. Two models are presented [35] of cross-correlation for a convoy scenario on a highway: one using a joint path-loss model for all communication links between all vehicles, including LOS and obstructed LOS (OLOS) cases [17], and another using a specific path-loss model for each communication link, as well as LOS and NLOSv separately [18]. The auto- and cross-correlations do not affect the average value of the received power if the data ensemble is big enough. However, the correlations will cause the system to experience longer large-scale fading dip durations compared to the uncorrelated case. This is especially important for VANET safety applications, where the consecutive packet error rate is a critical factor. The findings in [35] regarding path-loss models, autocorrelation behavior, and the cross-correlation of large-scale fading processes stress the benefits of GBM for VANET simulators. It is significant that geometry-based models distinguish between LOS and OLOS communication and apply different path-loss models for the two cases. Otherwise the VANET simulator needs to consider the cross-correlation between different communication links, i.e. implement [17], to achieve results close to reality. On the other hand, when using a geometry-based model as an input to the VANET simulator, the cross correlation can actually be neglected and the implementation of [18] is not necessary. This is a very useful and practical result, since it makes it much easier to implement VANET simulators for multi-link scenarios. The computational complexity could easily become an issue if the cross-correlation between a large numbers of links had to be considered.

4.2.3.3 Fast-Fading Parameters

For <6 GHz, we propose to use the parameters described in [36]. For >6 GHz, we propose to use the shadow and fast-fading parameters presented in Section 4.2.3.3 and in more detail in [37].

Table 4.16 Summary of channel-modeling components to include in the framework in [29].

	V2V					
	Urban			Highway		
≤ 6 GHz	LOS	NLOSv	NLOSb	LOS	NLOSv	NLOSb
LOS_blockage	[BGX16]			[BGX16]		
PL	[32]	[32]	[32]	[32]	[32]	[32]
SF	[32, 34]	[32, 34]	[32, 34]	[32, 34]	[32, 34]	[32, 34]
FF	[36]	[36]	[36]	[36]	[36]	[36]
>6 GHz						
LOS_blockage	[BGX16]			[BGX16]		
PL	[38]			[38]		
SF	[38]			[38]		
FF	[37]			[37]		
	V2I (vehicle-to-base station: V2B)					
	Urban			Highway		
V2B ≤ 6 & > 6GHz V2RSU > 6GHz	LOS	NLOS		LOS	NLOS	
LOS_blockage PL SF FF	[29]					
	V2I (vehicle to road-side unit (RSU))					
	Urban			Highway		
V2RSU ≤ 6GHz	LOS	NLOSv	NLOSb	LOS	NLOSv	NLOSb
LOS_blockage	[29]			[29]		
PL	[32]			[32]		
SF	[33]			[33]		
FF	[29]			[29]		

4.2.3.4 Summary

Table 4.16 summarizes which models to use for each of the V2X channel-modeling components as part of the channel-modeling framework described in [29]. By and large, for V2I channels, the components in [29] are suitable for use. On the other hand, LOS blockage, path loss, shadow fading, and fast fading in [29] are not suitable for V2V channels; therefore, alternative parameterization for those components is proposed based on the existing literature. For any parameters not mentioned in Table 4.16, for initial evaluation, the existing parameters in [29] can be used, with the note that some of the parameters (e.g. elevation angles) need to be revisited.

4.2.4 Recent V2X Channel Measurements and Models

Recent V2X channel measurements and models are presented in the following subsections, describing V2V measurements in cmWave and mmWave, mmWave V2V channel modeling, and multi-link shadowing extensions.

4.2.4.1 V2V Measurements in cmWave and mmWave

Measurements at 60 GHz have been carried out with a dual-polarized, ultra-wideband, multi-channel sounder, which offers after back-to-back calibration a null-to-null bandwidth of 5.1 GHz. The spatial characterization of the environment was done by automatically rotating dual-polarized horn antennas with 30° HPBW in 30° steps, covering the whole azimuth range at Tx and Rx. On the other hand, a single elevation of 0° was measured at both sides. The scenario was a T intersection in an urban environment with parked vehicles, multi-story buildings, and lampposts, as shown in Figure 4.12a. Tx and Rx were located at 1.44 m height, emulating two vehicles 44 m apart communicating with each other with transceivers at vehicle rooftop level. Two parked vehicles were present during the measurements to increase the scattering effects, denominated as Position 0. A third vehicle, denominated the blocking vehicle (Mercedes Benz Sprinter – a delivery van), was located in 15 different positions (Positions 1–15 in Figure 4.12) emulating an overtaking situation. All the measurements were performed in as static scenario as possible by restricting access on the streets.

In Figure 4.12, (a) shows the measurement setup, with positions of the blocking vehicle marked with white circles and sequentially denominated Positions 1–15. Position 0 refers to the scenario without the blocking vehicle. Results for the following positions are shown: Position 0 (b), Position 4 (c), Position 5 (d), and Position 8 (e). In addition, (b)–(e) indicate the angular position of contributions from LOS, building A, and building B, which, depending on the position, are significant as in (b) or blocked, e.g. building A contribution in (c) or building B contribution in (d).

Figure 4.13 shows the power delay profile (PDP) for Position 6 in Figure 4.12, with the primary contributing components identified. Figure 4.13 shows that LOS cluster (including the LOS component and those close to it in delay, such as diffraction around the vehicle body), while attenuated, still exists. This is in contrast with a "classic" NLOS state (blockage by buildings), where LOS cluster is most often insignificant. In the case of NLOSv, LOS cluster often contributes non-negligible energy due to transmission through the vehicle, diffractions around the blocking vehicle, and ground reflection under it.

4.2.4.2 mmWave V2V (Sidelink) Channel Modeling

In the following subsections the mmWave sidelink channel modeling are presented, starting with the directional channel characterization, followed by path loss description, and ending with shadow and fast-fading parameter analysis.

Directional Channel Characterization

The previous measurements can be used for directional channel characterization of mmWave V2V channels. Different scatterers were identified using large resolution in the time domain and different scans in the angular domain. The interpolated and normalized

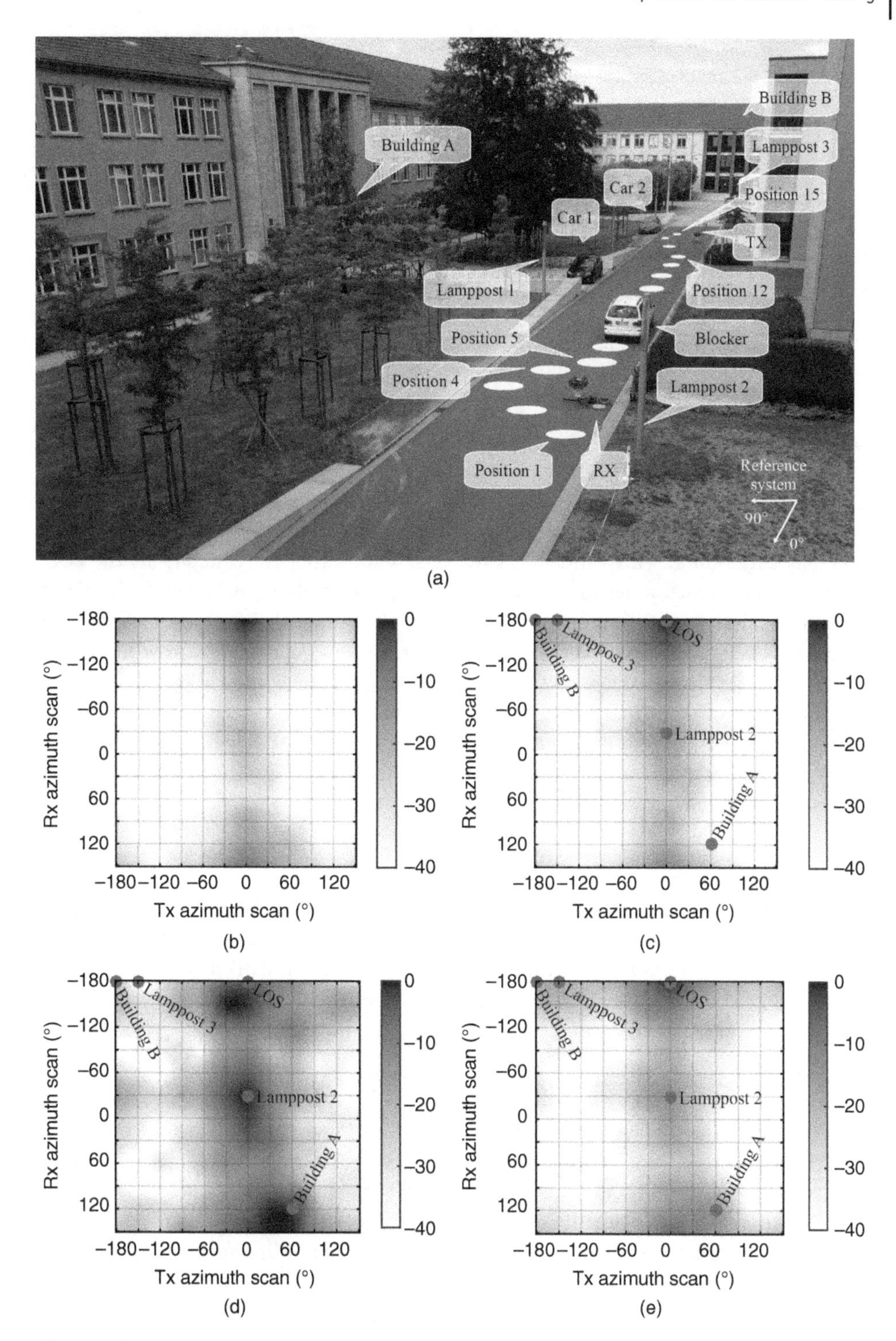

Figure 4.12 Measurements of mmWave V2V channels at 60 GHz and the results of directional channel characterizations. (a) Measurement setup, with positions of blocking vehicles marked with white circles and sequentially denominated Positions 1–15.

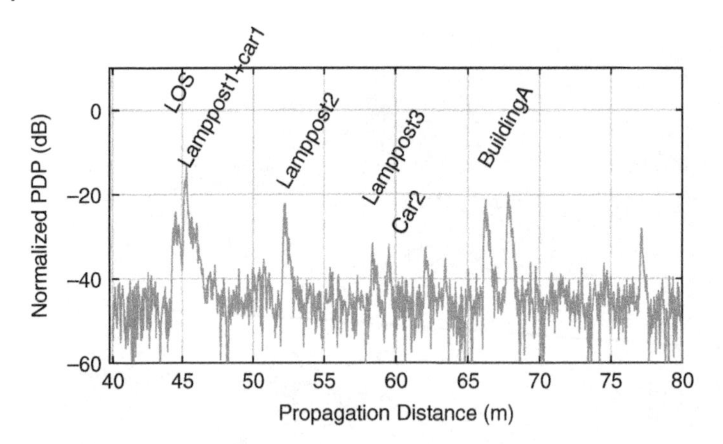

Figure 4.13 PDP for Position 6 normalized to the LOS component in scenario without a vehicle blocker.

per position power bi-azimuthal profiles are shown in Figure 4.12b–e for selected positions of the blocking vehicle. At Position 0 (Figure 4.12b), the strongest scatterers are the LOS, building A, and a back reflection in the T intersection in building B. When the blockage starts in Position 4 (Figure 4.12c), building A is blocked, and the number of relevant scatterers within the displayed dynamic range increases. In Position 5 (Figure 4.12d), the LOS is completely blocked. Finally, when the blocking vehicle is in the middle of the communicating vehicles, in Position 8 (Figure 4.12e), the strongest scatterer is building A; building B is still blocked, but scatterers at ±30° Tx azimuth become more relevant. The results of directional characterizations show that NLOSv V2V propagation significantly increases the angular spread (AS) of MPCs at 60 GHz.

Path Loss

The alpha-beta-gamma (ABG) path-loss (PL) model is applied in [38] for V2V channels in various urban and highway scenarios under LOS, and under building and vehicle blockage cases. Note that the ABG PL model is currently used in the 3GPP 3D model [29]. For the reason of aligning V2V PL modeling to [29], we propose to use the PL equations in [38].

The ABG PL model is given as:

$$\mathrm{PL}^{ABG}(f,d)\left[dB\right] = 10\alpha\log_{10}(d) + \beta + 10\gamma\log_{10}(f) + X_{\sigma}^{ABG}$$

where α captures how the PL increase as the transmit-receive distance (in meters) increases, β is the floating offset value in dB, γ captures the PL variation over the frequency f in GHz, and X_{σ}^{ABG} is the SF term in dB. This equation shows the fitting of the ABG PL model for various environments. We note that that the models in [38] are multi-frequency models, with the parameters in the model extracted from measurements in the above-6 GHz band (specifically, 6.75, 30, 60, and 73 GHz).

Table 4.17 presents the ABG model parameters in various scenarios for V2V links in LOS and NLOSb cases, and Table 4.18 presents the vehicle blockage effect model for V2V (i.e. the NLOSv case). Further details are available in [38].

Table 4.17 ABG model parameters for various scenarios with LOS and NLOSb.

Scenarios		ABG model parameters
Urban grid	**LOS**	$\alpha = 1.67$, $\beta = 38.77$, $\gamma = 1.82$, SF = 2.04 dB
	NLOSb	$\alpha = 2.38$, $\beta = 36.85$, $\gamma = 1.89$, SF = 3.05 dB
Highway	**LOS**	$\alpha = 1.58$, $\beta = 37.9$, $\gamma = 2$, SF = 3.13 dB
	NLOSb	$\alpha = 2.73$, $\beta = 25.98$, $\gamma = 2$, SF = 3.47 dB

Table 4.18 Additional path loss for NLOSv for various scenarios. Parameters for normally distributed loss: Mu (mean) and Sigma (standard deviation).

Scenarios	Separate for normal and large-size vehicle		Combined for normal and large-size vehicle
	Normal size (passenger car/ van)	Large size (truck/big van)	Passenger car/SUV/ truck/big van
Urban grid NLOSv	Mu = 5.86 Sigma = 3.08 dB	Mu = 10.43 dB Sigma = 4.48 dB	Mu = 8.95 dB Sigma = 4.61 dB
Highway NLOSv	Mu = 4.77 dB Sigma = 4.26 dB	Mu = 15.39 dB Sigma = 5.02 dB	Mu = 10.08 dB Sigma = 7.06 dB

Shadow and Fast-Fading Parameter Analysis

The dynamics of blocked scatterers have a significant impact on mmWave V2V channels. The mean and variance of the parameters have been calculated considering Positions 0–3 and 13–15 as LOS and Positions 4–12 as NLOSv. The synthetic omnidirectional characteristic of the channel has been utilized to calculate the DS, shadow fading, and Rice K-factor. This is computed by averaging the PDPs from the different directional scans. Angular spread (AS) has been calculated using marginal power angular profiles. The values are summarized in Table 4.19 (based on [37]), showing that the DS and angular spread increase under blockage. On the other hand, due to obstruction of the LOS component, the K-factor is reduced from approximately 10 dB to −5 dB. The mean vehicle blocking loss (mean additional attenuation due to vehicle blockage in Positions 4–12 compared to Position 0) is about 12 dB. For each of the states, Table 4.19 contains the parameter values extracted from the measurements explained in Section 4.2.4.1 and [37]. While the NLOSb state in an urban environment is due to building blockage, in a highway environment, it is due to blocking by foliage. In Table 4.19, for values in square brackets, parameters are taken from the UMi model in [29] (Tables 7.5 and 7.6). Where new measurement results do not contain a certain parameter for NLOSv, we reuse the value for the NLOS state from the UMi model in [29] (Tables 7.5 and 7.6) for both urban and highway scenarios. Note again that UMi model parameters are used for lack of a better solution. To that end, some of the parameters (e.g. elevation angles) might need to be further investigated and updated when appropriate measurements become available. Further details are available in [37].

Table 4.19 Fast-fading parameters for V2V sidelink.

Scenarios		Urban			Highway		
		LOS	NLOSb	NLOSv	LOS	NLOSb	NLOSv
Delay spread (DS) $\text{lgDS} = \log_{10}(\text{DS}/1\text{s})$	μ_{lgDS}	$-0.2 \log_{10}(1+f_c) - 5.7$	$-0.12 \log_{10}(1+f_c) - 6.3$	$-0.3 \log_{10}(1+f_c) - 4.5$	-8.16	-7.66	-8.04
	σ_{lgDS}	$0.17 \log_{10}(1+f_c) - 10$	$0.1 \log_{10}(1+f_c) - 9$	$-0.14 \log_{10}(1+f_c) - 6.7$	-8.53	-7.62	-8.08
Angle-of-departure (AOD) spread (ASD)	μ_{lgASD}	$-0.1 \log_{10}(1+f_c) + 2.4$	$-0.1 \log_{10}(1+f_c) + 2.2$	$-0.1 \log_{10}(1+f_c) + 2.9$	1.39	1.32	1.29
$\text{lgASD} = \log_{10}(\text{ASD}/1°)$	σ_{lgASD}	$-0.03 \log_{10}(1+f_c) + 1.1$	$-0.3 \log_{10}(1+f_c) + 4.3$	$-0.2 \log_{10}(1+f_c) + 3.5$	0.59	0.77	0.82
Angle-of-arrival (AOA) spread (ASA)	μ_{lgASA}	$-0.1 \log_{10}(1+f_c) + 2.4$	$-0.1 \log_{10}(1+f_c) + 2.2$	$-0.1 \log_{10}(1+f_c) + 2.9$	1.39	1.32	1.29
$\text{lgASA} = \log_{10}(\text{ASA}/1°)$	σ_{lgASA}	$-0.03 \log_{10}(1+f_c) + 1.1$	$-0.3 \log_{10}(1+f_c) + 4.3$	$-0.2 \log_{10}(1+f_c) + 3.5$	0.59	0.77	0.82
Zenith-of-arrival (ZOA) spread (ZSA)	μ_{lgZSA}	$[-0.1 \log_{10}(1+f_c) + 0.73]$	$[-0.04 \log_{10}(1+f_c) + 0.92]$	$[-0.04 \log_{10}(1+f_c) + 0.92]$	$[-0.1 \log_{10}(1+f_c) + 0.73]$	$[-0.04 \log_{10}(1+f_c) + 0.92]$	$[-0.04 \log_{10}(1+f_c) + 0.92]$
$\text{lgZSA} = \log_{10}(\text{ZSA}/1°)$	σ_{lgZSA}	$[-0.04 \log_{10}(1+f_c) + 0.34]$	$[-0.07 \log_{10}(1+f_c) + 0.41]$	$[-0.07 \log_{10}(1+f_c) + 0.41]$	$[-0.04 \log_{10}(1+f_c) + 0.34]$	$[-0.07 \log_{10}(1+f_c) + 0.41]$	$[-0.07 \log_{10}(1+f_c) + 0.41]$
Shadow fading (SF) [dB]	σ_{SF}	See Table 4.18	See Table 4.18	See Table 4.18	See Table 4.18	See Table 4.18	See Table 4.18
K-factor (K) [dB]	μ_K	3.48	N/A	N/A	[10]	N/A	N/A
	σ_K	1.71	N/A	N/A	[3.5]	N/A	N/A

Cross-Correlations	*ASD* vs *DS*	0.6	0.8	0.8	0.8	0.3	0.3
	ASA vs *DS*	0.6	0.8	0.8	0.8	0.3	0.3
	ASA vs *SF*	−0.3	−0.3	−0.4	[−0.4]	[−0.4]	[−0.4]
	ASD vs *SF*	−0.3	−0.3	−0.4	[−0.5]	[0]	[0]
	DS vs *SF*	−0.5	−0.5	−0.5	[−0.4]	[−0.7]	[−0.7]
	ASD vs *ASA*	0.4	0.1	0.1	0.8	0.3	0.3
	ASD vs *K*	−0.3	[N/A]	[N/A]	[−0.2]	[N/A]	[N/A]
	ASA vs *K*	−0.3	[N/A]	[N/A]	[−0.3]	[N/A]	[N/A]
	DS vs *K*	−0.2	[N/A]	[N/A]	[−0.7]	[N/A]	[N/A]
	SF vs *K*	0.1	[N/A]	[N/A]	[0.5]	[N/A]	[N/A]
Cross-correlations [1]	*ZSD* vs *SF*	[0]	[0]	[0]	[0]	[0]	[0]
	ZSA vs *SF*	[0]	[0]	[0]	[0]	[0]	[0]
	ZSD vs *K*	[0]	[N/A]	[N/A]	[0]	[N/A]	[N/A]
	ZSA vs *K*	[0]	[N/A]	[N/A]	[0]	[N/A]	[N/A]

Table 4.19 (Continued)

Scenarios		Urban			Highway		
		LOS	NLOSb	NLOSv	LOS	NLOSb	NLOSv
	ZSD vs *DS*	[0]	[−0.5]	[−0.5]	[0]	[−0.5]	[−0.5]
	ZSA vs *DS*	[0.2]	[0]	[0]	[0.2]	[0]	[0]
	ZSD vs *ASD*	[0.5]	[0.5]	[0.5]	[0.5]	[0.5]	[0.5]
	ZSA vs *ASD*	[0.3]	[0.5]	[0.5]	[0.3]	[0.5]	[0.5]
	ZSD vs *ASA*	[0]	[0]	[0]	[0]	[0]	[0]
	ZSA vs *ASA*	[0]	[0.2]	[0.2]	[0]	[0.2]	[0.2]
	ZSD vs *ZSA*	[0]	[0]	[0]	[0]	[0]	[0]
Delay scaling parameter r_τ		Delay scaling parameter r_τ	[2.1]	[2.1]	[4]	[2.1]	[2.1]
XPR [dB]	μ_{XPR}	[10]	[8.0]	[8.0]	[10]	[8.0]	[8.0]
	σ_{XPR}	[4]	[4]	[4]	[4]	[4]	[4]
Number of clusters N		Number of clusters N	[20]	[20]	[13]	[20]	[20]
Number of rays per cluster M		Number of rays per cluster M	[21]	[21]	[21]	[21]	[21]
Cluster DS (c_{DS}) in [ns]		Cluster DS (c_{DS}) in [ns]	[N/A]	[N/A]	[N/A]	[N/A]	[N/A]
Cluster ASD (c_{ASD}) in [deg]		Cluster ASD (c_{ASD}) in [deg]	[11]	[11]	[4]	[11]	[11]

Cluster *ASA* (c_{ASA}) in [deg]		Cluster *ASA* (c_{ASA}) in [deg]	[23]	[23]	[18]	[23]	[23]
Cluster *ZSA* (c_{ZSA}) in [deg]		Cluster *ZSA* (c_{ZSA}) in [deg]	[8]	[8]	[8]	[8]	[8]
Per cluster shadowing std. ζ [dB]		Per cluster shadowing std. ζ [dB]	4	4	4	4	4
Correlation distance in the horizontal plane [m]	*DS*	[8]	[11]	[11]	[8]	[11]	[11]
	ASD	[9]	[11]	[11]	[9]	[11]	[11]
	ASA	[9]	[10]	[10]	[9]	[10]	[10]
	SF	[11]	[14]	[14]	[11]	[14]	[14]
	K	[16]	[N/A]	[N/A]	[16]	[N/A]	[N/A]
	ZSA	[13]	[11]	[11]	[13]	[11]	[11]
	ZSD	[13]	[11]	[11]	[13]	[11]	[11]

4.2.4.3 Multi-Link Shadowing Extensions

For realistic performance evaluation of V2V communication systems, it is crucial that the channel models used for system simulations are realistic and cover relevant details. There are knowledge gaps and no model to date covering all significant aspects of the NLOS case in urban intersections. In [39], the authors fill the knowledge gap by providing path-loss and fading parameters for vehicles of various kinds and sizes, addressing multilink shadowing effects in urban intersections. In addition, auto-correlation properties of a single link and cross-correlation properties of large-scale fading between different links are analyzed in [39]. An NLOS channel-gain model that considers that a communication link can be obstructed by other vehicles (NLOSv) and that is reciprocal is described as

$$G\left(d_t, d_r\right)|\mathrm{dB} = \underbrace{10\log_{10}\left(m^2\right)}_{\text{Offset}} + \Psi_\sigma + 10\log_{10}\left(\underbrace{\left(g1\frac{\lambda}{4\pi\left(d_t+d_r\right)}\right)^2}_{\text{Single interactions}} + \underbrace{\left(g_2^N\frac{\lambda}{4\pi\left(d_t+d_r\right)}\right)^2}_{\text{Multiple interactions}}\right)$$

where

$$N = \max\left\{2\sqrt{\frac{d_t d_r}{w_t w_r}} - 1, 0\right\}$$

where the widths of the streets at which the transmitting and receiving vehicles are located are denoted by w_t, w_r,; and d_t, d_r are the distances to the intersection center. The term Ψ_σ represents large-scale fading and is modeled as a non-zero mean Gaussian process (which is not the common case) for each communication link and iteration (repetitive measurement runs in the same intersection). The mean of Ψ_σ has a Gaussian distribution that represents the differences in the particular traffic situation and gain of the involved antennas for the particular communication link during that specific iteration. Measurements were performed in [39] to capture the joint (simultaneous) behavior of the link gain between six vehicles with one or two antennas on each, in total 41 links, in an urban environment. The six vehicles were divided into two groups driven in the following order: (1) XC90 Silver, Truck Blue, and XC90 Black; and (2) S60, Truck Gold, and V60. The two groups were driven on two orthogonal streets with the purpose of meeting at various intersections. Estimated parameters of g_1, g_2, m, and σ for two different intersections, Yngve (T-crossing with buildings in two corners) and Xerxes (X-crossing with buildings in four corners), are presented in Table 4.20, and a schematic description of the channel-gain model is shown in Figure 4.14.

The solid lines in Figure 4.14 represent single interactions from the center area of the intersection, which are only seen by Truck Blue and S60. XC90 Black is blocked by Truck Blue and has NLOSv toward the intersection center; therefore it has no single interactions toward S60. Dashed lines represent multiple interactions between S60 and Truck Blue and between S60 and XC90 Black, respectively.

Table 4.20 Estimated parameters for the NLOS model.

Link type	Yngve g1[dB]	g2[dB]	M[dB]	σ[dB]	Xerxes g1[dB]	g2[dB]	M[dB]	σ[dB]
Truck-to-truck	-6.12	-1.05	-3.31	4.92	-58.02	-1.96	4.55	4.34
Truck-to-car	-9.63	-2.00	1.03	7.38	-14.18	-3.36	8.78	5.07
Car-to-car	-9.08	-1.45	-2.47	4.39	-47.41	-2.05	3.67	4.27
Truck-to-obstructed car	-51.07	-1.01	-5.33	5.57	-12.17	-3.40	6.42	4.24
Car-to-obstructed car	-61.51	-1.62	-3.61	5.11	-15.68	-3.34	6.81	4.72

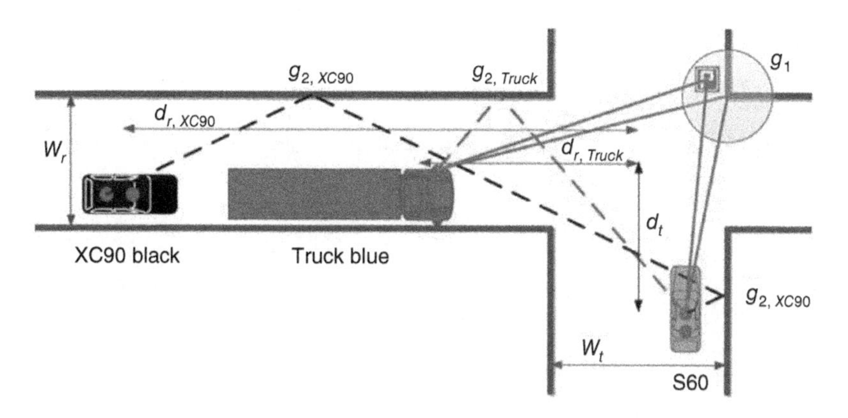

Figure 4.14 Description of the channel gain model for a typical NLOS communication link between vehicles in an urban intersection.

The large-scale fading, Ψ_σ, is achieved by subtracting the distance-dependent mean from the overall channel gain. Then, the spatial autocorrelation of the large-scale fading can be written as

$$r\left(\Delta d_i\right) = E\left\{\Psi_\sigma\left(d_i\right)\Psi_\sigma\left(d_i + \Delta d_i\right)\right\},$$

where d_i is the Manhattan-distance between Tx and Rx. The autocorrelation of the shadowing process can be approximated by a well-known model proposed by Gudmundson [40], based on a negative exponential function,

$$r\left(\Delta d_i\right) = \sigma^2 e^{-\left|\Delta d_i\right|/d_c} = \sigma^2 \rho\left(\Delta d_i\right)$$

Figure 4.15 shows the histograms of the decorrelation distance for the two intersections. By using the proposed channel gain model [39], the analysis shows that the cross-correlations between different links are small, even for communication links with antennas located at the same vehicle (see Figure 4.16). The possibility of neglecting cross-correlation makes the implementation of realistic models in VANET simulations much easier.

The X-axis for the first three graphs in Figure 4.16 represents the link between XC90 Silver's roof antenna (XC90SR) and S60's roof antenna (S60R), and the Y-axis represents

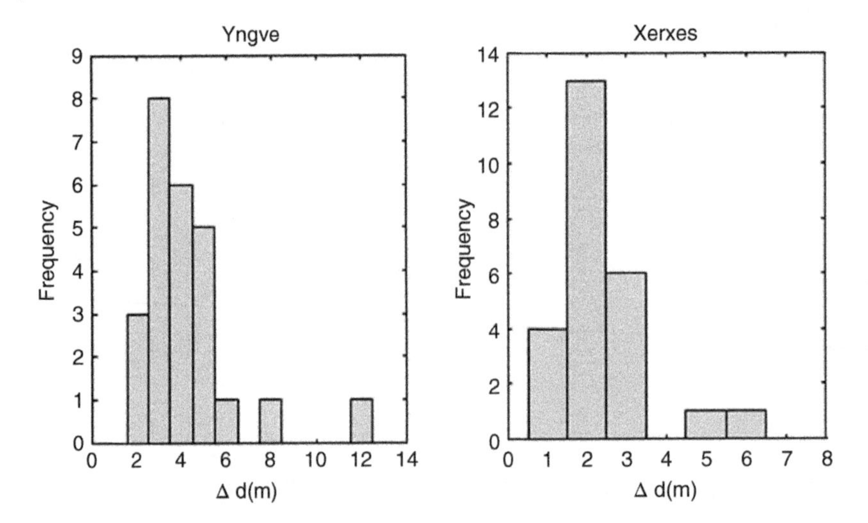

Figure 4.15 Histograms of the de-correlation distances of all NLOS communication links in intersections Yngve and Xerxes.

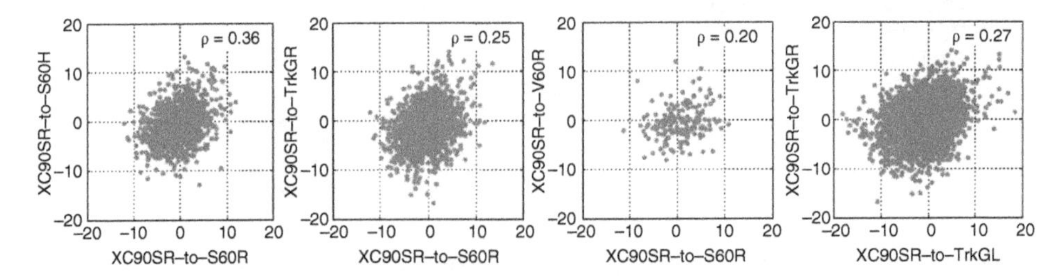

Figure 4.16 Estimated multilink shadowing correlation ρ between different links for the associated scatter plots in intersection Yngve.

communication between XC90 Silver's roof and one antenna on each vehicle in the other convoy, that is S60's hat rack antenna (S60H) and V60's roof antenna (V60R). The last graph shows the cross-correlation of the links between XC90 Silver's roof and the left (TrkGL) and right (TrkGR) antennas on Truck Gold. Note that both the first from left and last scatter plots are cross-correlations when using different antennas on the same vehicle.

4.2.5 Summary

Section 4.2 has discussed state-of-the-art channel models and related measurements for V2X communications and also described some recent extensions to existing models standardized mainly in 3GPP. Many of the results presented in this chapter have already been discussed in 3GPP technical reports, but they are covered here for completeness. This section highlights open issues in 3GPP channel models and also describes recent results in the literature that will be helpful in resolving those remaining issues. Most of these recent results and solutions discussed here are very well aligned with the channel-modeling approach adopted previously in 3GPP, which makes them suitable candidates to be adopted as standards in the future by 3GPP.

References

1 5GAA. (2020). Working group standards and spectrum study of spectrum needs for safety related intelligent transportation systems – day 1 and advanced use cases. TR S-200137, version 1.0.
2 European Commission. (2008). Commission decision of 5 August 2008 on the harmonised use of radio spectrum in the 5 875-5 905 MHz frequency band for safety-related applications of Intelligent Transport Systems (ITS). C(2008) 4145), 2008/671/EC.
3 CEPT ECC. (2008). Use of the band 5855-5875 MHz for Intelligent Transport Systems (ITS). ECC Recommendation (08)01.
4 CEPT ECC. (2008). The harmonised use of the 5875-5925 MHz frequency band for Intelligent Transport Systems (ITS). ECC Decision (08)01.
5 ETSI. (2014). Radiocommunications equipment operating in the 5 855 MHz to 5 925 MHz frequency band; harmonised standard covering the essential requirements of article 3.2 of Directive 2014/53/EU. v2.1.1.
6 CEPT ECC. (2016). Harmonised use of the 63-64 GHz frequency band for Intelligent Transport Systems (ITS). Approached 13 March 2009, Amended 04 March 2016.
7 3GPP. (2018). Evolved universal terrestrial radio access (E-UTRA); user equipment (UE) radio transmission and reception. TS 36.101 v14.6.0.
8 5G Americas. (2017). Spectrum landscape for mobile services.
9 Labib, M., Marojevic, V., Reed, J.H., and Zaghloul, A.I. (2017). Extending LTE into the unlicensed spectrum: technical analysis of the proposed variants. *IEEE Communications Standards Magazine* 1 (4): 31–39.
10 ARIB. (2001). Dedicated short-range communication system. ARIB STD-T75, version 1.0.
11 ACMA. (2017). Radiocommunications (intelligent transport systems) class licence.
12 Autorià per le Garanzie nelle Comunicazioni. (2018). Delibera N. 231/18/CONS.
13 Ministero dello sviluppo economico. (2019). Bando 5G per l'assegnazione di diritti d'uso delle frequenze.
14 Bell, P. (2018). Italian 5G auction sees high price tags, raised eyebrows. TeleGeography.
15 European 5G Observatory. (2018). 700 MHz auction results in Sweden.
16 Bundesnetzagentur. (2019). Frequenzauktion 2019 - frequenzen für 5G.
17 Bundesnetzagentur. (2018). Konsultationsentwurf einer Entscheidung der Präsidentenkammer der Bundesnetza-gentur.
18 Ministerio de Economía y Empresa. (2018). Nota informativa sobre la subasta del espectro banda 3600–3800 MHz.
19 Ministry of Energy, Tourism and Digital Agenda. (2018). Spain's 5G national plan 2018–2020.
20 L'Agence nationale des fréquences (ANFR). (2018). Préparer les bandes actuelles de la téléphonie mobile pour la 5G.
21 Federal Communications Commission (FCC). (2019). The FCC's 5G FAST plan.
22 Federal Communications Commission (FCC). (2019). Auction 102: spectrum frontiers – 24 GHz.
23 Tomas, J.P. (2018). China issues spectrum for nationwide 5G trials. RCR Wireless News.
24 European 5G Observatory. (2019). Japan assigns 5G spectrum to four operators.
25 GSMA. (2019). Auction best practice: GSMA public policy position.

26 Abbas, T. (2014). Measurement based channel characterization and modeling for vehicle-to-vehicle communications. Department of Electrical and Information Technology, Lund University.

27 Boban, M., Vinhoza, T.T.V., Ferreira, M. et al. (2011). Impact of vehicles as obstacles in vehicular ad hoc networks. *IEEE Journal on Selected Areas in Communications* 29 (1): 15–28.

28 Boban, M., Viriyasitavat, W., and Tonguz, O.K. (2013). Modeling vehicle-to-vehicle line of sight channels and its impact on application-layer performance. In: *Proceeding of the 10th ACM International Workshop on Vehicular Inter-networking, Systems, and Applications (VANET 13)*, 91–94. Taipei, Taiwan.

29 3GPP. (2017). 3rd generation partnership project; technical specification group radio access network; study on channel model for frequencies from 0.5 to 100 GHz (Release 14). TR 38.802 V14.2.0.

30 Karedal, J., Czink, N., Paier, A. et al. (2011). Path loss modeling for vehicle-to-vehicle communications. *IEEE Transactions on Vehicular Technology* 60 (1): 323–328.

31 ITU-R. (2013). Propagation by diffraction. Recommendation P.526.

32 Boban, M., Barros, J., and Tonguz, O. (2014). Geometry-based vehicle-to-vehicle channel modeling for large-scale simulation. *IEEE Transactions on Vehicular Technology* 63 (9): 4146–4164.

33 Aygun, B., Boban, M., Vilela, J.P., and Wyglinski, A.M. (2016). Geometry-based propagation modeling and simulation of vehicle-to-infrastructure links. In: *Vehicular Technology Conference (VTC Spring)*, vol. 83, 1, 5. IEEE IEEE, May 2016.

34 Abbas, T., Sjöberg, K., Karedal, J., and Tufvesson, F. (2015). Measurement based shadow fading model for vehicle-to-vehicle network simulations. *International Journal of Antennas and Propagation.*

35 Nilsson, M., Gustafson, C., Abbas, T., and Tufvesson, F. (2017). A measurement based multilink shadowing model for V2V network simulations of highway scenarios. *IEEE Transactions on Vehicular Technology* 66 (10): 8632–8643.

36 Acosta-Marum, G. and Ingram, M.A. (2007). Six time- and frequency- selective empirical channel models for vehicular wireless LANs. *IEEE Vehicular Technology Magazine* 2 (4): 4–11.

37 3GPP. (2018). V2X fast fading model. 3GPP TSG-RAN1 Meeting #92, Athens. R1-1802721.

38 3GPP. (2018). V2X path loss and shadowing. 3GPP TSG-RAN1 Meeting #92, Athens. R1-1802720.

39 Nilsson, M.G., Gustafson, C., Abbas, T., and Tufvesson, F. (2018). A path loss and shadowing model for multilink vehicle-to-vehicle channels in urban intersections. *MDPI Sensors* 18 (12): 4433.

40 Gudmundson, M. (1991). Correlation model for shadow fading in mobile radio systems. *Electronics Letters* 27 (23): 2145–2146.

41 Wiquist, W. (2020). FCC modernizes 5.9 GHz band for Wi-Fi and auto safety. *FCC News.*

5

V2X Radio Interface

Malte Schellmann[1], Jian Luo[1], Erik Ström[2], Tommy Svensson[2], Gabor Fodor[3,4], Hieu Do[3], Keerthi Kumar Nagalapur[3], Zexian Li[5], Hanwen Cao[1], and Konstantinos Manolakis[1]

[1]*Huawei German Research Center, Germany*
[2]*Chalmers University of Technology, Sweden*
[3]*Ericsson Research, Sweden*
[4]*KTH Royal Institute of Technology, Sweden*
[5]*Nokia Bell Labs, Finland*

Communication among vehicles, road infrastructure, and the network plays a pivotal role for obtaining a comprehensive view of the environment for any road user, and as such, it paves the way toward automated and autonomous driving. To this end, communication becomes a fundamental component of a safety-critical system. Hence, it has to fulfill high demands on reliability and low latency: radio links must show stable behavior and need to be robust against typical distortions that may occur in the radio environment, and specific delay bounds have to be met with high reliability. If these requirements are met, communication will become a key enabler for connected automated driving (CAD) and connected road user (CRU) service applications. The latest mobile radio generation, 5G new radio (NR), offers the radio service of ultra-reliable low latency communication (URLLC), which provides fundamental system functionalities for addressing those requirements. These URLLC system functionalities thus form the basis for the vehicle-to-everything (V2X) radio interface, while further extensions need to be designed to properly match the particular needs of V2X communication and to beneficially exploit properties inferred from environmental and propagation conditions prevailing in typical V2X communication scenarios.

Similar to mobile radio networks like 4G long term evolution (LTE), vehicles can communicate with the network through base stations (BSs), thus establishing a vehicle-to-network (V2N) connection. These links exhibit propagation conditions similar to those known from mobile networks; however, for V2X communication, they must be capable of providing stable connections even at very high vehicular speeds. Considering the latest beamforming techniques with narrow beams, which are used especially at higher carrier frequencies in

* With contributions from Hao Guo, Behrooz Makki, and Tapisha Soni.

Cellular V2X for Connected Automated Driving, First Edition.
Edited by Mikael Fallgren, Markus Dillinger, Toktam Mahmoodi, and Tommy Svensson.

5G NR, beam tracking and adjustment need to be carried out reliably in a fast manner, calling for the development of novel schemes. Correspondingly, the first section of this chapter, Section 5.1, is devoted to this topic.

Moreover, roads may be equipped with special infrastructure elements that vehicles can directly communicate with, such as elements of traffic control systems or dynamically changing traffic signs. A wireless connection to these infrastructure elements is called a vehicle-to-infrastructure (V2I) link. Propagation conditions for these links are, to a large extent, comparable to V2N links, though the height of the infrastructure element is significantly lower in general, and the coverage range is therefore much more delimited. Nevertheless, technologies developed for V2N communication are usually also applicable for V2I communication.

Finally, vehicles need to be able to communicate with each other by establishing a vehicle-to-vehicle (V2V) link, which may be done under the control of the network (i.e. requiring the vehicles to be in network coverage with an established V2N link) or in a fully autonomous fashion. For direct V2V communication, the connection between any pair of vehicles is called a *sidelink* (SL), which has specific properties rendering it sufficiently different from V2N and V2I links: antennas of transmitters and receivers are typically placed at a much lower level above the ground than in cellular communications, low levels of transmit powers are used, the small coverage range allows only short-range communication, and a high probability of blocking is experienced due to other cars in the vicinity. Moreover, the speed meant to be supported for V2V links amounts to twice the maximum vehicle speed of 250 km/h, since two communicating vehicles may be moving toward each other, yielding speed support of up to 500 km/h. These substantially different properties require a tailored redesign of core elements of the air interface for the sidelink, embracing the lowest two layers of the system protocol stack: the physical (PHY) layer and medium access control (MAC) layer. Consequently, PHY and MAC extensions are the focus of the second section of this chapter, Section 5.2.

In addition, thanks to the short-range communication enabled by the sidelink, novel communication concepts are facilitated for V2X communication. As such, cooperation between multiple user equipments (UEs) can be established for opening up additional communication links, which can be exploited for improving overall communication reliability. Also, the technology of full duplex becomes feasible for V2V communication, which allows for efficient use of the available spectrum and reduction of communication latencies. The application of these novel communication concepts in the context of V2X communication is elaborated in the third section of this chapter, Section 5.3, and the last section of this chapter, Section 5.4, contains a summary.

5.1 Beamforming Techniques for V2X Communication in the mm-Wave Spectrum

A key technology component in the 5G NR system is the application of beamforming techniques with narrow beams in the mm-wave spectrum to overcome the dominant path-loss (PL) effect. Highly mobile users pose the requirement of fast beam tracking and adjustment when using this technology. In this section, three different beamforming schemes are presented for this case, where each of these schemes is devoted to one of the three casting types: unicast, multicast, and broadcast.

5.1.1 Beam Refinement for Mobile Multi-User Scenarios

Due to power limitations and high path loss of omnidirectional transmissions at mm-wave frequencies, the coverage range is typically small, so highly directional transmissions are required. On the other hand, the physical size of the antenna elements at the mm-wave band is relatively small. Thus, there is a potential for employing large antenna arrays such that large-scale beamforming can be performed, provided that sufficient and fast beam-alignment can be achieved. In this section, the focus is on unicast scenarios, in which no common data of interest can be assumed among the vehicles; in the subsequent two sections, techniques are described that can harness the benefits of using multicasting or broadcasting whenever there are common data of interest for groups of vehicles or for all vehicles, respectively.

As proposed in [1], after a basic connection is established, during the initial access (IA) procedure, BSs and users can begin exchanging messages and implement a beam-refinement procedure to further improve beam directions. Generally, IA beamforming at mm-wave frequencies is different from cm-wave since it is hard to acquire the channel state information (CSI) at mm-wave frequencies. For this reason, codebook-based beamforming has been proposed as an efficient method to reduce the dependency on CSI estimation/feedback. Several works have been presented on both physical layer and procedural algorithms of codebook-based beamforming. The basic steps of the beam-refinement approach are:

1) Selecting a precoding matrix at the BS while selecting a combining matrix at the receiver side out of predefined codebooks
2) Sending test signals
3) Updating the selection results based on the users' feedback about their performance metrics

However, in those works, either the algorithms are designed for special metrics, precoding/combining schemes, and channel models, or the implementation complexity grows significantly with an increasing number of BSs/users. Moreover, the running delay of the algorithm has rarely been considered in performance evaluation. On the other hand, machine learning–based schemes have been recently proposed for IA [2], which can be effectively applied for different channel models with acceptable implementation complexity, enabling low delays.

This section introduces a beam-refinement scheme to include beamforming at both the transmitter and receiver in mm-wave multi-user multiple input multiple output (MU-MIMO) networks with moving vehicle users. The goal is to perform a system-level rate optimization by proposing and analyzing a genetic algorithm (GA)-based scheme with the knowledge of user positions. As will be shown, user mobility can also be handled by beam refinement. In the moving scenario, the beam-refinement process can keep tracking the beams by exploiting spatial correlations so that the computational delay can

In each time slot with instantaneous channel realization $\boldsymbol{H} \in \mathscr{C}^{N\times M}$, do the following

(I) Initialization: Consider L, e.g. $L = 10$, sets of precoding matrices $\boldsymbol{V}_l$ and combining matrices $\boldsymbol{U}_l$, $l = 1, \ldots, L$, randomly selected from the pre-defined codebook $\boldsymbol{W}_{\mathrm{T}}$ and $\boldsymbol{W}_{\mathrm{R}}$.

(II) Selection: For each $\boldsymbol{V}_l$ and $\boldsymbol{U}_l$, evaluate the instantaneous value of the objective metric R_l, $l = 1, \ldots, L$, for example end-to-end throughput (5). Find the best beamforming matrix which results in the best value of the considered metric, named as the *Queen*, e.g. $\boldsymbol{V}_q$ and $\boldsymbol{U}_q$ satisfies $R(\boldsymbol{V}_l, \boldsymbol{U}_l) \leq R(\boldsymbol{V}_q, \boldsymbol{U}_q), \forall l = 1, \ldots, L$ if the end-to-end throughput is the objective function.

(III) Save the Queen: $\boldsymbol{V}_1 \leftarrow \boldsymbol{V}_q, \boldsymbol{U}_1 \leftarrow \boldsymbol{U}_q$

(IV) Genetic operation I-Crossover: Create $S < L$, e.g., $S = 5$, beamforming matrices $\boldsymbol{V}_{\mathrm{s}}^{\mathrm{new}}$and $\boldsymbol{U}_{\mathrm{s}}^{\mathrm{new}}$, $s = 1, \ldots, S$, around the Queen $\boldsymbol{V}_1$ and $\boldsymbol{U}_1$. These sets are generated by making small changes in the Queen $\boldsymbol{V}_q$ and $\boldsymbol{U}_q$.

(V) $\boldsymbol{V}_{s+1} \leftarrow \boldsymbol{V}_{\mathrm{s}}^{\mathrm{new}}, \boldsymbol{U}_{s+1} \leftarrow \boldsymbol{U}_{\mathrm{s}}^{\mathrm{new}}, s = 1, \ldots, S.$

(VI) Genetic operation II-Mutation: Regenerate the remaining sets $\boldsymbol{V}_s$ and $\boldsymbol{U}_s$, $s = S + 2, \ldots, L$, randomly with the same procedure as in Step (I).

(VII) Go back to Step (II) and run for N_{it} iterations,. N_{it} is a fixed number decided by designer. Return the final Queen as the beam selection rule for the current time slot.

Figure 5.1 GA-based beam-refinement algorithm. Source: [3].

be substantially reduced. Cooperative side links, i.e. V2V links that allow users to share their received message to further increase throughput, are also shown to be beneficial.

5.1.1.1 Algorithm Description

The GA-based algorithm starts by creating L possible beam-selection sets at the transmitter, $\boldsymbol{V}_l$, and receiver, $\boldsymbol{U}_l$, respectively, i.e. L submatrices of each predefined codebook $\boldsymbol{W}_{\mathrm{T}}$, $\boldsymbol{W}_{\mathrm{R}}$ at transmitter and receiver, respectively, which then form L pairs $\{\boldsymbol{V}_l, \boldsymbol{U}_l\}$. During each iteration, the best set $\{\boldsymbol{V}_l, \boldsymbol{U}_l\}$, referred to as the *Queen*, is chosen based on a suitable performance metric, e.g. rate R_l. Next, the Queen is kept, and $S < L$ similar sets around the Queen are regenerated by making small changes to the Queen (typically 10% of the Queen columns are replaced randomly without loss of generality, as adopted here). Finally, the other $L - S - 1$ beamforming matrices are selected randomly from the transmit and receive codebooks $\boldsymbol{W}_{\mathrm{T}}$ and $\boldsymbol{W}_{\mathrm{R}}$, respectively, to avoid the algorithm from being trapped in a local minimum. Note that increasing S for a given L can increase the chance of being trapped. After N_{it} iterations (set by the designer as a trade-off between complexity and desired closeness to asymptotic performance), the Queen is returned as the result of the beam selection in the current time slot. The algorithm is summarized in Figure 5.1, and further details can be found in [3].

5.1.1.2 Illustrative Performance Results

Here the results of beam-tracking performance for moving vehicles and the performance gain of cooperative users (CUs) are presented, in which they share channel knowledge e.g. over the sidelink to perform joint decoding. For illustration purposes, consider a single-cell MU-MIMO system in which four vehicles are served on the same time-frequency resource, as shown in Figure 5.2. Moreover, performance gains are shown for deploying CUs that are forming a virtual antenna array among the vehicles, and this is compared with the case of non-cooperative users (NCUs).

The performance of the GA-based beam-refinement algorithm is evaluated and compared with related state-of-the-art schemes: link-by-link search [4], two-level search [5] and tabu search [6], in terms of service outage-constrained end-to-end throughput

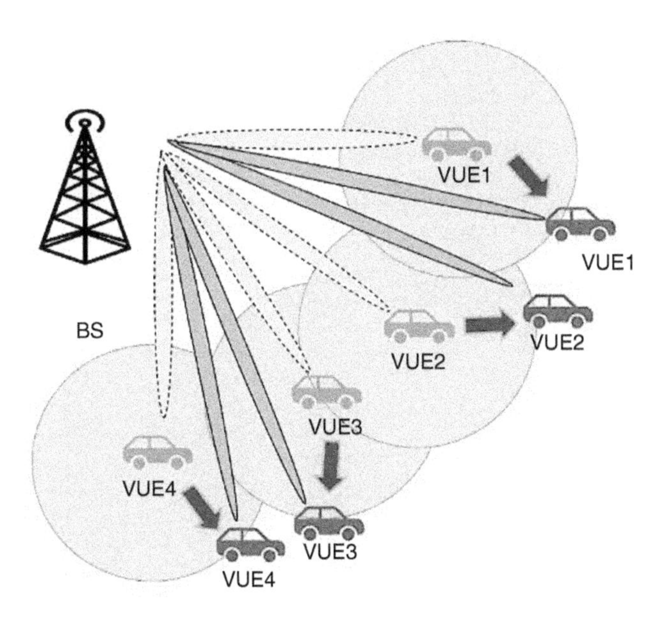

Figure 5.2 Beam-tracking network with proposed GA-based scheme in a MU-MIMO system. Circles indicate potential movement range for vehicle user equipment (VUE) during a particular time slot.

in bit-per-channel-use (bpcu), as defined in [3]. In the link-by-link search, the beam design of the users is not optimized simultaneously. Instead, with a greedy approach, the beamforming solution is settled sequentially user-by-user by considering the interference from the previous assignment step. System performance improves in successive iterations. In the first level of the two-level search, the BS transmits messages over wider sectors, while in the second level, it searches the optimal solution within the best such sector by using narrower beams [7]. Tabu search can be regarded as a restricted GA-based scheme, with the constraint that while updating the Queen, each updated precoding matrix is restricted to differ in only one column.

Figure 5.3a shows the effect of user mobility on the beam-refinement delay for the considered algorithms, calculated as the number of iterations required for a given moving speed times the calculation time for each iteration. The beam-refinement delay is evaluated (where it is assumed that the calculation time for each iteration amounts to 10^{-4} of the considered moving time slot) of each algorithm in two cases to check the suitability of these algorithms for mobile users. Here, the number of transmit and receive antennas is 32 and 8, respectively (2 antennas for each user), and the signal-to-noise ratio (SNR), defined as transmitted power to receiver noise, is 32 dB, while the moving time slot is 1 ms. *Case 1* refers to the situation that the algorithm uses random guesses at the starting point, while *Case 2* means the algorithm starts with the optimal sets from the previous time slot by exploiting the spatial correlation. The algorithm running delays in Case 1 and Case 2 for each method are all presented in the plot.

Comparing the performance for the two cases in the figure, both the GA-based algorithm and the tabu-based algorithm can significantly reduce the beam-refinement delay for a broad range of user speeds, since they benefit from using the beam-refinement solution

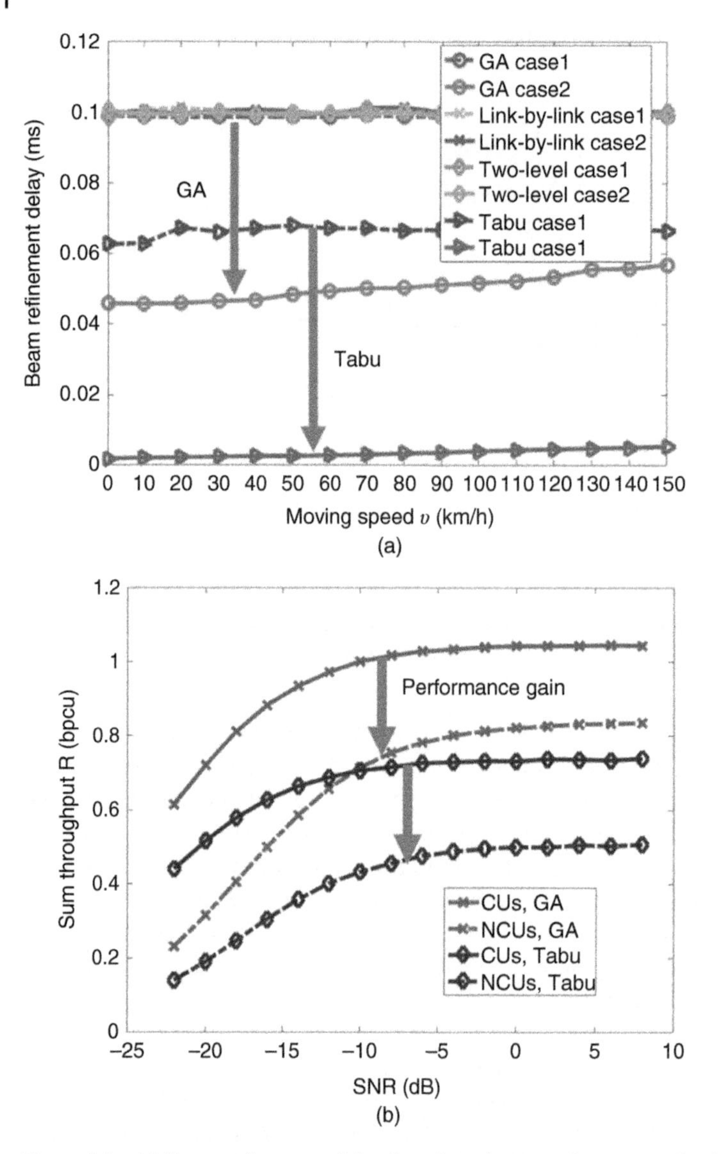

Figure 5.3 (a) Beam-refinement delay for a broad range of user speeds with the considered methods. The vertical arrows refer to the reduction from Case 1 to Case 2 for GA and tabu search. (b) The optimized service outage-constrained end-to-end sum throughput of all users with cooperative and non-cooperative cases for GA and tabu. Case 1, using 1000 iterations (sufficient for essentially asymptotic performance in all cases). Source: [3].

from the previous transmission as the initial guess when the moving distance is not large. Note that tabu search has the lowest delay in both cases since it simply changes the Queen to its neighbors, which takes full advantage of spatial correlations. However, as user speed increases, the beam-refinement delay increases slightly, intuitively because the spatial correlation between the positions in successive time slots decreases. Moreover, the link-by-link

Table 5.1 Average number of required iterations in different cases. Source: [3].

M/N	GA, CUs	GA, NCUs	Tabu, CUs	Tabu, NCUs
32/12	502	1	498	1
32/8	500	1	501	1
32/4	488	1	502	1

M is the number of transmit antennas, while N is the total number of receive antennas, with $N/4$ being the number of antennas per user.

search and two-level-based search do not show noticeable performance gains when comparing the two cases.

Figure 5.3b shows the effect of user collaboration on the end-to-end throughput for the GA and tabu methods. As seen in the figure, performance of the GA-based algorithm and tabu-based restricted GA algorithm is substantially improved in the case of CUs. These improvements decrease somewhat as the SNR increases. The GA-based algorithm is shown to be able to better scan the overall beam space than the tabu-restricted GA search, showing the value of fewer restrictions while updating the Queen in the GA algorithm.

Table 5.1 presents the average number of required iterations for both the GA and tabu-restricted GA search in the cases of CUs and NCUs. In contrast to Figure 5.3b, in the table, it can be seen that the NCUs case requires a much shorter iteration time compared to the CUs case for the considered system configurations. Thus, the presented performance gains in the CA case should be regarded as estimates of the potential performance improvements with CUs over NCUs, i.e. allowing vehicles to form a virtual antenna array. However, for practical implementations, further studies on how to best confine the GA search in this case would be needed to reduce the required number of iterations. Thus, the presented performance in Figure 5.3b should be regarded as an upper bound for any such algorithm. Further details and performance results can be found in [3].

To summarize, GA-based beam-refinement schemes for codebook-based beam refinement during initial access can substantially improve the delay-constrained throughput of V2N links by exploiting spatial correlation, and the performance gain remains at high speed. Spatial correlation can be utilized by setting the Queen, i.e. the optimal beam set, of the previous position as the initial guess for the current position. Vehicle cooperation using sidelinks to form a virtual antenna array shows substantial performance gains for further throughput increase. However, this comes at the expense of computational complexity and sidelink communication overhead; therefore, further study is needed to make such an approach practical for implementation.

5.1.2 Beamformed Multicasting

In Section 5.1.1, initial access and beam selection were briefly discussed. During initial access, a beamformed broadcast is applied. In this section, beamformed multicast (BM) is addressed. It is well known that multicast transmissions can significantly improve the resource efficiency of V2I and V2N links compared to that of unicast transmission whenever there is common data to transmit. Relying on multicast capability, the common data of interest is transmitted only once to the concerned vehicles instead of addressing every vehicle separately by sending the same message over multiple dedicated radio channels. The multicast transmission mode thus enhances the downlink capacity in the context of downlink re-distribution of cooperative awareness messages (CAMs) and decentralized environmental notification messages (DENMs), which are characterized by modest packet sizes [8]. Even greater gains can be expected in scenarios where vehicles receive high-data-rate V2I transmissions supporting applications such as high-definition map acquisition and infotainment services requiring Gbps transmission rates [9].

To this end, a beamformed multicast technique is presented, which builds on adaptive and robust beam-management techniques. This technique is especially suitable for mm-wave bands, where large antenna arrays can be deployed at BSs and road-side units (RSUs).

For ease of presentation, a single-cell communication system is considered, in which an infrastructure node such as a BS or RSU (commonly referred to as a transmission/reception point [TRP]) provides reliable multicast transmission to UEs within cell coverage. The key idea is to exploit the fact that UEs geographically close to each other can be served by the same beam containing common messages for such UEs. Furthermore, the TRP exploits feedback from UE acknowledgment (ACK) or negative-acknowledgment (NACK) schemes, which are designed to trigger retransmissions for enhancing the reliability of the multicasting service. The proposed BM transmission technique consists of the following control plane and signaling support:

- Signaling design for initial transmission
- Signaling design for hybrid automatic repeat request (HARQ) feedback
- Signaling design for retransmission

For initial transmission, a novel physical channel is implemented, called the physical broadcast multicast channel (PBMCH), which is designed for data transmission only, as illustrated as option 1 in Figure 5.4. The associated control signaling is supported by existing physical channels such as the physical broadcast channel (PBCH) carrying system information blocks defined in legacy cellular systems [10]. Additionally, another frame structure is considered, in which control signaling and data transmission are incorporated in the multicast/broadcast physical layer downlink control channel and the broadcast/multicast physical layer downlink shared channel, as illustrated as option 2 in Figure 5.4. While option 1 provides more compatibility with the existing standard, option 2 allows more flexibility.

The necessary downlink control information associated with the PBMCH transmission enables the receiving UEs to decode the PBMCH messages. This control information includes information regarding the applied modulation and coding scheme (MCS), as well as ACK/NACK indication. The resources for these pieces of information are either pre-allocated or dynamically allocated using the standardized physical downlink control channel (PDCCH). In this way, the overhead of the initial transmission is reduced.

For HARQ feedback, different timing indications of ACK/NACK feedback are supported. Specifically, HARQ feedback can be sent either immediately after each beam or after all

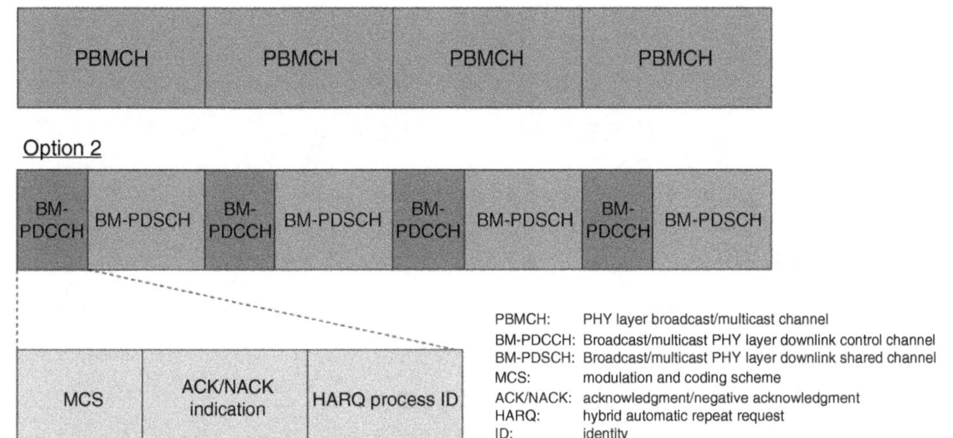

Figure 5.4 Illustration of the two options for 5GCAR PBMCH for initial transmission. Source: © 2019 IEEE. Reprinted, with permission, from [13].

beams have scanned their corresponding cell area. In addition, since geographically close UEs usually belong to a single group, these UEs may receive signals from adjacent scanning beams, on which different levels of received signal power may be experienced. To improve the spatial reuse gain, UEs belonging to the same group are enabled to combine signals either from all beams they received signals from or on the single beam with the highest receive power (the "best" beam). The selection of the "best" beam can be realized, for example, by linking to the synchronization of the "best" beam from the received synchronization signal block or the received channel state information reference signals (CSI-RSs) utilizing channel reciprocity.

Finally, for retransmissions, a similar beam-selection scheme is utilized, as illustrated in Figure 5.5. Specifically, the packets on retransmission links are identical within each beam or group of beams, if HARQ feedback is transmitted in all beam directions (upper part of Figure 5.5). This option enhances the chance of successful retransmission. Alternatively, if HARQ feedback is transmitted only along the "best" beam (lower part of Figure 5.5), the packets on the retransmission links are specific to each beam/group and supported by beam/group-specific PDCCH indications. This option allows high efficiency of resource usage, since remaining beams can be used for the next BM initial transmission.

The performance of the proposed BM scheme is compared with two benchmark schemes: broadcast schemes in LTE-V (where V stands for V2X) [11] (without beamforming and retransmission) and LTE-V+HARQ (with retransmission, but without beamforming), in terms of achievable data rate and latency. The simulation parameters, which are in line with [12], are summarized in Table 5.2.

In the simulator, a maximum of 16 beams (where the minimum beamwidth equals 22.5°) can be simultaneously exploited to cover the entire angular space. The number of simultaneous beams can be any divisor of 16 (1, 2, 4, 8, or 16). Further, simultaneous transmissions to different groups of vehicles along different beams are supported at the TRP. Other details

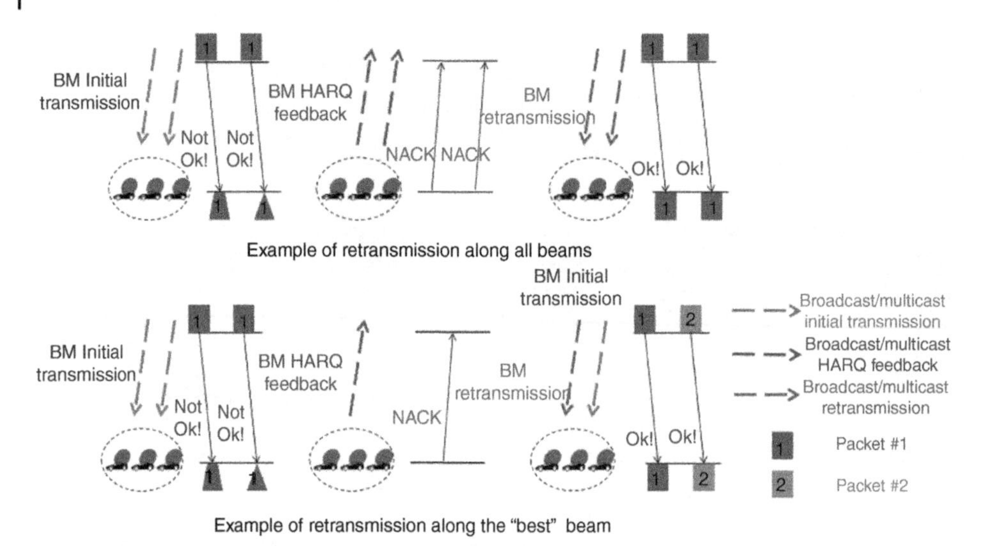

Figure 5.5 Illustration of HARQ feedback and retransmission. Source: © 2019 IEEE. Reprinted, with permission, from [13].

Table 5.2 Simulation setup for beamformed broadcast/multicast design.

Parameter	Value
Carrier frequency	30 GHz
Bandwidth	200 MHz
Maximum transmit power	43 dBm
Noise figure	7 dB
Subframe length	1 ms
Number of OFDM symbols	14

of the simulation setup, including the network layout, channel model, and TRP antenna configurations, are in line with those in [14]. To obtain numerical results, the simulation samples are averaged over 1000 independent snapshots.

The results in Figure 5.6a show that the proposed scheme outperforms the benchmark broadcast/multicast schemes in terms of achievable data rate, due to the redundant copies of data received on adjacent beams by each UE. Receiving redundant copies enables the usage of high-order MCS that improves the achievable data rate. In particular, a small performance gain for the proposed scheme as compared with LTE-V and LTE-V+HARQ is observed in the low SNR range. The main reason for this behavior is that a low SNR leads to transmission failures invoking retransmissions, for which conservative MCS is applied in all schemes. This, in turn, implies that the proposed scheme benefits less from

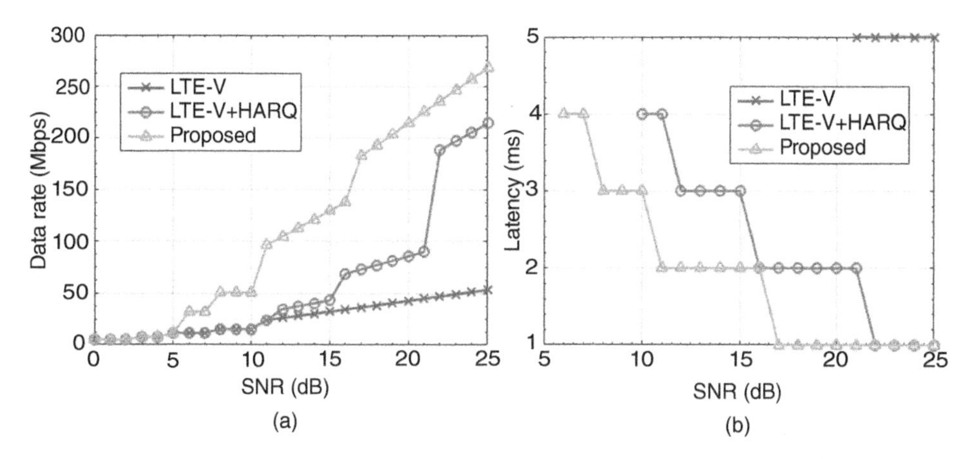

Figure 5.6 Performance comparison of different broadcast/multicast schemes vs. SNR in terms of (a) data rate and (b) latency. LTE-V denotes LTE-V2X without beamforming and retransmission; while LTE-V+HARQ denotes LTE-V2X with retransmission, no beamforming. Source: © 2019 IEEE. Reprinted, with permission, from [13].

the capability of using high-order MCS in the low SNR range. In contrast, the proposed BM scheme achieves significantly higher data rates in the medium and high SNR ranges. This gain achieved by the proposed method comes at the cost of a moderate complexity increase as compared with the LTE-V and LTE-V+HARQ schemes. This complexity increase is due to computing the beamforming weights inherent in the proposed scheme.

In Figure 5.6b, the performance of different broadcast/multicast schemes is compared in terms of latency, defined as the time after a UE is turned on until it successfully decodes the broadcast information. As can be observed from this figure, the proposed scheme achieves the lowest latency, owing to the redundant copies received on adjacent beams. Note that in the case of LTE-V, at some SNR values (especially in the low SNR range), the corresponding latency value is infinite, which means the transmitted information cannot be successfully decoded by the receiver within the maximum number of retransmissions (set to 5, whereas for LTE-V, there is no retransmission); therefore, the latency value is not plotted in the figure.

5.1.3 Beam-Based Broadcasting

Broadcast transmission is essential for the initial access of UEs. Further, certain V2X-related information, e.g. traffic-related information, can be transmitted via broadcasting mode from TRPs to vehicles. Compared to the multicast transmission in the previous section, broadcast transmission does not require any feedback from the UEs. While the mm-wave spectrum provides a large amount of bandwidth, which is suitable for broadcast with a large amount of data for V2X applications, mm-wave transmissions usually rely on beamforming [15]. This complicates the broadcast transmission at high frequencies for V2X

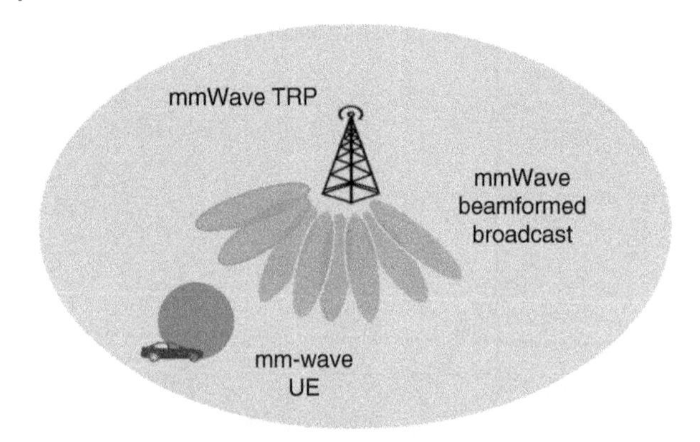

Figure 5.7 System model for studying mm-wave beam-based broadcasting for V2X. Source: © 2018 IEEE. Reprinted, with permission, from [16].

communications and requires proper design considering different beam patterns, beam configurations, multiplexing schemes, etc. This section describes a beam-scan scheme, which depends on the targeted performance metric. Specifically, by properly selecting the beam-scan scheme, a trade-off can be managed between latency and signaling overhead, the latter being defined as the percentage of resources used for the broadcast signaling among the overall used resources.

Without loss of generality, downlink transmission in a single-cell standalone network is considered with one mm-wave TRP at its center of radius r = 100 m. In general, UEs are assumed to be randomly dropped in the network. Figure 5.7 shows an example of a single-cell standalone mm-wave network with one TRP.

It is assumed that the TRP employs directional steerable antenna arrays and can perform both 2D and 3D beamforming. The UEs are assumed to be able to synthesize a quasi-omni antenna pattern for signal reception. Consider a basic frame structure with length T consisting of a system control interval, a broadcast transmission interval, and a unicast transmission interval, as illustrated in Figure 5.8. In the broadcast transmission interval, the TRP broadcasts information via beam scan over different beam slots. The entire cell is covered by N beam-scan areas, where TRP forms M simultaneous beams – limited by the number of radio frequency (RF) chains at the TRP – to successively scan these areas, as illustrated in the bottom part of Figure 5.8. Within the duration t of each slot, the formed beam(s) deliver broadcast information to UEs located in the corresponding areas. Note that these slots are separated by guard intervals (GIs) maintained for duration t_{GI}, reserved for beam switching in the case of hybrid or analog beamforming. The TRP periodically scans the cell via angular probing in the broadcast transmission interval within each frame, and maintains the scanning beam order in each frame: the beam(s) scan the same area of the cell during the same slot of each frame.

The basic frame structure can be further extended as follows (illustrated in Figure 5.9):

- Each frame contains multiple broadcast intervals.
- There exists a unicast-only frame.
- One broadcast interval is distributed into several frames.

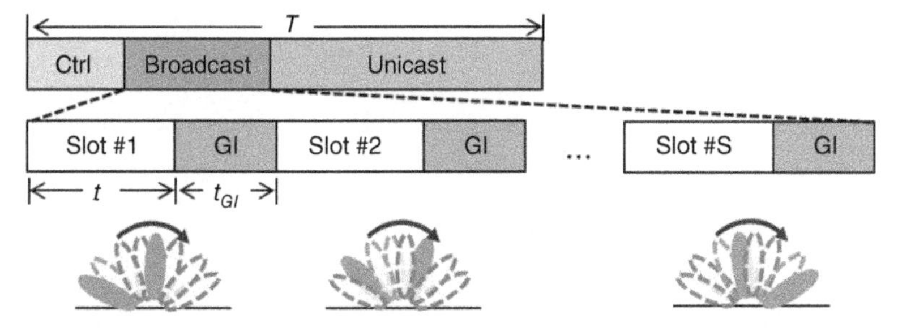

Figure 5.8 Illustration of a basic frame structure for broadcast transmission, where t is the slot duration and t_{GI} is the guard interval duration. Source: © 2018 IEEE. Reprinted, with permission, from [16].

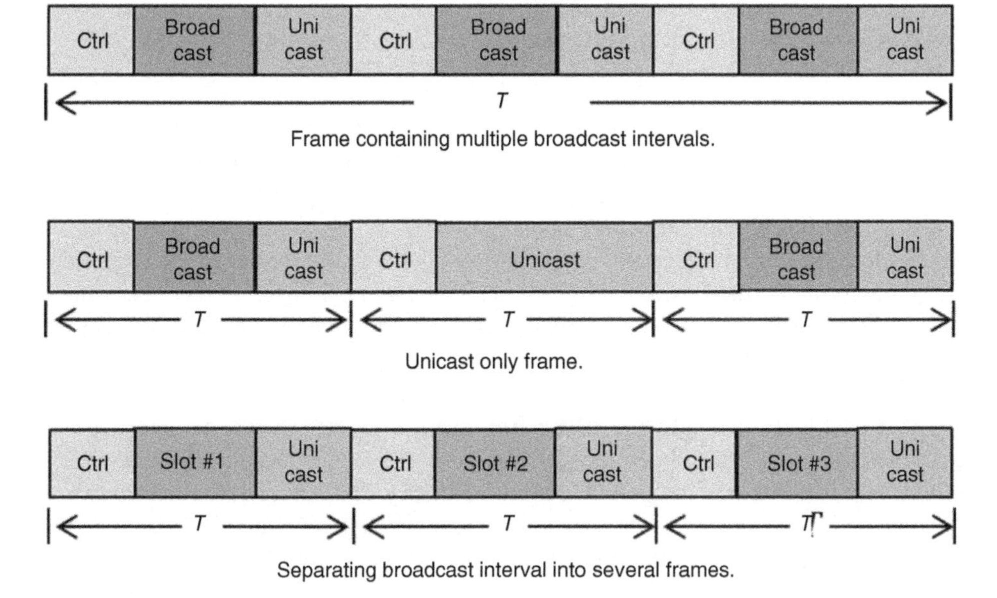

Figure 5.9 Illustration of different frame structure extensions, where T is the frame duration. Source: © 2018 IEEE. Reprinted, with permission, from [16].

To harvest the multiplexing capability, different broadcasting schemes are investigated to enable the TRP to broadcast information. The considered schemes are described as follows [17]:

- *Single-beam exhaustive scan*: TRP scans the entire cell area with a single beam at a time in time-division (TD) multiplexing manner. This scheme is referred to as the *baseline* design, which is especially suitable for transmitters with a single RF chain.
- *Multi-beam exhaustive scan*: If multiple RF chains are available at the transmitter, more than one beam can be formed simultaneously. In contrast to the baseline, this scheme exploits multiple simultaneous beams multiplexed in the frequency domain. Here, three multiplexing schemes are applied: frequency-division (FD), code-division (CD), and space-division.

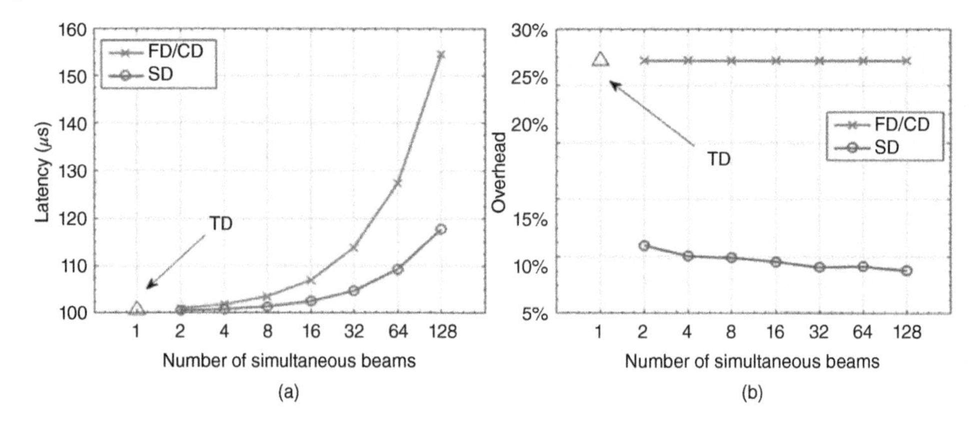

Figure 5.10 Performance of different beamformed broadcast schemes for V2X applications in terms of (a) latency and (b) overhead. TD: time division; FD: frequency division; CD: code division; SD: spatial division. Source: © 2018 IEEE. Reprinted, with permission, from [16].

By evaluating the performance metrics, including broadcast latency and overhead, based on the system assumptions, the following insights are obtained for the beam-based broadcast design [16].

How wide should the beam be? Latency is optimized when the thinnest beam is formed, since the thinner the beam, the higher the SNR, and the shorter the slot duration t [17]. Interestingly, the beamforming architecture has no impact on latency, which renders the latency performance of analog/hybrid beamforming the same as that of digital beamforming. In contrast, thinner beams result in higher overhead, as shown in Figure 5.10b.

Is it beneficial to exploit multi-beam simultaneous scan? As depicted in Figure 5.11a, a multi-beam simultaneous scan leads to a latency penalty. A single-beam exhaustive scan is found to be optimal in terms of latency. This is reversed when considering overhead, where the single-beam exhaustive scan, as well as frequency-division/code-division multi-beam scan, suffer from high overhead, as shown in Figure 5.11b. The spatial-division multi-beam scan, however, achieves the lowest overhead. In a word, by configuring the number of simultaneous beams, latency can be traded with overhead or vice versa.

How does frame structure affect the broadcast performance? It is straightforward that a frame containing multiple broadcast intervals leads to lower latency at the price of higher overhead. On the other hand, unicast-only frames and/or separating one broadcast interval into several frames are recommended to achieve lower overhead with relatively high latency.

What is the impact of block error rate (BLER)? BLER could affect performance, since the higher the BLER, the higher the chance of decoding errors, even when the UE has captured the best beam. Thus, the UE may need to wait several beam-scanning periods until the broadcast information is successfully decoded. Higher BLER typically occurs if a higher MCS is used; then fewer resources are consumed by broadcast signaling, corresponding to a

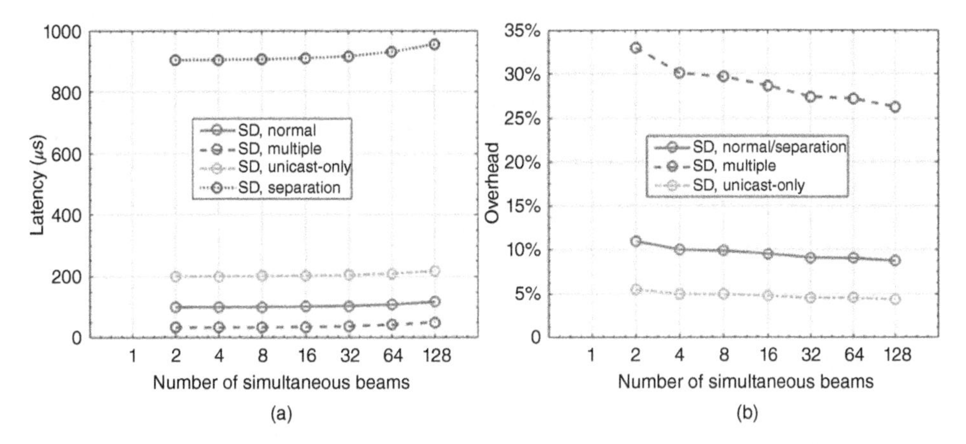

Figure 5.11 Performance of different frame structure designs for V2X applications compared in terms of (a) latency and (b) overhead. Source: © 2018 IEEE. Reprinted, with permission, from [16].

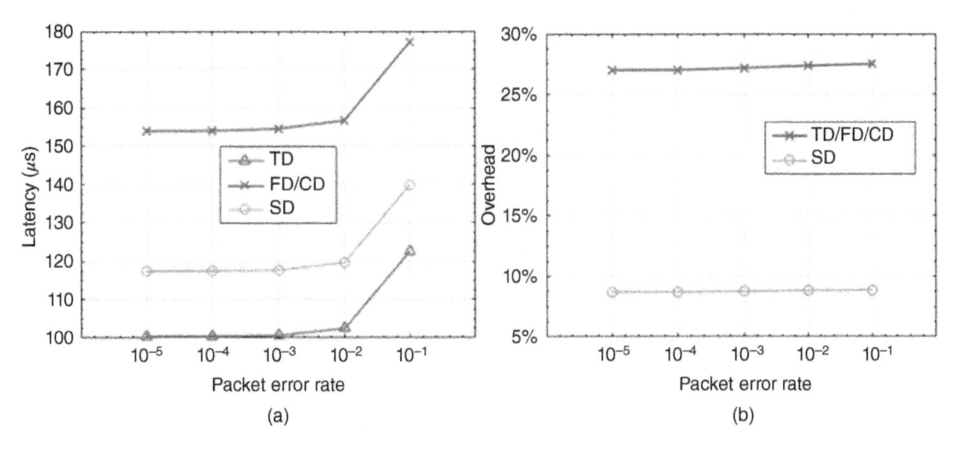

Figure 5.12 Performance of different block error rates for V2X applications compared in terms of (a) latency and (b) overhead. Source: © 2018 IEEE. Reprinted, with permission, from [16].

shorter duration t and thus lower overhead. In contrast, the lower the BLER, the higher the chance of correct decoding once a UE captures the best beam. Lower BLER can be achieved by lower MCS, which translates to more resources being used for broadcast signaling and thus to a longer duration t, finally yielding higher overhead. It has been demonstrated in Figure 5.12 that latency and overhead are relatively insensitive to extremely low packet error rates (10^{-5}). Therefore, the relatively low packet error rate (10^{-3}) would be sufficient for V2X beam-based broadcast, unless an extreme coding scheme is desired to achieve better performance.

Overall, beam-based broadcast schemes allow a flexible configuration to manage the trade-off between latency and overhead. By leveraging the analysis and simulation results presented, insights into the selection of beam-scan scheme can be obtained: e.g. when the information needs to be broadcasted to all vehicles with the lowest latency (e.g. a RSU on the highway), the thinnest beam should be used and the beam space should be sequentially scanned, i.e. without multiple beams in parallel. When the overhead of the broadcast transmission should be minimized, while latency is not a critical issue (e.g. on roads with strict speed limits), then multiple beams should be transmitted in parallel, and the number of simultaneous beams can be carefully selected to achieve a trade-off between latency and overhead.

5.2 PHY and MAC Layer Extensions

Direct V2V communication between vehicles in the neighborhood can be enabled through a sidelink, which may be either controlled by the BS or operated fully autonomously. This section focuses on extensions of the air interface in the PHY and MAC layer to address the challenges for V2V communication among a large number of distributed vehicles in highly mobile environments. First, a theoretical analysis of the trade-off between pilot overhead and spectral efficiency for a cellular MU-MIMO system under high mobility is presented, along with a corresponding robust receiver design. Then, the latest advances for V2V sidelink design are introduced, covering reference signal design, synchronization, scheduling, and power control.

5.2.1 Channel State Information Acquisition and MU-MIMO Receiver Design

In wireless systems in general, and orthogonal frequency division multiplexing (OFDM) and MU-MIMO systems in particular, spectral and energy-efficient operation relies heavily on the acquisition of accurate CSI at the transmitters and receivers. This section examines the inherent trade-off between allocating resources to CSI acquisition and data transmission and proposes a novel MU-MIMO receiver that is robust to errors in the CSI. Numerical results indicate that tuning the pilot and data power levels and employing a robust and CSI-error-aware receiver achieve higher spectral efficiency than that of legacy systems.

5.2.1.1 The Importance and Challenges of Channel State Information Acquisition in MU-MIMO Systems

CSI acquisition and channel-estimation methods in MU-MIMO systems have been studied extensively, and a large number of schemes, including blind, data-aided, and decision-directed non-blind techniques, have been proposed in the literature [18, 19].

In the uplink (UL) of cellular MU-MIMO systems, the BS typically acquires the CSI of the uplink by means of an uplink pilot or reference signals (RSs) that are orthogonal in the code domain. UEs in LTE and NR systems, for example, use cyclically shifted Zadoff-Chu sequences to form demodulation reference signals (DMRSs), allowing the BS to acquire channel state information at the receiver (CSIR), which is necessary for uplink data reception [20, 21]. In contrast, to acquire channel state information at the transmitter (CSIT), BSs either rely on downlink pilots and quantized information fed back by UEs [22] or assume channel reciprocity [23].

It has been pointed out by several related works that in systems employing pilot-aided channel estimation, the number of pilot symbols and the pilot-to-data power ratio play a crucial role in optimizing the inherent trade-off of sharing the available resources between pilot and data symbols [23–25].

The seminal work in [24] determined lower and upper bounds on the difference between mutual information when the receiver has an estimate of the CSI and when it has perfect CSI. Subsequently, the results in [25] showed how pilot-based channel estimation affects the capacity of the fading channel. The insights in these works indicate that training imposes a substantial information-theoretic penalty, which can be severe when the coherence interval T (measured in terms of the number of symbols available for pilot and data transmission) is only slightly larger than the number of transmit antennas M, or when the SNR is low. In these regimes, learning the entire channel is highly suboptimal. Conversely, if the SNR is high and T is much larger than M, pilot-based CSI acquisition schemes can come close to achieving capacity. Therefore, the power that should be spent on training and data transmission depends on the relation between T and M. Specifically in MIMO OFDM systems that employ minimum mean squared error (MMSE) channel estimation, [26] and [27] computed lower bounds on the system capacity. It was also shown that the optimal pilot and data power setting that maximize this lower bound or minimize the average symbol error rate significantly increase the capacity compared with a system that uses a suboptimal power setting. More recently, specifically for MU-MIMO systems, the trade-off between pilot and data symbols was analyzed in [28].

Since one of the main characteristics of vehicular communication systems is the fast time-variation and short coherence time of the fading channel, CSI acquisition in these systems is problematic. Recognizing that fast channel variations make accurate CSIT and CSIR acquisition difficult, previous works suggest that the effect of CSI feedback delay can be reduced by employing channel-prediction schemes when the channel exhibits temporal correlation [29]. Unfortunately, the temporal correlation may be weak in high-mobility scenarios, in which case channel prediction is not a viable option. Indeed, at high vehicular speeds, the effect of the Doppler shift and Doppler spread becomes severe since the channel coherence time is inversely proportional to the Doppler shift, which depends on the carrier frequency and the speed of the vehicular UE.

When a vehicular UE in a 6 GHz carrier frequency system travels at 500 km/h, for example, the maximum Doppler shift is 2.78 kHz, and the coherence time is approximately 152 μs, while at a speed of 30 km/h, the coherence time is approximately 2.5 ms. Thus, CSI acquisition becomes challenging at high speeds because of the short coherence time. In typical traffic situations, the velocity of the UE changes over time, leading to time-varying Doppler effects and thereby non-stationary fading coefficients. Such non-stationary fading effects and the fast-changing propagation environment make accurate CSI acquisition difficult in vehicular communication systems [29, 30].

5.2.1.2 Interplay Between CSIR Acquisition and MU-MIMO Receiver Design

Due to the importance of CSI acquisition for data transmission and reception, designing reference (pilot) signals and employing accurate channel-estimation techniques are of particular importance in cellular MIMO and vehicular communication systems. In addition, designing MU-MIMO receivers that are robust to CSI errors complements CSI acquisition techniques and helps to achieve spectral and energy-efficient wireless systems.

In the LTE and NR uplink, demodulation reference (pilot) signals (DMRS) are used to facilitate channel estimation for the coherent demodulation of the physical uplink shared and control channels. The DMRS occupy specific OFDM symbols within the uplink subframe according to the block type arrangement and support a large number of UEs by utilizing cyclic extensions of the well-known Zadoff-Chu sequences. Because of its importance in practical systems, this section elaborates on the interplay between reference signal design and channel estimation in cellular MIMO systems, and refer to related works for further details.

To illustrate the interplay between CSIR acquisition and MU-MIMO receiver design, let us consider the uplink transmission in a multi-antenna single-cell wireless system, in which UEs (vehicles) are scheduled on orthogonal frequency channels. In cellular MU-MIMO systems, UEs employ orthogonal pilot sequences in the code domain, so that no interference between pilots within any given cell is present in the system. (Note that in multi-cell systems, due to pilot reuse across neighbor cells, pilot contamination still causes pilot interference.) This is a common practice in massive MU-MIMO systems in which a single UE may have a single antenna.

The BS estimates the channel $\boldsymbol{h}$ (column vector of dimension N_{r}, where N_{r} is the number of receive antennas at the BS) by either least squares (LSs) or MMSE channel estimation to initialize an MMSE equalizer for uplink data reception. Since orthogonal pilot sequences are employed to facilitate CSIR acquisition at the serving BS, the channel-estimation process can be assumed independent for each UE within any given cell of the cellular system. Let us consider a time-frequency resource of T time slots in the channel coherence time, and F subcarriers in the coherence bandwidth, with a total number of symbols $\tau = F \times T$. Denote by τ_{p} the number of symbols allocated to pilots, and by τ_{d} the number of symbols allocated to data ($\tau_{\mathrm{p}} + \tau_{\mathrm{d}} = \tau$). Furthermore, assume a transmission power level P_{p} and P for each pilot and data symbol, respectively. For illustration, the comb-type pilot symbol-allocation method is used (Figure 5.13); note that in practice, both block-type and comb-type pilot symbol allocations, as well as a combination of these, are used to construct uplink and downlink reference signals [31].

In the comb-type pilot arrangement, a certain number of subcarriers are allocated to pilot symbols continuously in time. This approach is a suitable strategy for time-varying channels. Given F subcarriers in the coherence bandwidth, F_{p} subcarriers are allocated to the pilot, and $F_{\mathrm{d}} = F - F_{\mathrm{p}}$ subcarriers are allocated to the data symbols. Each UE transmits at a constant total power P_{tot}; however, the transmission power can be distributed unequally in each subcarrier. In particular, if transmitted power P_{p} is considered for each pilot symbol and P for each data symbol transmission, the following constraint is enforced:

$$F_{\mathrm{p}} P_{\mathrm{p}} + F_{\mathrm{d}} P = P_{\mathrm{tot}}.$$

The total number of symbols for pilots is $\tau_{\mathrm{p}} = TF_{\mathrm{p}}$ and for data is $\tau_{\mathrm{d}} = TF_{\mathrm{d}}$. However, with a comb-type pilot arrangement, the trade-off between pilot and data signals includes the trade-offs between the number of frequency channels and between the transmit power levels, which is an additional degree of freedom compared with the block-type arrangement.

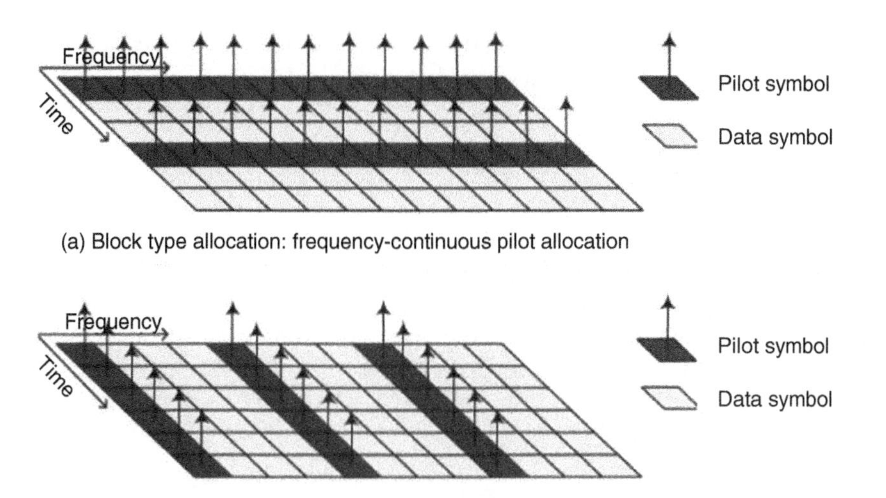

Figure 5.13 Illustration of block and comb-type pilot symbol allocations. Pilot symbols are mapped to pilot subcarriers carefully chosen among the subcarriers in the time-frequency domains.

Consider a vehicular UE that transmits an orthogonal pilot sequence $\boldsymbol{s} = [s_1, \ldots, s_{\tau_p}]^T$, where each symbol is scaled as $|s_i|^2 = 1$, for $i = 1, \ldots, \tau_p$; further, let $[.]^T$, $[.]^*$, and $[.]^H$ denote the transpose, conjugate, and conjugate transpose, respectively. Thus, the $N_r \times \tau_p$ matrix of the received pilot signal at the multiple-antenna BS from the UE is:

$$\boldsymbol{Y}^p = \alpha\sqrt{P_p}\boldsymbol{h}\boldsymbol{s}^T + \boldsymbol{N},$$

where $\boldsymbol{h}$ denotes the vector of random complex channel coefficients and α denotes the path loss. For illustration purposes, let us assume that $\boldsymbol{h}$ is a circular symmetric complex normal-distributed random vector with mean vector $\boldsymbol{0}$ and complex covariance matrix $\boldsymbol{C}$ (of size N_r), denoted as $\boldsymbol{h} \sim \mathcal{CN}(\boldsymbol{0}, \boldsymbol{C})$, α accounts for the propagation loss, and $\boldsymbol{N} \in \mathcal{C}^{N_r \times \tau_p}$ is the spatially and temporally additive white Gaussian noise (AWGN) with element-wise variance σ^2. In the numerical evaluations, the 3GPP-recommended channel models for vehicular scenarios are used. By way of example, when the BS uses conventional LS channel estimation, it relies on correlating the received signal with the known pilot sequence and estimates the channel vector as:

$$\hat{\boldsymbol{h}}_{LS} = \frac{1}{\alpha\sqrt{P_p}}\boldsymbol{Y}^p\boldsymbol{s}^*\left(\boldsymbol{s}^T\boldsymbol{s}^*\right)^{-1} = \boldsymbol{h} + \frac{1}{\alpha\tau_p\sqrt{P_p}}\boldsymbol{N}\boldsymbol{s}^* = \boldsymbol{h} + \tilde{\boldsymbol{h}}_{LS},$$

where $\tilde{\boldsymbol{h}}_{LS}$ denotes the channel-estimation error vector when employing LS channel estimation, and

$$\boldsymbol{N}\boldsymbol{s}^* = \left[\sum_{i=1}^{\tau_p} s_i^* n_{i,1}, \ldots, \sum_{i=1}^{\tau_p} s_i^* n_{i,N_r}\right]^T,$$

and

$$\boldsymbol{N}\boldsymbol{s}^* \sim \mathcal{CN}\left(\boldsymbol{0}, \tau_p\sigma_p^2\boldsymbol{I}_{N_r}\right).$$

By considering $\boldsymbol{h} \sim \mathcal{CN}$ **(0, C)**, it follows that the LS estimated channel of the tagged user l, which is $\hat{\boldsymbol{h}}_{\text{LS}}$, is a circular symmetric complex normal-distributed vector $\hat{\boldsymbol{h}}_{\text{LS}} \sim \mathcal{CN}\left(\boldsymbol{0}, \boldsymbol{R}_l\right)$ with covariance matrix

$$\boldsymbol{R}_l = E\left(\hat{\boldsymbol{h}}_l \hat{\boldsymbol{h}}_l^H\right) = \boldsymbol{C}_l + \frac{\sigma_{\text{p}}^2}{\alpha_l^2 P_{\text{p}} \tau_{\text{p}}} \boldsymbol{I}_{N_{\text{r}}}.$$

It is worth noticing that the covariance matrix of the estimated channel is close to the covariance matrix of the actual vector channel ($\boldsymbol{C}_l$) if either the pilot power (P_{p}) is high or the number of pilot symbols (τ_{p}) is large.

The objective of the MU-MIMO receiver is to estimate the transmitted data symbol based on the received MU-MIMO signal, which can be written as

$$\boldsymbol{y} = \alpha_l \boldsymbol{h}_l \sqrt{P_l} x_l + \sum_{k \neq l}^{K} \alpha_k \boldsymbol{h}_k \sqrt{P_k} x_k + \boldsymbol{n}_{\text{d}},$$

where x_k is the data symbol transmitted by the kth user and $\boldsymbol{n}_{\text{d}}$ is the thermal noise on the received data signal. The BS applies a linear receiver $\boldsymbol{G}_l \in C^{1 \times N_{\text{r}}}$ to estimate the transmitted data symbols of user l. In the case of perfect knowledge of the channel gains, the mean square error (MSE) of the received data symbols averaged over the transmitted symbols and the thermal noise can be written as the expectation of the squared error of the estimated symbol ($\boldsymbol{G}_l \boldsymbol{y}$) with respect to the transmitted symbol x_l as follows:

$$MSE\left(\boldsymbol{G}_l, \boldsymbol{H}\right) = E_{x, \boldsymbol{n}_d}\left(\left|\boldsymbol{G}_l \boldsymbol{y} - x_l\right|^2\right).$$

Unfortunately, since perfect knowledge of the channel gains is not available at the receiver, the linear receiver denoted by G has access only to the estimated channel vector ($\hat{\boldsymbol{h}}$), rather than to the actual channel vector ($\boldsymbol{h}$). The interplay between the channel-estimation process and the MU-MIMO receiver design can therefore be thought of as two aspects of data-symbol estimation, whose quality depends on both the quality of the channel estimation – affected by the trade-off between pilot and data resources – and the quality of the receiver algorithm using the estimated channel vector as input. In particular, as proposed in the next section, a MU-MIMO receiver that takes into account the statistics of the channel-estimation error outperforms naive receivers that treat the erroneous channel estimates as if they were the true channel vectors.

5.2.1.3 Novel Approaches to Near-Optimal MU-MIMO Linear Receiver Design and the Impact of CSIR Errors

In this section, a linear MMSE receiver is developed that employs noisy instantaneous channel estimates to minimize the MSE, and the dependence of receiver performance on the pilot-to-data power ratio is highlighted. A key observation in MU-MIMO systems is that the performance of the receiver of a particular user (user l) depends critically not only on the estimated channel of user l, but also on the estimated channel of other users. Previous works on MU-MIMO receiver design often either assumed the availability of perfect CSI or treated the estimated channel in the receiver as if it was the actual channel [32, 33]. An exception, in which a MU-MIMO receiver was proposed, is [34], in which the receiver actively takes into account the statistics of the estimated channel of the intended user (user l) and also the second-order statistics of the interfering users. However, the performance of that MU-MIMO receiver is limited by relying on second-order statistics of interfering users rather than utilizing the estimated channels of each (tagged and interfering) user. In [35], a MU-MIMO receiver has been proposed, which utilizes each channel estimate in the multi-user system. Specifically, it was shown that assuming Gaussian channels and Gaussian signaling, the receiver that minimizes the MSE of the uplink-received data symbols of the intended user in the presence of errors in the channel estimates of all users (where the estimation errors depend on the respective user's pilot power) can be expressed as

$$G_l^* = \alpha_l \sqrt{P_l} \hat{h}_l^H D_l^H J^{-1},$$

where

$$J = \sum_{k=1}^{K} \alpha_k^2 P_k \left(D_k \hat{h}_k \hat{h}_k^H D_k^H + Q_k \right) + \sigma_d^2 I_{N_r},$$

where σ_d^2 is the antenna element-wise thermal noise power on the received data signal, and the D_k and Q_k matrices express the quality of the channel estimation in terms of the statistics of the estimated channel (referred to as the *channel-estimation noise*) as follows:

$$\left(h_l \mid \hat{h}_l \right) \sim D_l \hat{h}_l + \underbrace{CN\left(0, Q_l\right)}_{\text{channel est.noise}},$$

where, in the case of least squares channel estimation,

$$D_l = C_l R_l^{-1} \text{ and } Q_l = C_l - C_l R_l^{-1} C_l,$$

where R_l is the covariance matrix of the estimated channel: that is, $\hat{h}_l \sim \mathcal{CN}\left(0, R_l\right)$.

5.2.1.4 Performance Modeling and Numerical Results in Multi-Antenna Cellular Vehicle Scenarios

The combined effect in terms of the MSE of the uplink received data symbols of the two intertwined technology components – a novel and robust MU-MIMO receiver design [35] and the proposed decentralized pilot-to-data power ratio setting [36–38] – is illustrated in Figures 5.14 and 5.15. In these numerical examples, seven UEs are multiplexed spatially, each of which uses a pilot sequence seven symbols long to achieve code-domain orthogonality. The proposed receiver minimizes the MSE of the uplink-received data symbols by using the instantaneous channel estimates of the desired and interfering users and compensates for channel-estimation noise by employing the proposed G_l^* MMSE receiver for each user from earlier (details in [35]). In contrast to the state-of-the-art receiver developed in [39] and [40], the proposed receiver thus makes use of all instantaneous channel estimates as an input to the MU-MIMO receiver and minimizes the average squared error of the uplink-received data symbols in the presence of channel-estimation errors. As the figure shows, this CSI-error-aware receiver ("With CSI error compensation") has lower MSE performance than the state-of-the-art receiver. To fully take advantage of the potential of this receiver, the pilot-to-data power ratio must be tuned at the transmitter (that is, UE) side such that the MSE is minimized. Minimizing the MSE for each user – here illustrated in Figure 5.14 in an urban macro setting where the vehicle speed is 40 km/h – translates to achieving high sum spectral efficiency (SSE), shown in Figure 5.15.

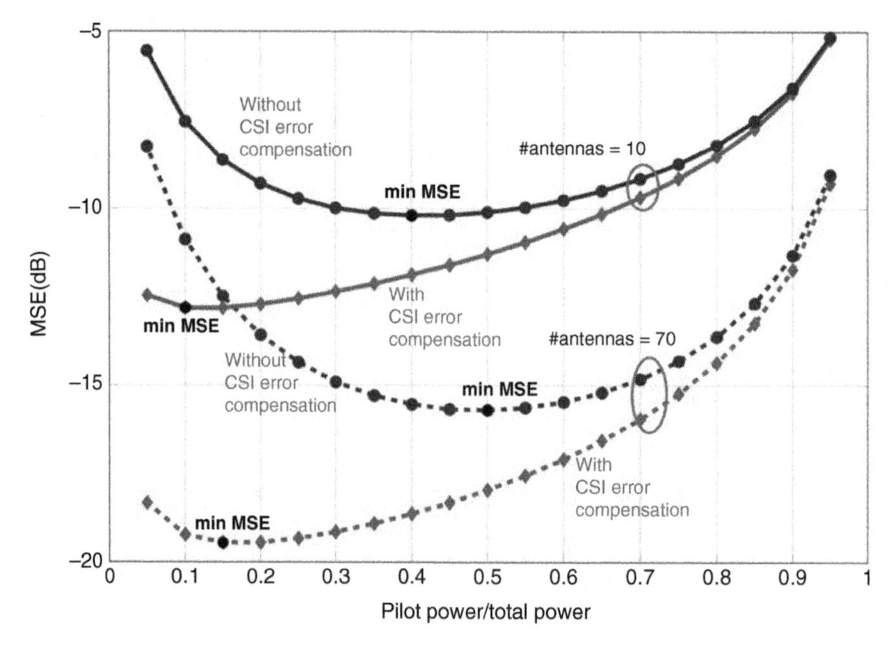

Figure 5.14 Performance of the proposed MU-MIMO receiver (referred to as the CSI-error-aware robust receiver) in terms of the mean squared error (MSE) of the uplink received data symbols at the BS. The curve labeled "with CSI error compensation" is obtained using a MU-MIMO receiver that uses the instantaneous channel estimates of each user and compensates for channel-estimation errors.

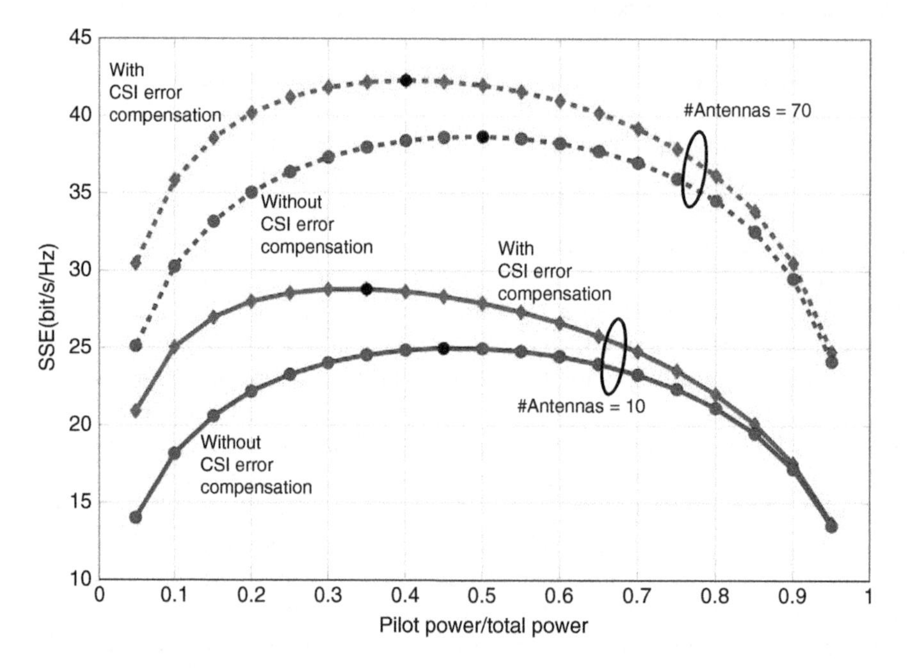

Figure 5.15 Performance of the proposed MU-MIMO receiver (referred to as the CSI-error-aware robust receiver) in terms of the achieved sum spectral efficiency (SSE) as a function of the pilot power. The curve labeled "with CSI error compensation" is obtained using the proposed MU-MIMO receiver that uses the instantaneous channel estimates of each user and compensates for channel-estimation noise.

5.2.2 Reference Signal Design

In wireless communications, it is often desirable that the two ends of a communication link – the transmitter and the receiver, respectively – know the propagation channel between them, since such knowledge enables efficient utilization of the available resources. From the transmitter's perspective, knowing the channel helps in adapting the transmission techniques, e.g. via precoding[1] the signal to be sent or via link adaptation, to best suit the channel condition. From the receiver perspective, as discussed in detail in Section 5.1.3, knowing the channel allows for effective recovery of the transmitted signal from the received signal.

To facilitate the acquisition of channel knowledge, commonly known as CSI, pilot signals or reference signals (RS) are widely used in wireless communications. These RS are typically known beforehand by both the transmitter and the receiver and are designed to have special properties for the receiver to exploit.

In 3GPP cellular technologies, several RS are used for estimating the channel (i.e. for obtaining CSIR), most notably cell-specific reference signals (CRSs) and DMRS. These RS are inserted in certain predefined resource elements (REs) of a transmission and are used by the receiver to estimate the channel at data-bearing resource elements of the same transmission, enabling the recovery of the desired data.

For obtaining CSIT, the CSI-RS or sounding reference signals (SRSs) are commonly used. In contrast to DMRS, CSI-RS are normally transmitted without precoding and allow the measurement of the non-precoded channel at a receiver. The transmitter sends out the RS, and the receiver performs some measurements based on the received RS. The receiver then reports to the transmitter either the measured channel or indicators derived from the measured channel. The transmitter can use the report (commonly termed *feedback*) to select the best precoder or adapt the MCS to maximize certain measures of performance: for example, the data rate.

5.2.2.1 Challenges to CSI Acquisition in V2V Sidelink Communication

The wireless channel experienced by a receiver influences CSI acquisition or channel estimation to a great extent. In V2V sidelink communication, in contrast to uplink and downlink cellular communications (hereafter called a cellular link), both the transmitter and receiver antennas can be mounted at similar heights and surrounded by both fixed and mobile scatterers. Such environments may result in many multipath components and the blocking of the line-of-sight (LOS) component. Therefore, the propagation in V2X scenarios can vary from LOS to non-line-of-sight (NLOS) due to static blockers such as buildings and vehicular-NLOS (NLOS-V) due to vehicular blockers (see Chapter 4). These blocking conditions appear in all highway, rural, and urban traffic scenarios. In NLOS scenarios, the delay spread can be substantial, depending on the distribution of the vehicles and other scatterers. Furthermore, the mobility of the objects in the propagation environment also has a significant impact on the wireless channel. Specifically, the transmitter, receiver, and scatterers can be moving at high speeds (as high as 250 km/h in some parts of the world),

---

1 *Precoding* refers to the process of applying different amplitude scaling and phase shifts to the signals to be transmitted from different antennas.

resulting in potentially very large Doppler spreads. The large Doppler and delay spreads result in a highly time- and frequency-varying channel. Therefore, the mechanisms for CSI acquisition for V2X must be robust in highly time- and frequency-varying channels.

One of the defining characteristics of 5G (and beyond 5G) wireless communication systems is the utilization of high-frequency spectrum, most notably the mm-wave bands. The use of this spectrum provides many benefits, including large bandwidth for high data rates, which is very beneficial for many V2X use cases. However, at higher carrier frequencies, carrier frequency offset and phase noise become more detrimental [41], and their effects need to be alleviated with the support of suitably designed RS. Note that the phase noise in an OFDM system introduces inter-carrier interference (ICI) and a phase error that is the same for all subcarriers, often referred to as common phase error (CPE). While the impact of ICI can be alleviated by large subcarrier spacing, the estimation of CPE often requires dedicated reference signals. In the 5G NR interface, phase-tracking reference signals (PT-RSs) were introduced for this purpose.

In addition to the challenges posed by the propagation environment, the strict requirements of current and future automotive use cases, such as cooperative maneuvering or autonomous driving, add extra demands to CSI acquisition. In particular, the very stringent requirements in latency (a few milliseconds) and reliability (99.99% and higher) of those use cases partly translate into the need for a robust RS structure to allow for correct decoding of a transmission while reducing the need for retransmissions.

5.2.2.2 Reference Signal Design for V2V Sidelink

Driven by the aforementioned challenges in CSI acquisition for V2V sidelink as well as by other aspects of V2X communications, an effective design of reference signals for 5G V2V sidelink should fulfill the following targets:

- Enable robust channel estimation in adverse propagation conditions as presented earlier, namely the severe Doppler and delay spreads as well as frequency and phase error, especially at high vehicle speeds and high carrier frequencies.
- Facilitate CSIT acquisition, if necessary, for enabling transmission link adaptation and precoding for multi-antenna transmission.
- Minimize the overhead of reference signals. Unlike the cellular link, the sidelink generally requires resources to account for the automatic gain control (AGC) settling and the guard period (GP), as described shortly. As a result, these overheads reduce the amount of available resources for data and reference signals.
- Allow for effective separation of reference signals at the receiver. Note that for the case of sidelink transmissions not being scheduled by a network node (i.e. a central coordinator is lacking), transmissions from different nodes can happen in the same slot, thereby interfering with each other. In such a scenario, it is important for a receiver to be able to separate and estimate the channels corresponding to the individual transmitters.

The first two design targets listed here necessitate at least three types of RS for V2V sidelink: an RS for CSIR acquisition, an RS for CSIT acquisition, and an RS for correcting the CPE. Adopting the NR convention, these signals are called DMRS, CSI-RS, and PT-RS, respectively. In the following, these RS are described in detail, with a focus on the design of the DMRS.

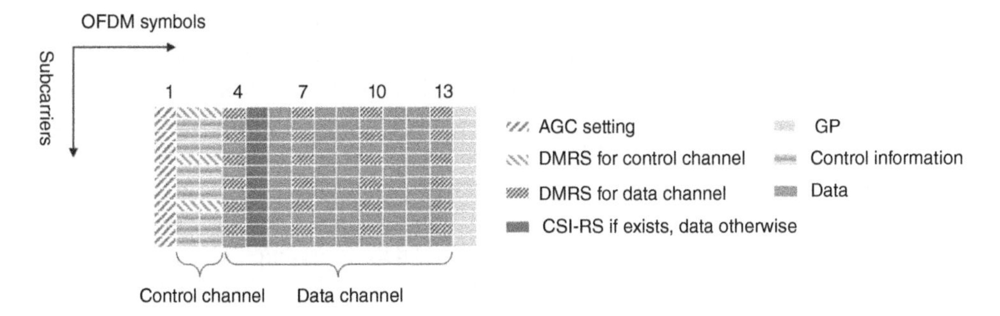

Figure 5.16 A realization of a DMRS and CSI-RS design for V2V sidelink.

The basic physical-layer structure considered for the V2V sidelink RS design is the NR sidelink slot, which is based on the OFDM waveform with a cyclic prefix [42] and reuses the NR slot structure [43] specified for cellular communications to a large extent. A sidelink slot consists of 14 OFDM symbols (OS) and has a duration of 1, 0.5, 0.25, and 0.125 ms for 15, 30, 60, and 120 kHz subcarrier spacing, respectively. A sidelink slot typically starts with an OS for AGC settling, followed by a few symbols for the control channel, which in turn is followed by symbols for the data channel. Finally, the slot ends with a symbol for the GP. The control channel carries information useful for medium access mechanisms and information required to decode the data channel. The AGC settling symbol is needed because the received power in the sidelink can change dramatically due to the movement of vehicles and/or different transmitters transmitting in different slots. The GP is needed for a device to switch its circuit from transmission mode to reception mode and also to ensure minimal interference from sidelink to cellular link due to imperfect synchronization between the links.

An example of the sidelink slot structure is given in Figure 5.16. In this figure, the control channel is assumed to occupy two OS, followed by 10 OS for the data channel. The AGC settling and the GP occupy one OS each. One resource block (RB), corresponding to 12 subcarriers, is shown in the frequency direction. The figure also shows a realization of the structure of the DMRS and CSI-RS according to our design, with details of the design described in the succeeding sections.

DMRS Design

The focus is on a DMRS design that delivers good channel-estimation performance with non-iterative channel-estimation algorithms.[2] The channel estimates at the data resource elements (REs), where an RE corresponds to a subcarrier in an OFDM symbol, are obtained using interpolation, extrapolation, and filtering of the estimated channel at the RS REs. Therefore, sufficient and well-located RS are necessary to sample the time- and frequency-varying channel.

For the control channel, since it carries a small amount of information, only a few OS (typically one to three OS) are needed. On the other hand, due to its essential role in medium access control and in scheduling the data channel, the control channel requires very high

2 Channel-estimation algorithms using decision-feedback and turbo-equalization provide improved estimation performance at the cost of increased latency in decoding the data.

channel-estimation performance. As a result, it is desirable to have DMRS REs in every OS of the control channel, allowing for a good sampling of the channel and high-quality channel estimation. Figure 5.16 shows a configuration where the DMRS is mapped in a comb-4 fashion with REs containing control information in every OS of the control channel.

Another design target for the DMRS of the control channel is the possibility to decode at least one of the control channels when multiple of them overlap in time and frequency. The overlapping is a consequence of the need to limit the allowed locations of the control channel in the time-frequency grid to reduce the blind decoding complexity of the receiver. Therefore, it is necessary to orthogonalize or randomize the DMRS of different control channels, should they overlap. This can be achieved by defining a small set of orthogonal cover codes (OCCs) to be applied to the same DMRS sequence and/or defining a small set of DMRS sequences that behave noise-like to each other. A transmitter can randomly pick one of the DMRS sequences in the set for transmitting the control channel, and the receiver blindly detects one of the sequences.

For the data channel, up to four DMRS symbols are needed to track the time variations of the channel at very high vehicular speeds, as illustrated in Figure 5.16. In some cases, e.g. at reasonably low vehicle speeds and/or high subcarrier spacings, fewer than four DMRS symbols can give satisfactory performance. Therefore, in our design, multiple DMRS configurations are allowed to maximize resource utilization. The transmitter can select one of the allowed DMRS configurations based on one or multiple conditions, e.g. the operating subcarrier spacing, the transmitter's absolute speed, the relative speed to the intended receiver, or the intended MCS used for the transmission. For example, in a 15 kHz subcarrier spacing system and at very high speed, where the coherence time of the channel is comparable to the OS duration, it is more suitable to use four DMRS symbols in a slot. In contrast, in a 120 kHz subcarrier spacing system and at low speed, two DMRS symbols in a slot suffice to track the channel's time variations. Additionally, early decoding of data using the first DMRS is also supported by our design by placing a DMRS near the start of the data resources.

In an OS carrying the DMRS, the DMRS subcarriers are comb-interleaved with the data subcarriers. A comb of one RS RE in every two subcarriers, as shown in Figure 5.16 and referred to as comb-2 mapping, in an OS allows tracking of a highly frequency-varying channel. Thanks to the comb structure, when performing multi-layer transmissions, DMRS for multiple layers can be transmitted on non-overlapping combs for orthogonalization. Furthermore, orthogonal DMRS for multiple layers can be transmitted by using OCC on a single comb. Using two non-overlapping combs with comb-2 mapping and two-dimensional OCC on each comb, a maximum of four orthogonal DMRS for the transmission of four layers can be supported.

As with the control channel, for the data channel, it is desirable to randomize the interference from DMRS of overlapping transmissions. This is to avoid the situation where two overlapping transmissions utilizing the same DMRS sequence prevent the receiver from performing proper channel estimation. Such a target can be achieved by using pseudo-random sequences for the DMRS of the data channel and ensuring that each transmission uses a DMRS sequence that is generated in a transmitter-specific and radio-resource-specific manner.

As mentioned earlier, the AGC settling symbol is a particular requirement of the sidelink. While a dedicated AGC training sequence can be transmitted during this symbol, a transmission of normal data is also possible. This latter solution increases the resource utilization efficiency because it allows extra data to be sent when the received power does not change too much across sidelink slots (i.e. the receiver can both adapt its AGC circuit and decode the data). This is particularly desirable when AGC operation requires a duration smaller than an OS. A smarter and more resource-efficient slot can be designed by imposing a special resource mapping structure in the AGC OS. In this structure, every other subcarrier in the first OS is left empty, resulting in two identical copies of the signal in the time domain. Using such a design, the first half of the OS is used for AGC, while the second half is used to demodulate the symbols carried by the non-empty subcarriers. Moreover, both data and DMRS can be mapped to the non-empty subcarriers, facilitating the channel estimation of the AGC OS. This is particularly important when there is a gap of several OS, due to the control channel, between the AGC and other data-bearing OS. In such situations, the receiver has to rely on DMRS in the AGC OS to estimate the channel of this symbol, rather than using the DMRS in other OS of the data channel. Due to space limitations, the resource-efficient designs of the AGC symbol are omitted from the illustration in Figure 5.16.

CSI-RS Design

In our design, the CSI-RS is only transmitted when the transmitter needs it, e.g. when the transmitter wants to perform link adaptation or precoding for multi-antenna transmission in unicast or multicast transmissions. This design avoids always-on CSI-RS transmission, thereby improving resource utilization efficiency.

When transmitted, a single CSI-RS can be placed in one of the OS, multiplexed with data or with DMRS, as depicted in Figure 5.16. The presence of CSI-RS and the request for CSI-report is indicated in the sidelink control information carried by the control channel.

In some cases, when the DMRS pattern is dense, the transmitter can also choose to mute one of the DMRS and use it for transmitting the CSI-RS.

PT-RS Design

As presented in Section 5.2.2.2, at high carrier frequencies, typically at mm-wave frequencies and above, it is necessary to track the changes in the carrier phase resulting from the phase noise of the oscillators. To enable such tracking, the PT-RS, which are more densely distributed in time (e.g. in every OS or every second OS) compared to the DMRS and can be placed only on selected subcarriers, is used. The sparse structure of the PT-RS in the frequency domain is due to the fact that the CPE is the same for all subcarriers. The PT-RS is used in combination with the DMRS in channel estimation at the receiver. Other details of the PT-RS are the same as that of the PT-RS for cellular communications; see [43] for details. Note that Figure 5.16 does not show the PT-RS.

5.2.2.3 Performance Evaluation

Here the influence of the RS design on performance is shown. The focus is on the DMRS used for the channel-estimation necessary for demodulating and decoding the data.

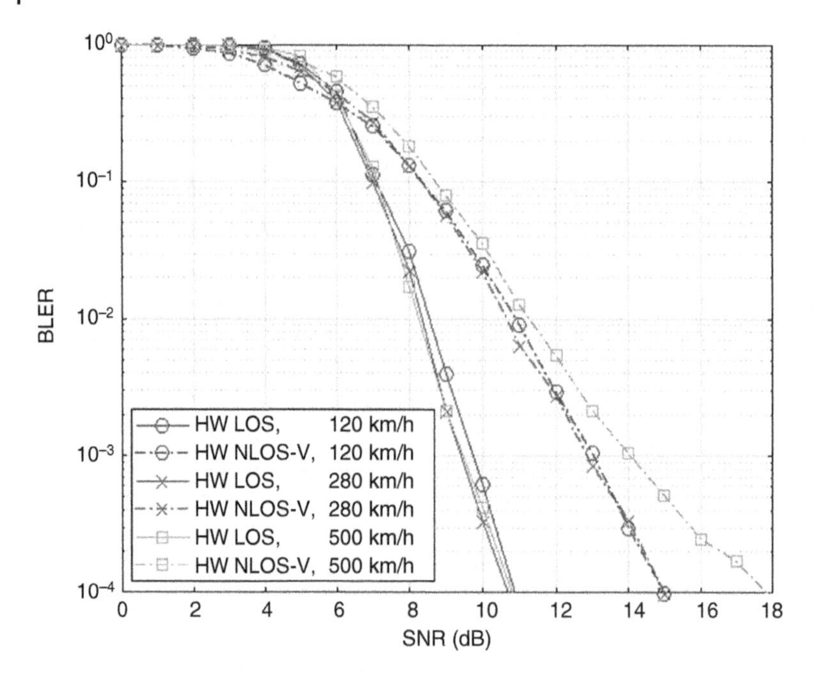

Figure 5.17 Link-level evaluation of a DMRS design at different relative vehicular speeds in highway line-of-sight (LOS) and non-line-of-sight due to vehicle blocking (NLOS-V) scenarios; 6 GHz carrier frequency, 30 kHz subcarrier spacing, 16-QAM, code rate 0.8. Single-layer transmission with one transmit and four receive (two dual-polarized) antennas.

Figure 5.17 shows BLER as a function of the transmit SNR (E_S/N_0) of the four-DMRS configuration shown in Figure 5.16. The evaluation uses the channel models developed in Chapter 4. Two channel models, Highway LOS and Highway NLOS-V, specified in [12], have been considered for evaluation. Relative vehicular velocities of 120, 280, and 500 km/h are evaluated. A carrier frequency of 6 GHz and a subcarrier spacing of 30 kHz are used. A bandwidth of 12 RBs is allocated. Single-layer transmission with one transmit and four receive (two dual-polarized) antennas is performed. The MCS uses a low-density parity check (LDPC) channel code with rate 0.8 and a quadrature amplitude modulation (QAM) format of 16-QAM. A practical linear MMSE based channel-estimation algorithm is used. As seen in the figure, the RS design with four DMRS delivers low BLER, and hence high reliability, for large relative vehicular speeds of 500 km/h and high MCS. Furthermore, to achieve a given BLER (for BLER < 0.3), a higher SNR is needed in the case of the NLOS-V channel compared to the LOS channel. This is due to the NLOS propagation and loss in received SNR caused by vehicle-blocking in the Highway NLOS-V model.

5.2.3 Synchronization

In a typical V2X communication scenario, as shown in Figure 5.18, mobile users who are in or out of cellular coverage, and possibly connected to different and non-synchronized base

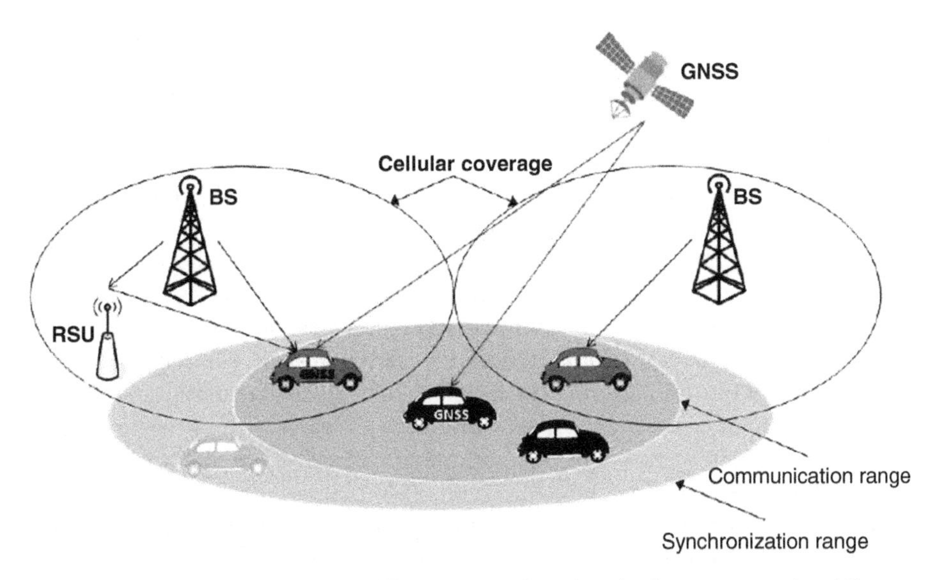

Figure 5.18 Cellular network with different types of synchronization sources and mobile users with and without cellular coverage and/or GNSS. Source: © 2017 IEEE. Reprinted, with permission, from [44].

stations, need to communicate and share the available sidelink frequency band. Moreover, a subset of users may be equipped and have access to a global navigation satellite system (GNSS), whereas others may not. To enable multiple sidelink connections that coexist in the same frequency band and potentially also share a band with the uplink, users communicating via sidelink need to be mutually synchronized by means of having a common time reference.

For in-coverage UEs and UEs with GNSS, it is feasible to obtain a common timing, if the time offset(s) between different synchronization sources (base station, GNSS, etc.) or their offsets to a predefined time reference such as coordinated universal time (UTC) are known to the user. In 5G NR (Release 15), the synchronization source type and/or time offset information are provided by the PBCH in the downlink. It is noted that the priority for selection of the synchronization source for the sidelink is configured by the network and is common for all UEs: so-called "GNSS-based" and "BS-based" synchronization priority tables are defined.

The situation is more challenging for users without cellular coverage or GNSS, which in the absence of high-priority synchronization sources need to rely on the sidelink synchronization signals (SLSSs) sent by other users to obtain their own time reference for sidelink transmission. Multiple SLSS are sent by users originally synchronized to different sources, either directly or indirectly, over one or more sidelink hops. According to Release 16, the priority of a user serving as a source is defined by the type of its own synchronization source and number of hops to the receiving user.

In the current version of the 3GPP Release 16 NR V2X standard, priority-related information is included in the physical sidelink broadcast channel (PSBCH). As a consequence,

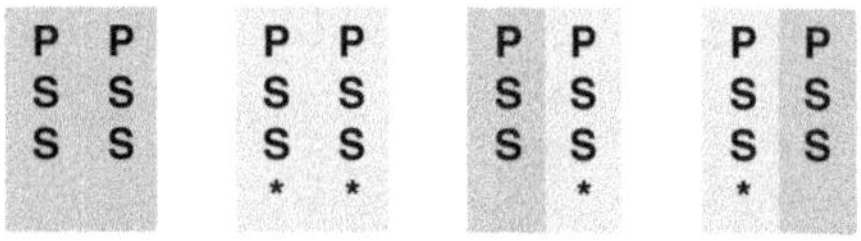

Figure 5.19 Example for a two-segment sequence. With $n = 2$, up to four different cases can be distinguished. Source: © 2017 IEEE. Reprinted, with permission, from [44].

a receiver needs to detect multiple SLSS, decode the corresponding PSBCH to obtain the source priority information, and finally select for synchronization the SLSS sent by the highest-priority source. The benefit of conveying priority information by the SLSS is that it allows the receiver to select earlier the SLSS sent by the user with the highest priority. This approach avoids the receiver performing unnecessary SLSS detection and PSBCH decoding for all SLSS to find the highest-priority source. If a user is already synchronized through the sidelink, using prioritized SLSS reduces the search for sequences to the SLSS exhibiting higher priority than the one it is already synchronized to. Clearly, such a scheme reduces computational complexity, power, and buffering and simplifies the overall synchronization procedure and the time it may require.

In Release 16 NR sidelink, the SLSS includes two consecutive symbols for the sidelink primary synchronization signal (S-PSS) and two consecutive symbols for the sidelink secondary synchronization signal (S-SSS). Currently, two different sequences are defined, but the same sequence is repeated in the two symbols for both S-PSS and S-SSS. Without any additional resources, by supporting combinations of different sequences in the two symbols of S-PSS, four different combinations are defined in total, which can be mapped to different synchronization source priorities. Two different sequences are already used for S-PSS, which means a standard receiver would not need to perform any additional search for new sequences, avoiding complexity overhead. The same may also be applied for S-SSS, yielding an overall increase in the number of sequence combinations from 4 to 16.

In what follows, an example is presented using Zadoff-Chu sequences, as in the LTE V2X standard. As a design approach for the two different sequences, the root indices are selected in such a way that the sequences are complex conjugates while remaining mutually orthogonal. The cross-correlation of the received signal with local copies of S-PSS at the receiver is implemented by calculating separately the terms involving real and imaginary parts of S-PSS. In terms of implementation, complex-conjugate S-PSS sequences have the benefit that the terms involving the real part need to be calculated only once, thus saving computational overhead. As a concrete design example, consider a sequence L, with $L = 63$ as in LTE, and root index u_1 (u_1 being relatively prime to L) for the first sequence and $u_2 = L - u_1$ for the second sequence. It is noted that PSS1 and PSS2 remain mutually orthogonal. Figure 5.19 shows an example with two different sequences and two symbols for S-PSS, which distinguishes four different cases.

Figure 5.20 shows an exemplary receiver implementation with a two-step sequence detection. In the first step, sequences are detected, and time synchronization is performed by estimating the beginning of frame (BOF) and beginning of symbol (BOS). In addition to the cross-correlation-based operations, which have already been described, the blue blocks are also included to improve performance. There, the phase shifts due to time/frequency offsets

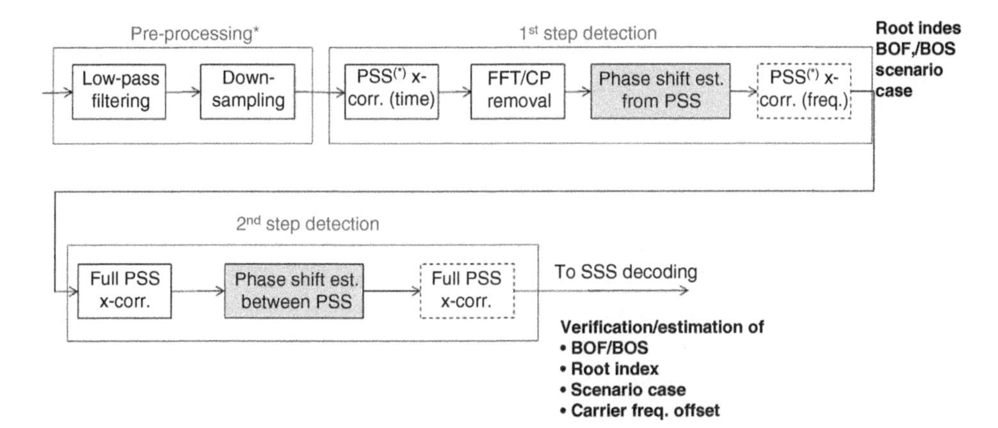

Figure 5.20 Implementation of a two-step detection procedure at the receiver, providing prioritized SLSS detection, time, and frequency synchronization. Source: © 2017 IEEE. Reprinted, with permission, from [44].

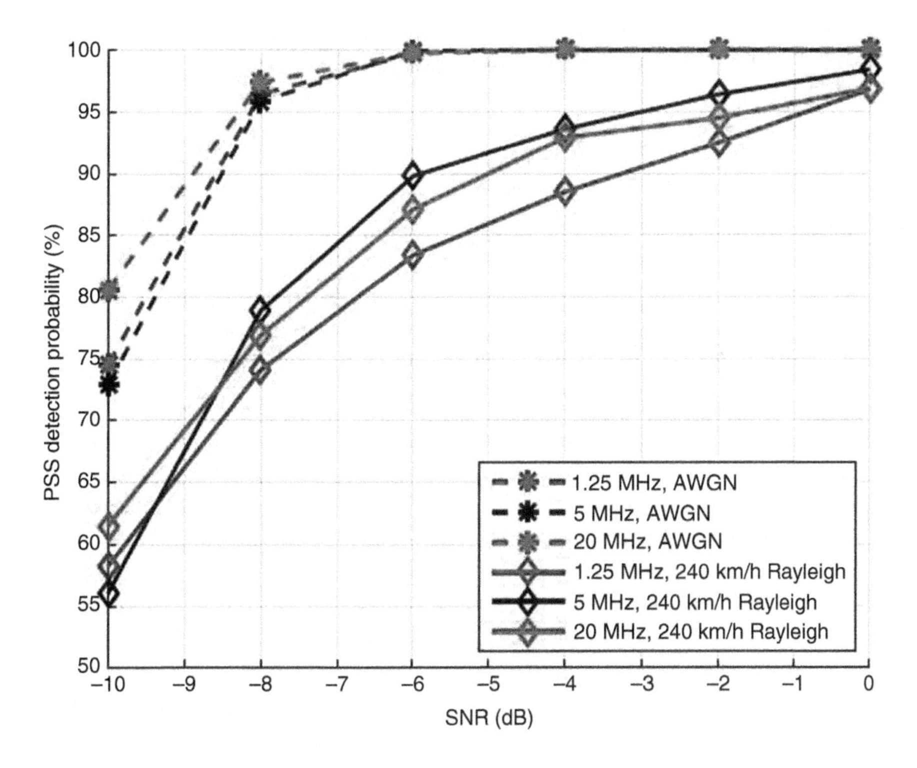

Figure 5.21 Probability of correct sequence detection vs. SNR. Source: © 2017 IEEE. Reprinted, with permission, from [44].

are estimated and compensated, and S-PSS detection is refreshed (blocks with dashed lines). In the second step, results from the first step are verified and refined, and frequency synchronization is performed, including detection and compensation of the carrier frequency offset.

Figure 5.21 shows the sequence detection probability versus the SNR for the first step of the procedure shown in Figure 5.20. Here, cross-correlation with both S-PSS sequences is applied to the received signal in the time domain, and a correlation peak is considered valid if it exceeds a threshold depending on the noise level. If two significant peaks are detected, and the actual transmitted signal is correctly identified, the S-PSS detection is considered correct. The reference curves are shown for the AWGN channel, while 1000 independent, identical distributed (i.i.d.) realizations of a multipath channel with continuous Rayleigh fading over the length of the received signal block for a speed of 240 km/h are considered as a realistic and challenging scenario. For system bandwidths above 1.25 MHz, time-domain low-pass filtering and down-sampling are applied, resulting in similar performance for all system bandwidths. The differences in performance are only due to the properties of the filters. In general, results show that for SNR values around 0 dB, the S-PSS detection probability approaches 100%, even under fast-fading channel conditions. The confirmation of correct sequence detection of the first step by the second step has been found to be close to 100%. Therefore, the role of the second step is mainly seen in the refinement of the residual time and carrier frequency estimates and improvement of overall synchronization accuracy. For more details on this work, the interested reader is referred to [44].

5.2.4 Scheduling and Power Control

Scheduling and power control, collectively referred to as radio resource management (RRM) in this section, are crucial to control packet errors and latency. A significant cause for packet errors is interference (caused by the reuse of radio resources). Moreover, scheduling naturally affects latency. When performance guarantees are to be provided, it is a significant advantage to use centralized scheduling and power control. In our setting, the central controller could be a base station, cloud or edge computing device, or specially elected vehicle in the network. We will limit the discussion here to centralized RRM solutions, although decentralized RRM algorithms are of interest and should not be excluded as candidates when selecting the RRM approach in a practical system.

Many traffic safety and traffic efficiency applications are enabled by the local broadcast of periodic status messages (e.g. CAM messages) or event-triggered warning messages (e.g. DENM messages). There are usually requirements for timeliness and reliability. Timeliness can be measured in terms of latency or age-of-information, and reliability is measured as the probability that the timeliness metric (or a function of it) is less than a pre-described threshold. For example, suppose the end-to-end latency τ is desired to be less or equal to $\tau_{\max}$ with reliability $1-\varepsilon_{\max}$; then these requirements are fulfilled if

$$\Pr\left\{\tau \leq \tau_{\max}\right\} \geq 1-\varepsilon_{\max}$$

or, equivalently, if

$$\Pr\left\{\tau > \tau_{\max}\right\} < \varepsilon_{\max}.$$

In the latter formulation, the left-hand side can be interpreted as the outage probability. We use the convention that $\tau = \infty$ whenever the transmitted packet is not received or is received with errors in the payload. This convention allows characterizing both late packets and erroneous packets with the same condition on the latency, namely that $\tau > \tau_{max}$.

To formalize the multicast nature of the communication and how to measure performance, a number of definitions are made. Consider a network consisting of N vehicular UEs. The set of all UEs is denoted $\mathcal{N} \triangleq \{0, 1, \ldots, N-1\}$, and the set of intended receivers to UE $i \in \mathcal{N}$ is denoted $\mathcal{R}_i \subset \mathcal{N}$: that is, UE i wants to transmit a packet to UE j if and only if $j \in \mathcal{R}_i$. The model allows for unicast $|\mathcal{R}_i| = 1$, multicast $1 < |\mathcal{R}_i| < N$, and broadcast $|\mathcal{R}_i| = N$. Finally, the set of all intended links (i, j) in the network is denoted $\mathcal{L} = \{(i,j) : i \in \mathcal{N}, j \in \mathcal{R}_i\}$. It is assumed that all UEs generate a single packet to be transmitted at the beginning of the scheduling interval, and it is assumed that the scheduling interval duration is equal to the latency constraint τ_{max}. The goal with the scheduling and power control is to allocate radio resources to the N UEs during the scheduling interval to satisfy the latency and reliability constraint for as many as possible of the links in $\mathcal{L}$ (which consists of all intended unicast, multicast, and broadcast links). To this end, define the performance metric $Z_{i,j}$ such that $Z_{i,j} = 1$ if the packet from UE i is received at UE j with the required latency and reliability, and $Z_{i,j} = 0$ otherwise. When $Z_{i,j} = 1$, we say that UE i is *connected* to UE j. We can measure the number of intended receivers that UE i is connected to as

$$Z_i = \sum_{j \in \mathcal{R}_i} Z_{i,j},$$

and the average number of intended receivers a UE in the network is connected to as

$$Z = \frac{1}{N} \sum_{i \in \mathcal{N}} \sum_{j \in \mathcal{R}_i} Z_{i,j} = \frac{1}{N} \sum_{i \in \mathcal{N}} Z_i.$$

Note that we only consider the intended receivers in these performance metrics. Hence, even if packets can be decoded by non-intended receivers, this is not considered when measuring the performance of the scheduling and power-control algorithm.

It should be noted that centralized scheduling and power control comes at the price of control signaling, e.g. measurement and reporting of CSI to the central controller and dissemination of the schedule and transmit power settings from the central controller to the network of transmitters and receivers. Due to the potentially highly time-varying channels in a V2V setting, measuring and reporting instantaneous CSI, i.e. the channel gain, including small-scale fading (SSF), can be infeasible due to the volume of control signaling, but also since the RRM will be based on outdated CSI. In highly dynamic environments, it is therefore more attractive to study RRM schemes that are based on slowly varying CSI, i.e. the combined effect of path loss and large-scale fading (LSF). In many cases, the variation of path loss and LSF over the scheduling interval can be neglected. Indeed, this assumption is made for the remainder of this section. Moreover, it is assumed that the slow CSI does not change over the system bandwidth. With these two assumptions, it is sufficient to measure and report the average power gain $H_{i,j}$ between UE i and UE j once per scheduling interval. If there are N UEs in the network, this corresponds to $N(N-1)/2$ measurements per scheduling interval (since it is assumed that the channel gains are reciprocal). For the rest of this section, it is assumed that the slow CSI is available to the RRM algorithm and that the schedule and power settings are disseminated to all N vehicles without errors.

The RRM approach adopted here starts by dividing the available bandwidth into F frequency slots and the scheduling interval into T timeslots. A frequency-time slot (f, t), where $f \in \{0, 1, \ldots, F-1\} \triangleq \mathcal{F}$ and $t \in \{0, 1, \ldots, T-1\} \triangleq \mathcal{T}$, is called a resource block (RB). The output of an RRM algorithm is the transmit powers $P_{i,f,t}$ and scheduling variables $X_{i,f,t}$ for all UEs, $i \in \mathcal{N}$, and all RBs $(f, t), f \in \mathcal{F}, t \in \mathcal{T}$. The scheduling variable $X_{i,f,t} = 1$ if UE i is scheduled in RB (f, t) and $X_{i,f,t} = 0$ otherwise. The transmit power is subject to a max power constraint in a timeslot (due to device limitations and regulations)

$$\sum_{f \in \mathcal{F}} P_{i,f,t} \leq P_{\max},$$

and the transmit power is also constrained by the scheduling variable as

$$0 \leq P_{i,f,t} \leq P_{\max} X_{i,f,t}, i \in \mathcal{N}, f \in \mathcal{F}, t \in \mathcal{T}.$$

The bandwidth and time duration of an RB is dimensioned such that an RB can carry a complete packet with a packet error probability that is at most $\varepsilon_{\max}$ for an SNR in the desired range. Hence, the RB dimension is, in general, different from the LTE RB dimension. It is assumed that the packet error probability is less or equal to $\varepsilon_{\max}$ if the average received signal-to-interference-plus-noise ratio (SINR) is equal to or above a certain threshold γ_{T}. To be precise, a transmission from UE i to UE j in RB (f, t) will be received with packet error probability less or equal to $\varepsilon_{\max}$ if the SINR

$$\frac{S_{i,j,f,t}}{I_{i,j,f,t} + \sigma^2} \geq \gamma_{\mathrm{T}} \Longleftrightarrow S_{i,j,f,t} \geq \gamma_{\mathrm{T}} \left(I_{j,f,t} + \sigma^2 \right),$$

where $S_{i,j,f,t}$ is the desired signal power, $I_{i,j,f,t}$ is the interference power, and σ^2 is the noise power in an RB. Note that the desired and interference powers are the powers averaged over the SSF distribution. It should be clear that γ_{T} depends on the packet size, coding and modulation, and statistic of the SSF.

To compute our main performance metrics $Z_{i,j}$, we need to evaluate the SINR constraint for all transmissions in the network. Note that, depending on which transmitter-receiver pair is considered, a particular transmission will be either a desired signal or an interfering signal. Recall that UE i transmits its packet in RB (f, t) with transmit power $P_{i,f,t}$. From the perspective of an intended receiver UE $j \in \mathcal{R}_i$, the desired signal power when decoding the packet from UE i is

$$S_{i,j,f,t} = P_{i,f,t} H_{i,j}.$$

Now consider a transmission from UE k in RB (f', t'), which is recalled to have transmit power $P_{k,f',t'}$. This transmission can potentially cause interference. Consider two types of interference: co-channel interference (CCI) and adjacent channel interference (ACI). If $t' \neq t$, then there will be no interference, since it is assumed that RBs are orthogonal in time. However, if $t' = t$, there will CCI if $f' = f$ and ACI if $f' \neq f$. In general, the interference power due to the transmission from UE k at receiving UE j is $P_{k,f',t'} H_{k,j} \lambda_{|f-f'|}$, where λ_r is the adjacent channel interference ratio (ACIR). The ACIR λ_r describes the relative amount of power that spills over from the scheduled RB (f', t) into the RBs $(f' \pm r, t)$. The power leakage is mainly due to power amplifier nonlinearities [43]. We use the convention that $\lambda_0 = 1$ to quantify CCI and ACI with the same expression. Hence, the total (desired plus interfering) received power at UE j in RB (f, t) is

$$R_{j,f,t} = \sum_{f' \in \mathcal{F}} \sum_{k \in \mathcal{N}} P_{k,f',t} H_{k,j} \lambda_{|f-f'|}.$$

Hence, one can write $I_{i,j,f,t} = R_{j,f,t} - S_{i,j,f,t}$, and the SINR constraint can be reformulated as

$$S_{i,j,f,t} \geq \frac{\gamma_{\mathrm{T}}}{1+\gamma_{\mathrm{T}}} \left(\sigma^2 + R_{j,f,t} \right),$$

and $Z_{i,j} = 1$ if the SINR constraint is fulfilled for at least one RB in the scheduling interval. Formally,

$$Z_{i,j} = \bigvee_{f \in \mathcal{F}} \bigvee_{t \in \mathcal{T}} \left\{ P_{i,j,f,t} H_{i,j} \geq \frac{\gamma_{\mathrm{T}}}{1+\gamma_{\mathrm{T}}} \left(\sigma^2 + \sum_{f' \in \mathcal{F}} \sum_{k \in \mathcal{N}} P_{k,f',t} H_{k,j} \lambda_{|f-f'|} \right) \right\},$$

where $\bigvee_{m=0}^{M-1} c_m$ is the logical OR of the conditions $c_0, c_1, \ldots, c_{M-1}$. From this equation, the essence of the RRM problem can be deduced:

Given the slow CSI, noise variance, SINR threshold, and ACIR function: $\{H_{i,j} : i,j \in \mathcal{N}\}, \sigma^2, \gamma_{\mathrm{T}}, \{\lambda_r : r = 0, 1, \ldots, F\}$

Compute the transmit powers $\{P_{i,f,t} : i \in \mathcal{N}, f \in \mathcal{F}, t \in \mathcal{T}\}$, subject to the max power constraint, such that the latency and reliability requirements are satisfied, $Z_{i,j} = 1$, for all desired links $(i,j) \in \mathcal{L}$

A few remarks are in order:

- The schedule variable is implicit in this formulation. Indeed, $X_{i,f,t} = 0$ if and only if $P_{i,f,t} = 0$, or, equivalently, $X_{i,f,t} = 1$ if and only if $P_{i,f,t} > 0$.
- It might not be feasible (due to adverse channel conditions or limited radio resources) to satisfy the latency and reliability requirement for all desired links $(i,j) \in \mathcal{L}$. In this case, it is reasonable to maximize the number of satisfied links in the network, Z, subject to some fairness criterion, e.g. that Z_i is greater or equal to some fixed number.
- The formulation allows for both half-duplex and full-duplex operation. Indeed, the self-interference is controlled by the self-power gain $H_{i,i}$ and the ACIR function λ_r. Ideal full-duplex operation is allowed by setting $H_{i,i} = 0$. For the reasonable assumption that $\lambda_r > 0$ for all r, half-duplex is enforced by setting $H_{i,i}$ to a sufficiently large number. For intermediate values of $H_{i,i}$, it might be possible to decode some messages while transmitting, which constitutes a hybrid between half-duplex and full-duplex operation.
- In the more practical setting of half-duplex operation, it may be beneficial to transmit the same packet in more than one time slot. Indeed, suppose that UE i is transmitting its packet in RB (f, t); then no UE that is transmitting in the same time slot, i.e. in some RB (f', t), will be able to decode this packet.
- The optimization problem is combinatorial in nature, which implies that computational complexity scales poorly with the number of UEs N. In practice, one can use relaxations and approximations to yield more computationally efficient algorithms or partition the network into smaller subnetworks that can be individually scheduled and power-controlled. In the latter solution, a method for controlling the interference between subnetworks is required.

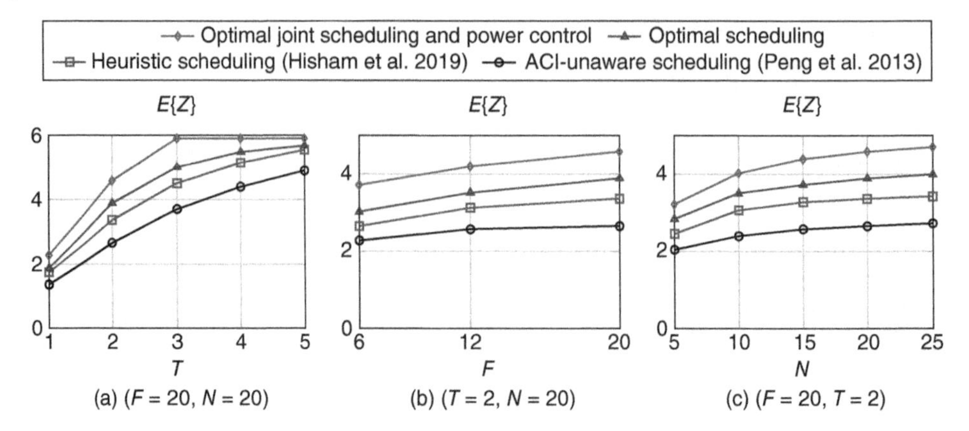

Figure 5.22 Average number of connected UEs (Z) versus the number of timeslots (T), number of frequency slots (F), and number of UEs in the network (N). The expectation of Z is over many network realizations.

- The ACI is significant in a so-called near-far situation, i.e. when $P_{k,f',t}H_{k,j} \gg P_{i,f,t}H_{i,j}$. This can indeed occur, e.g. when there are blocking vehicles between UE i and j and no blocking vehicles between V-EU k and j, which results in $H_{k,j} \gg H_{i,j}$. Note that transmit powers tend to be determined by the worst-case channel gain among the intended receivers. Hence, in a homogenous network, the transmit powers tend to be similar for all transmissions, and power control will therefore not remove near-far situations. It can therefore be concluded that ACI should not be neglected in these cases.

To quantify and exemplify the importance of scheduling and power control, a vehicular network in a highway scenario has been simulated. The details about the simulation setup and parameters are found in [45]. In Figure 5.22, we plot Z, i.e. the average number of connected vehicles per UE, where the expectation $E\{Z\}$ is over many network realizations (UE geographical positions and LSF realizations). In the simulation, the performance of four algorithms is compared: (i) optimal joint scheduling and power, i.e. the solution to the optimization problem described previously; (ii) optimal scheduling, i.e. the solution when the transmit powers are fixed to $P_{\max}$; (iii) the heuristic algorithm from [45]; and (iv) the algorithm proposed in [46], which is ACI-unaware and has been modified to fit our simulation setup. It should be noted that all algorithms except (iv) take ACI into account. The results in Figure 5.22 indicate that taking ACI into account noticeably improves performance. Both scheduling and power control are important, but most of the gain from algorithm (iv) is due to scheduling.

5.3 Technology Features Enabled by Vehicular Sidelink

Short-range communication established by the V2V sidelink enables the facilitation of new communication concepts for V2X: neighboring UEs in a close vicinity can cooperate in uplink and downlink transmissions with a BS to make use of additional link diversity, which can be beneficially exploited to improve overall reliability at a reasonable system cost. Moreover, the appealing concepts of full duplex, allowing UEs to transmit and receive at the same time, become feasible for short-range communications and thus offer new possibilities for V2X. This section will elaborate on these novel concepts in the V2X context and highlight their promising gains and benefits.

5.3.1 UE Cooperation for Enhancing Reliability

Improving the reliability of data transmission requires increasing the amount of system bandwidth, resulting in fundamental trade-offs between reliability, latency, and throughput [47]. Moreover, to maintain a reasonable bandwidth efficiency, it is of crucial importance to utilize all the diversity that can be made available to the system [48]. In V2X communications, many spatial diversity paths usually exist, due to other communication-capable vehicles in the direct neighborhood, and this source of diversity can be made available through sidelink communication. The exploitation of spatial diversity for URLLC has been elaborated on in [49], where the impact on the required SNR to achieve a desired packet error rate has been analyzed. In another recent work [50], the authors investigated sidelink-assisted retransmissions by a single relay node and the impact on system reliability and latency, characterized in terms of the BLER and the number of required retransmissions, respectively. Inspired by this work, parts of its analytical framework are reused here to investigate a resource-efficient, cooperative retransmission scheme for the V2X sidelink involving several neighboring vehicles, aiming at high reliability. In particular, we analyze the impact of the number of cooperating UEs on the communication reliability attainable in a V2X network, and the further improvement yield by additional retransmission based on automatic repeat request (ARQ) and HARQ schemes.

5.3.1.1 Communication Scenario

Consider a V2X communication scenario where a BS maintains a V2N communication link with a target vehicular UE (car A) to provide URLLC services (see Figure 5.23). Car A is surrounded by other neighbor UEs, to which it communicates through V2V communication via the sidelink. The V2N link to the target UE may be obstructed by a bypassing trailer, interrupting the ongoing URLLC communication via V2N link. If the neighbor UEs have overheard the URLLC packets dedicated to car A, they can support the URLLC transmission by retransmitting those packets via sidelink to car A. To enable neighbor UEs to overhear URLLC data communication, the BS needs to transmit URLLC packets dedicated to car A in a multicast fashion to the whole group of neighbor UEs (including car A), which needs to be formed in advance. Once car A does not receive a URLLC packet from the BS, it can request a retransmission from its neighbor UEs via sidelink. The neighbor UEs that have successfully received the URLLC packet from the prior multicast transmission can then retransmit it to the target UE.

Since the links between neighbor UEs and car A can be assumed to be independently fading, a large degree of spatial diversity can be made available. This can be exploited by using transmit diversity schemes applied in a distributed fashion, such as distributed space-time block codes (STBC) [51, 52] or cyclic delay diversity (CDD) [53]. Those schemes enable a constructive addition of signals that propagated via individual communication paths at the receiver without sacrificing resources for the redundant data transmission. Additional resources may need to be spent only on the pilot signals required to estimate the effective

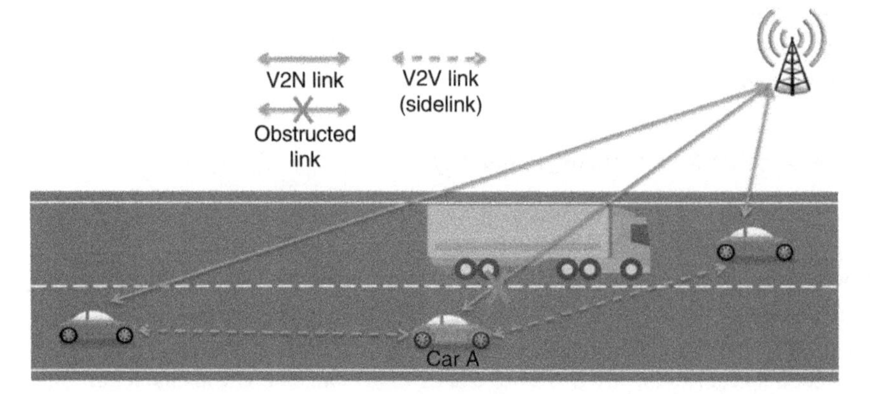

Figure 5.23 V2X communication scenario: link to car A obstructed by a bypassing trailer. Source: © 2019 IEEE. Reprinted, with permission, from [54].

channel being composed of the constituting V2V links. Using distributed STBC or CDD in this scenario is thus an efficient solution for resource-sparing retransmission of URLLC traffic attaining high reliability.

5.3.1.2 Reliability Analysis – Channels with Equal Power

This section analyzes the reliability gains attainable by a number of neighbor UEs cooperatively retransmitting missed packets via sidelink. Typical V2V channels are likely to exhibit a strong LOS component and may be subject to rich scattering. However, to ease analytical treatment, the LOS component is ignored in this analysis, and all V2V channels are assumed to be Rayleigh fading with an equal average power of unity. Sidelink transmissions of different UEs are assumed to be fully synchronous on the symbol level.

The power yield by the constructive addition of independently Rayleigh fading channel coefficients can be characterized by the Gamma distribution $\Gamma(k, \theta)$, with k denoted as the shape and θ as the scale parameter. While the shape parameter k reflects the number of complex Rayleigh fading variables stemming from statistically independent links, θ reflects the inverse of the scaling factor used to scale the transmit power per each link. If k users are assumed to be transmitting the same signal by using a spatial diversity transmission scheme with equal power-sharing and a constant sum power of unity, the distribution of the accumulated channel gain can be characterized by $\Gamma(k, k)$.

Let the probability distribution function (pdf) of the Gamma distribution be given as $f(x; k, \theta)$. Then the probability of the accumulated channel gain lying below a target value ρ is given by the cumulative distribution function (CDF)

$$\int_0^{\rho} f(x; k, \theta)\, dx = \int_0^{\theta\rho} f(x; k, 1)\, dx =: q_k (x \leq \theta\rho).$$

Since the parameter θ scales the overall distribution of x, it may also be used to scale the target value ρ while using the unscaled Gamma function $f(x; k, 1)$ instead, as shown on the right side of the equation. The probability expressed by the CDF is denoted as $q_k(x \leq \theta\rho)$, representing the measure of channel outage.

Let p_0 be the probability that a candidate UE for cooperative retransmission has successfully received the packet dedicated for car A through multicast transmission from the

BS, and let $\overline{p_0} = 1 - p_0$ be the probability of the opposite event (i.e. multicast packet not received). Assuming that probability p_0 is equal for all candidate UEs, the error event for a single-shot cooperative transmission of n users to car A can be calculated as

$$q_{1tx}(n,\theta) = \left(\overline{p_0}\right)^n + \sum_{k=1}^{n} p_0^k \left(\overline{p_0}\right)^{n-k} \binom{n}{k} q_k (x \le \theta\rho).$$

Equal power-sharing between the cooperatively transmitting UEs with a constant sum power of unity is considered; hence, each of the n UEs is assigned a constant transmit power of $1/n$, and thus $\theta = n$ holds. Since a user transmits only if it has successfully received the packet from the BS, the sum term covers all possible user combinations for $k \in \{1, \ldots, n\}$ simultaneously transmitting users, whereas the first term represents the case that none of the n users have successfully received the packet multicast from the BS (i.e. $k = 0$).

If the latency budget allows supporting more than one retransmission via sidelink, ARQ and HARQ mechanisms can be applied. Multiple retransmissions allow grouping UEs into different sets and letting them perform their cooperative retransmissions successively. While in the ARQ scheme, a reception signal from a single retransmission that cannot be decoded is discarded, the HARQ scheme combines signals from different retransmission such that the overall diversity contained in all the retransmissions can be utilized. First, the n UEs are divided into two disjoint subsets with size n_1 and n_2, respectively, while maintaining the equal power allocation per UE. Then the probability for a channel outage amounts to

$$q_{n_1}\left(x \le \frac{n}{2}\rho\right) \cdot q_{n_2}\left(x \le \frac{n}{2}\rho\right) \quad \text{for ARQ}$$

and

$$q_{n_1+n_2}\left(x \le \frac{n}{2}\rho\right) \quad \text{for HARQ.}$$

Since two retransmissions are considered now, the total available power is accumulated over two time slots, and thus the equal power allocation amounts to $2/n$ per each UE, yielding $\theta = n/2$. To determine the overall reliability for two retransmissions with ARQ and HARQ schemes, it should be considered that the UEs in the second set n_2 can listen to the UEs in the first set n_1 during the first retransmission, and hence they can improve their probability of having obtained the original packet for car A. Given that k UEs retransmitted the packet in the first attempt, and accounting for this "listening during retransmission," one obtains for the n_2 UEs in the second set the probability of successful reception of the original packet as

$$p_1(k) = 1 - \overline{p_0} \cdot q_k\left(x \le \frac{n}{2}\rho\right)$$

and $\overline{p_1} = 1 - p_1$accordingly. The probability that the packet cannot be decoded after the second retransmission, given there has been a first retransmission by k users, then yields

$$p_{\text{arq}} = \left(\overline{p_1}\right)^{n_2} + \sum_{r=1}^{n_2} p_1^r \left(\overline{p_1}\right)^{n_2-r} \binom{n_2}{k} \cdot q_r\left(x \le \frac{n}{2}\rho\right),$$

$$p_{\text{harq}}(k) = \left(\overline{p_1}\right)^{n_2} \cdot q_k\left(x \le \frac{n}{2}\rho\right) + \sum_{r=1}^{n_2} p_1^r \left(\overline{p_1}\right)^{n_2-r} \binom{n_2}{k} \cdot q_{r+k}\left(x \le \frac{n}{2}\rho\right).$$

These two expressions follow the structure of q_{1tx} presented earlier. For HARQ, though, it additionally considers in the first term the outage during the first retransmission carried out by the k UEs.

Given there was no first retransmission at all (i.e. $k = 0$), the probability that the packet cannot be decoded after the second retransmission is given by $q_{1tx}(n_2, n/2)$. Together with the previous expressions, one can finally calculate the overall error event for two successive cooperative retransmissions of $n = n_1 + n_2$ UEs by building the sum over all $k \in \{0, \ldots, n_1\}$ UEs transmitting in the time slot reserved for the first retransmission as

$$q_{2tx} = (\overline{p_0})^{n_1} \cdot q_{1tx}\left(n_2, \frac{n}{2}\right) + \sum_{k=1}^{n_1} p_0^k (\overline{p_0})^{n_1-k} \binom{n_1}{k} \cdot \begin{cases} q_k\left(x \leq \frac{n}{2}\rho\right) \cdot p_{\text{arq}} & \text{for ARQ} \\ p_{\text{harq}}(k) & \text{for HARQ} \end{cases}.$$

As a reference system, the single-frequency network (SFN) transmission mode is used, which is commonly used in broadcast transmissions. In this mode, the UEs transmit the identical signal in the same resource with equally shared transmit power. Since there is no further preprocessing of the transmit signals, signals from different UEs add up incoherently at the receiver, so no diversity gains can be obtained due to a lack of cooperation between the UEs. The SFN scheme thus cannot attain high performance, but it is simple. Following a similar analysis, one can obtain the corresponding analytical expressions for the SFN case.

5.3.1.3 Evaluation

As the aim is on high reliability, the BS should choose a coding scheme yielding a low bit error rate, and hence we set $p_0 = 0.95$ for the probability that the UEs successfully receive the multicast packet from the base station. The bit error rate relates to the channel outage, where for a single user in Rayleigh fading, an outage of $q_1(x \leq \rho) = 5\%$ translates to a channel gain of $\rho = -13$ dB, which is selected as the operation point for evaluation. Figure 5.24a shows the error event for one and two retransmissions, respectively, and for the SFN reference case. The SFN reference system shows a saturation behavior, as it cannot realize diversity gains from user cooperation. While the reliability for a single retransmission is poor, HARQ can improve it by roughly two orders of magnitude. The cooperative scheme, however, shows that the error event continuously scales with the number of cooperating users n, achieving one order of magnitude per user for the HARQ case. The curve for ARQ exhibits a slope that is less steep than that of HARQ, which reflects the reduced diversity order realized by ARQ compared to HARQ. Comparing to a single retransmission (rtx), 2 rtx HARQ provides a gain of one order of magnitude for $n = 4$ users, which increases to two orders of magnitude for $n = 7$ users.

Figure 5.24b demonstrates the error event for a single retransmission as well as two retransmissions with HARQ considering V2V channels with different power levels. The detailed analysis for this case has been omitted here due to space limitations; the interested reader is referred to [54]. The vector of power offsets (i.e., attenuation factors representing the path loss per V2V link) for $n = 8$ neighbor UEs is given by $[1, 1, 2, 2, 4, 4, 6, 6]^T$. Evaluation of these power offsets for a single retransmission (1 rtx) shows that these obviously cause clear performance degradations compared to the reference case of equal power per V2V channel, yielding a loss of more than two orders of magnitude for $n = 8$ cooperating users. The cooperative HARQ scheme is evaluated based on two types of clustering strategies: one where the n UEs are evenly distributed among the two subsets according to their power offsets; and another where one subset contains the stronger UEs (i.e. the $n/2$ UEs with smallest power offsets) and the other subset contains the weaker UEs (i.e.

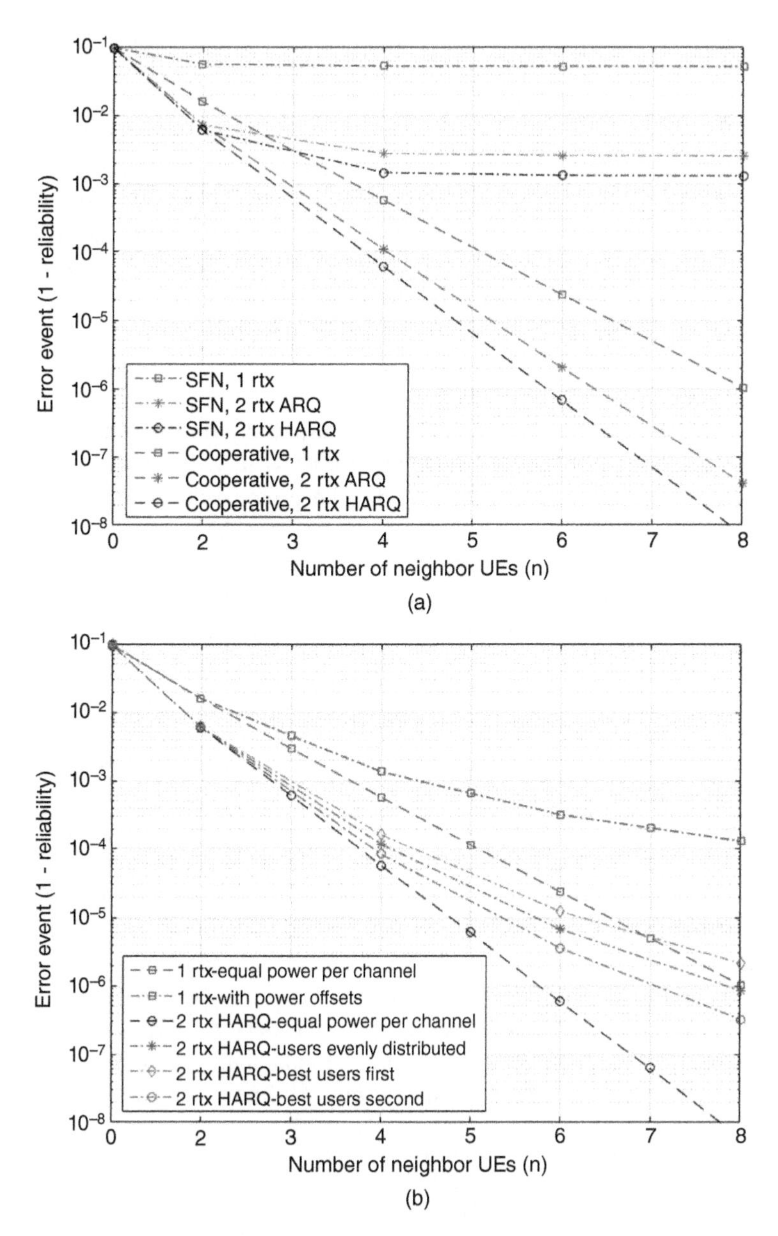

Figure 5.24 Improvement of reliability by the cooperative transmission of neighbor UEs, in (a) with equal average channel power, and in (b) with power offsets. Source: © 2019 IEEE. Reprinted, with permission, from [54].

the $n/2$ UEs with largest power offsets). Comparing the different cases of user clustering, it is observed that the best performance is attained if the strongest users transmit second. This can be attributed to the fact that the users in the second set have received the original packet with higher probability (since they were able to listen to the first retransmission), and hence their likelihood to retransmit is significantly increased. Since the strongest users experience better V2V channels than the others, the higher probability for transmission results in improved reliability. Finally, we note that the performance degradation for two HARQ retransmissions with this user grouping, compared to the reference case of equal power per V2V channel, amounts to significantly less than two orders of magnitude – and thus also exhibits a clear advantage over a single retransmission in this respect.

5.3.1.4 System Design Aspects

The previous sections have been focused on the link level, especially on the reliability enhancement due to joint operation of the sidelink (SL) and V2N link. In this section, the focus is on system aspects, covering an example of radio protocol design and also system-level simulation results.

Device discovery, which is used to determine the channel condition between devices and assess the potential of establishing direct sidelink communication, is one necessary step to enable the proposed concept of joint operation of the V2N interface and sidelink for reliability enhancement. Since our main focus is on communication aspects, in this section, it is assumed that the device-discovery phase is completed. In other words, the involved UEs have already discovered each other, and the link condition between the assisting UEs (UEs forwarding the received signal to the target UE or the BS) and the target UE is good enough to establish sidelink communication.

To efficiently support the cooperation of the V2N interface and SL interface on the system level, it is useful to look at the necessary enhancements of a radio access network (RAN)-level protocol. Figure 5.25 illustrates one example radio protocol stack for direct SL communication. In this example, only one direction of the communication link from the transmitting UE (Tx UE) to receiving UE (Rx UE) is illustrated. One thing worth pointing out is that the Rx UE should be able to differentiate two cases: (i) the packet received over the sidelink is for itself, and (ii) the packet received over SL needs to be forwarded further to BS once requested. Here it is assumed that the SL is used for enhancing communication reliability in the uplink direction. To be more specific, during the first transmission from

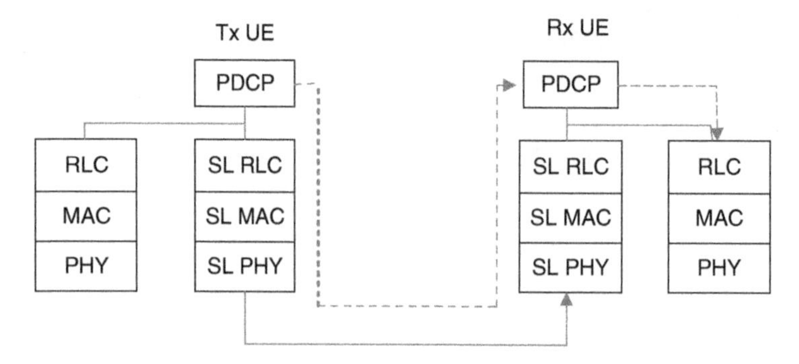

Figure 5.25 Example protocol stack (dashed line: packet delivery path).

the Tx UE, both the BS and Rx UE will try to decode the received data packet. In case the BS requests data packet retransmission due to the deep fading channel or strong interference, both the Tx UE and Rx UE can retransmit the same data packet to the BS. To distinguish the final target of the data packet, an indicator could be included at the PDCP layer. In this way, once the packet is received by a UE over sidelink, after successful decoding and processing at lower layers, at the PDCP layer it will learn that the packet needs to be forwarded to the BS if the BS requests retransmission from that UE.

Carrying indication at the PDCP layer is just one example. In principle, the indicator can be conveyed by other means as well: for example, via MAC control element (CE) or even physical layer signaling. At the RX UE side, depending on the detected indicator, the UE will determine whether to forward the received data packet toward upper layers or keep the decoded packet within the RAN and forward it to the BS once it is requested.

In the following, we discuss the procedure of applying SL-assisted reliable UL transmission with the example given in Figure 5.26, where it is assumed that UE1 and UE2 have already discovered each other and the channel condition is good enough to establish direct sidelink communication. Furthermore, they are assigned one group ID by the BS, e.g. SL_RNTI (sidelink radio network temporary identifier), which is used to identify the device pair or the sidelink group in case more than two UEs are acting as assisting UEs.

In Figure 5.26, it is assumed that UE1 has a UL data packet for transmission. After sending the scheduling request (SR) to BS and indicating to BS that the available sidelink can be used for reliability enhancement, BS sends a UL resource grant. With the same transmission, both BS and UE2 will receive the data packet from UE. In the case of failed data detection, BS can allocate resources for retransmission to both UE1 and UE2. Afterward, both UE1 and UE2 can send the same data packet to BS. Clearly, the benefit is the reduced latency for retransmission in case more than one retransmission is needed, and increased reliability as elaborated in the previous sections, which is one important performance indicator for URLLC. The detailed procedure example, as shown in Figure 5.26, is given as follows:

1. UE1 sends a UL resource request (e.g. scheduling request: SR) in a normal way to BS. At the same time, it can indicate that the sidelink can help for UL retransmission for enhancing reliability. Of course, after the discovery process, BS may also select and configure UE2 to assist UE1 in advance, and in this case, the SR of UE1 does not need to indicate the required help from UE2.
2. BS allocates resources to UE1 over the downlink control channel with SL_RNTI as the target identification. This way, UE2 can decode the control message as well and learn about the physical resources to be used by UE1 for sending UL data.
3. UE1 sends the data packets in the allocated resources. Since the resource information is known to both BS and UE2, UE2 can receive and decode the same packet as well. Due to the shorter distance between UE1 and UE2, the probability of successful decoding of the data packet from UE1 is higher than for BS.
4. If BS is not able to decode the first transmission correctly, BS sends a feedback signal, i.e. HARQ-NACK, with the allocated HARQ-ACK resource known to both UE1 and UE2. The resource allocated to UE1 and UE2 can be the same or different, depending on the operational mode. For example, the retransmission from UE1 and UE2 can take the SFN type of UL transmission (i.e. UE1 and UE2 sharing the same resource as well as transmission format) as a resource-sparing approach, as discussed in [50]. The resources allocated

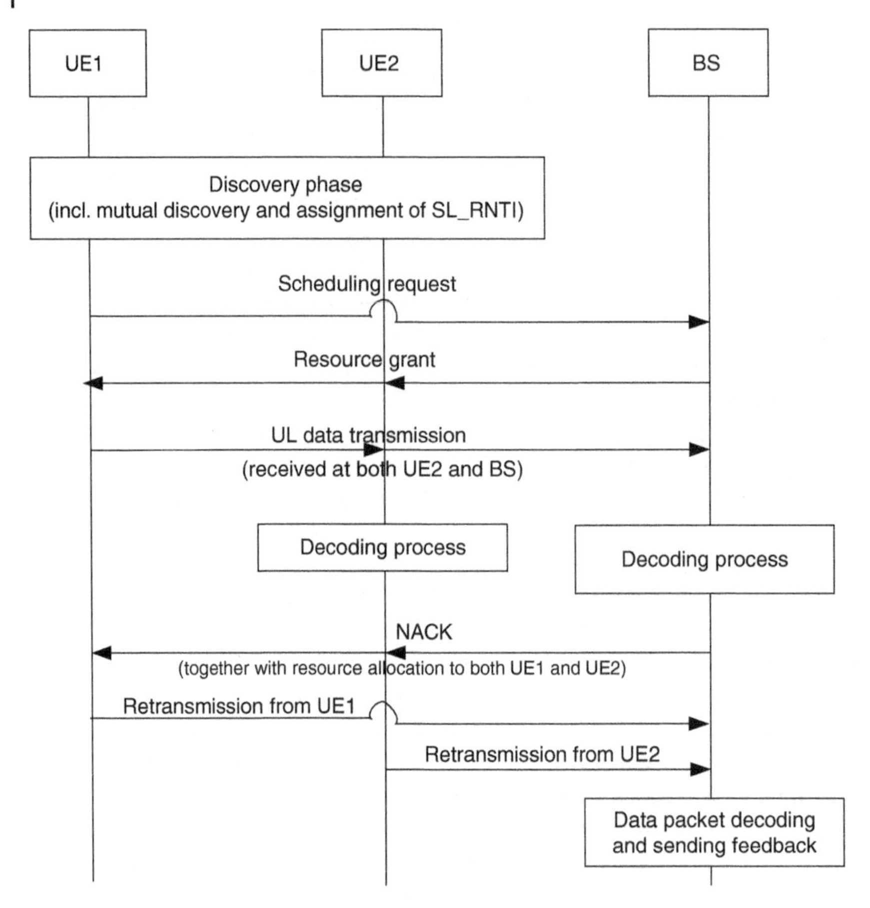

Figure 5.26 Example procedure for sidelink-assisted reliable communication.

to UE1 and UE2 may also be different, though. And in that case, the transmission format of UE1 and UE2 can be different.

5. Both UE1 and UE2 retransmit the identical data packet.
6. BS receives the packet retransmitted from both UE1 and UE2.

To evaluate the performance of the sidelink-assisted reliable communication in UL, a system-level simulation was carried out. Taking the highway scenario as described in [55], in our simulation, three BSs are deployed along a highway 2 km long. The traffic model is assumed to be periodic traffic with a packet length of 160 bytes, and the arrival time period is 100 ms. International Telecommunication Union (ITU)-Veh-A fading channel [11] is adopted in our simulation; system bandwidth is 10 MHz at 2 GHz center carrier frequency. Key simulation parameters are summarized in Table 5.3.

Taking the packet loss rate (i.e. packets that cannot be delivered correctly after two HARQ retransmission) as one example, Figure 5.27 illustrates the packet loss performance with different numbers of UEs, where the assisting UE is randomly selected among all UEs that can establish sidelink communication with the target UE. The comparison is between cases

Table 5.3 List of system-level simulation parameters.

Parameter	Value
Path-loss (PL) model	PL (dB) = A + B log(d); d = BS-UE distance A: 128.1 dB; B: 37.6 dB
Carrier frequency	2 GHz
Number of BSs	3
Number of UEs	Variable
Traffic model	Packet length 160 bytes Arrival time: every 100 ms
Bandwidth	10 MHz
Duplexing scheme	FDD
Subcarrier spacing	15 kHz
Fast-fading model	ITU-Veh-A
MIMO	1×2

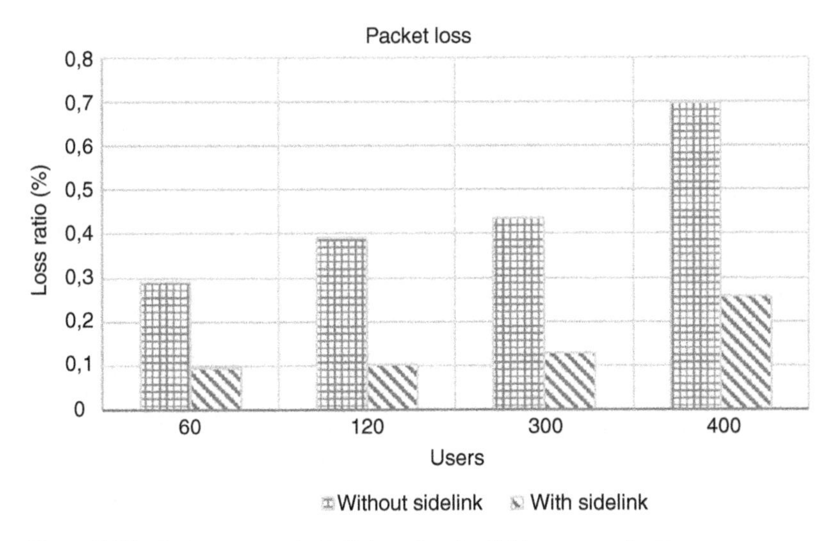

Figure 5.27 Performance of sidelink-assisted reliable communication.

with and without SL assistance, where the performance improvement due to SL assistance is by at least a factor of 3.

5.3.2 Full Duplex

Conventional wireless communication is subject to half-duplex constraints. Thus, a radio device cannot transmit and receive at the same time and in the same frequency band. This is because the transmitted signal's power is normally tens of decibel larger than the power of the received signal, which leads to very high self-interference, preventing the receiver circuits from "hearing" any signal sent by other nodes. During the past 10 years, there has been significant development in the self-interference cancelation (SIC) techniques, which opens the door for full-duplex radio enabling a wireless transceiver to transmit and receive simultaneously using the same frequency band. The different types of SIC techniques cover the entire range from antenna to baseband:

- *Passive RF isolation*: By physically separating the transmitting and receiving antennas by a certain distance, tens of decibel passive isolation can easily be created with the path loss between them. However, the modern cellular communication system normally requires the collocation of transmitting and receiving antennas due to size and cost constraints. Alternative passive isolation can rely on using different polarization of collocated antennas for transmission and reception, which could achieve 10–20 dB isolation. Furthermore, passive isolation can be further achieved using a tri-port circulator, which normally has 10–20 dB isolation, and an electrical balance duplexer, which can reach as high as 53 dB isolation in the 2.4 GHz ISM-Band (Industrial, Scientific and Medical Band) [56].
- *Active RF SIC*: After passive RF isolation, self-interference may still be too high, which can saturate the low-noise amplifier (LNA) in the receiver chain. A common solution is to create a copy of the transmitted RF signal with an extra auxiliary transmitter for further cancelation of the self-interference before it goes into the LNA. Since self-interference involves the channel response coming from the scattering environment around the antenna and the antenna itself, to achieve the best active RF SIC performance, the channel response should be taken into account for constructing the auxiliary cancelation signal. A closed-loop operation involving both baseband and RF is needed for the channel estimation, auxiliary signal construction, and cancelation process. The active RF SIC can normally provide 20–60 dB attenuation of self-interference.
- *Active digital SIC*: After RF SIC, residual self-interference is reduced to a level within the dynamic range of the analog to digital converter (ADC), and can be further canceled out with digital signal processing. Up to 30 dB of SIC can be achieved digitally.
- *Optical SIC*: There is a newer type of SIC based on converting the RF signal into an optical signal with e.g. a Mach-Zehnder modulator first, and then performing the SIC using optical signal processing [57]. This technique could support wider bandwidth, lower noise, and a higher tuning range for the carrier frequency. The company GenXComm is trying to commercialize it.

Prototypes of full-duplex radio have demonstrated that up to 110 dB overall cancelation of self-interference can be achieved by combining the SIC techniques. The direct benefit of full-duplex radio is that spectrum efficiency and data throughput can be doubled. In addition, some system overhead, such as the duplex gap in frequency-division duplex (FDD) and guard period in time-division duplex (TDD), can be avoided. There are many other advantages when applying full-duplex radio to cellular V2X (C-V2X) for both cellular and sidelink connections, which are introduced in the next section.

5.3.2.1 Advantages of Full-Duplex Radio for C-V2X

CAD aims to significantly improve road safety, traffic efficiency, and the comfort of driving. It requires C-V2X to achieve high reliability, high data throughput, and low latency with high user mobility and density. Full-duplex radio is a promising enabling technology for CAD, which has strong potential in the following aspects.

Low-Latency Feedback for Mobility Enhancement

In V2X communication, very fast channel fading is expected. With full duplex, ACK/NACK messages can be transmitted by reusing the receiving resource and without waiting for extra control channel resources for feedback. As a result, the round trip time (RTT) of the HARQ process and, subsequently, end-to-end latency can be significantly reduced. In an extreme case, full-duplex radio can enable the rateless type of channel coding and adaptation [57], where the receiver sends the acknowledgment instantaneously at the moment of successful decoding. In this way, not only lower latency but also higher spectrum efficiency and data throughput can be achieved.

Another aspect is that CSI feedback can be accelerated by full-duplex radio to mitigate the channel aging issue in fast fading, which is particularly helpful to 5G, where massive MIMO and beamforming are extensively used. A fast-moving, full-duplex UE can immediately feed back the CSI or transmit any other uplink/sidelink signal to assist its counterpart in quickly tuning its transmitting parameter (beam, resource block, MCS, etc.).

Optimal Collision Avoidance in Sidelink

C-V2X sidelink Mode 4 (LTE-V) and Mode 2 (NR-V2X) communication rely on sensing-based semi-persistent scheduling (SPS) based on orthogonal frequency division multiple access (OFDMA). In addition, IEEE 802.11p-based V2X relies on the listen-before-talk strategy based on carrier sensing multiple access with collision avoidance (CSMA/CA), as well as on the request to send (RTS) and clear to send (CTS) mechanisms. These designs are essentially based on sensing interference at the transmitter. However, what needs to be protected from interference is the receiver. As a result, the transmitter side sensing-based mechanism leads to the well-known hidden node problem: the nearby receiving node cannot be effectively detected, and signal collision occurs. The hidden node problem becomes particularly severe when nodes are dense and/or the channel environment is complex.

With full-duplex radio, this issue can be fundamentally solved. For example, a node that is receiving can simultaneously broadcast a beacon, reusing its receiving resource to tell its surrounding nodes to keep silent to protect its ongoing reception of data traffic with higher priority. Although the resource reservation in SPS and the RTS/CTS mechanisms may partially solve the hidden node problem, compared to the simultaneous approach enabled by full duplex, they are less flexible and less effective.

Low-Latency and High-Throughput Relay and Cooperation

A traditional repeater functions as an amplify-and-forward relay in a full-duplex manner, which solely relies on the strict isolation of the donor antenna and rebroadcast antenna via separated indoor/outdoor placement as well as careful design and deployment to minimize the front-to-back ratio of the antennas (i.e. the ratio of power gain between the front and rear of a directional antenna). With SIC capability, a relay node can forward the signal on a single shared transmitting/receiving antenna. Allowing for simultaneous transmission and reception, the data throughput and spectrum efficiency of the relay node can be doubled, no matter what forwarding technique is used.

Particularly for C-V2X sidelink, full-duplex radio enables the on-board unit (OBU) or RSU to cooperatively forward data from other nodes while avoiding disturbing its ongoing data reception. Combined with OFDMA, full-duplex relaying would be even more effective and flexible. For example, the relaying node can select its sub-channels for transmission and reception according to their gains in frequency-selective fading and also based on interference conditions and collision probabilities.

Cognitive Full-Duplex Communications for C-V2X

In a cognitive spectrum-sharing scenario, the full-duplex transceiver can enable a UE to continuously monitor the radio spectrum and immediately sense a transmission from a primary user that has higher priority or legacy rights for use of the spectrum. In this way, the cognitive UE could immediately suspend its transmission when needed, reducing the probability of interference imposed on the primary user. This scheme could enable sidelink communication to utilize uplink spectrum resources. In addition, with full-duplex operation, a dedicated sensing period can be replaced by uninterrupted spectrum sensing. As a result, a device can instantly vacate the spectrum when any other V2X transmission with higher priority or safety criticalness appears.

Interference-Free Adjacent Channel Operation

SIC can also enable simultaneous transmission and reception in adjacent frequency channels removing substantial out-of-band leakage. In this way, the large fixed-duplex gap in conventional FDD is avoided. For the cellular link, it can enable highly adaptive scheduling of downlink and uplink resources based on the flexible slot formats of 5G [9]. For the sidelink, it allows for better coexistence of different services that are assigned to dedicated frequency channels, e.g. in the 5.9 GHz band [58]. The products which are capable of SIC for adjacent channel have been developed by KUMU Networks.

Feasibility of Full-Duplex Radio for V2X

Although full-duplex radio technology is not yet sufficiently mature for large-scale commercialization, the practical issues can be largely mitigated in vehicular devices [58] in comparison with personal devices like smartphones:

- Compared to smartphones or IoT devices, the miniaturization of device and antenna is not a major concern. Significant passive isolation is feasible by placing antennas at different locations on the vehicle body.
- Cost constraints are relatively less stringent for vehicles [58], which allows for better SIC with high-quality RF components.
- Compared to smartphones or IoT device, OBU or RSU can host stronger signal processing with relaxed limitations on the power supply.

5.4 Summary

This chapter provided insights into key aspects of the radio interface design for V2X communications for CAD and CRU services. The major challenge for this design is providing stable, highly reliable communication links that must be maintained even with very high user mobility.

First, the focus was on the application of beamforming techniques with narrow beams in the mm-wave spectrum, where fast beam tracking and adjustment are required. Three different beamforming schemes were proposed, which are tailored for unicast, multicast, and broadcast transmissions, respectively. The unicast scheme represents a beam-refinement procedure for improving the beam directions under high mobility, facilitating fast beam tracking based on a genetic algorithm exploiting spatial correlations. The multicast scheme is built on a novel frame structure proposed for the application of adaptive and robust beam-management techniques, which enable the selection of beam sets for multicast transmission and reselection for retransmissions to a subset of the multicast users, fulfilling requirements for low latency and signaling overhead. For the broadcast scheme, distributing candidate beams for probing in the spatial, time, and frequency domains was investigated, and an important trade-off between the latency required for broadcast information acquisition and the overhead required for beam probing was found.

Second, extensions of the lower layers of the protocol stack were addressed to provide a tailored radio interface design for the V2V sidelink. A theoretical analysis of the trade-off between pilot overhead and spectral efficiency under high mobility was presented, along with a corresponding robust receiver design, which allow finding an optimum allocation of resources for pilot and data symbols for given mobility conditions. Thereafter, the latest advances in V2V sidelink design were detailed: For the design of reference signals for the V2V sidelink, an overview of the different types of signals required for 5G NR V2X was given, and it was explained how the overhead of these reference signals can be adjusted relative to user speed. For synchronizing UEs via the sidelink, the synchronization signals agreed on for 5G NR V2X were presented, along with potential extensions for further improvements that allow crucial information about priority to be available at a very early stage in the synchronization process. To address scheduling and power control for the direct communication of UEs via the sidelink, a centralized RRM scheme was introduced, aiming to find the optimum radio resource configuration for V2V communication to yield the maximum number of connected UEs.

Finally, novel communication concepts facilitated by short-range communication via the V2V sidelink were presented and elaborated. These concepts embrace cooperation of UEs, where neighboring UEs in close vicinity cooperatively forward data packets in uplink and downlink transmissions to attain substantial gains in reliability for transmissions at moderate system cost. Moreover, the appealing concept of full duplex was analyzed with respect to the latest state of the art in the field, and further elaboration showed that full duplex can be considered of high relevance for future applications in V2V communication.

References

1 mmMagic Project Deliverable 6.6. (2017). Final mmMAGIC system concept. Available at https://bscw.5g-mmmagic.eu/pub/bscw.cgi/d215278/mmMAGIC-D6.6.pdf

2 Guo, H., Makki, B., and Svensson, T. (2017). A genetic algorithm-based beamforming approach for delay-constrained networks. In: *Proceedings of IEEE Int'l Symposium on Modeling and Optimization in Mobile, Ad Hoc and Wireless Networks.*

3 Guo, H., Makki, B., and Svensson, T. (2018). Genetic-algorithm based beam refinement for initial access in millimeter-wave mobile networks. *Wiley-Hindawi Wireless Communications and Mobile Computing, Special Issue on Recent Advances in 5G Technologies: New Radio Access and Networking.*

4 Qiao, J., Shen, X., Mark, J. W., and He, Y. (2015). MAC-layer concurrent beamforming protocol for indoor millimeter-wave networks. *IEEE Transactions on Vehicular Technology* 64 (1): 327–338.

5 Hur, S., Kim, T., Love, D. J., et al. (2011). Multilevel millimeter wave beamforming for wireless backhaul. In: *Proceedings of IEEE Global Communications Conference*: 253–257.

6 Gao, X., Dai, L., Yuen, C., and Wang, Z. (2016). Turbo-like beamforming based on Tabu search algorithm for millimeter-wave massive MIMO systems. *IEEE Transactions on Vehicular Technology* 65 (7): 5731–5737.

7 Chen, L., Yang, Y., Chen, X., and Wang, W. (2011). Multi-stage beamforming codebook for 60GHz WPAN. In: *Proceedings of 6th Int'l ICST Conference on Communications and Networking in China (CHINACOM)*: 361–365.

8 Chen, S., Hu, J., Shi, Y., Peng, Y., Fang, J., Zhao, R., and Zhao, L. (2017). Vehicle-to-everything (v2x) services supported by LTE-based systems and 5G. *IEEE Communications Standards Magazine* 1 (2): 70–76.

9 3GPP. (2019). Physical layer procedures for control. TS38.213, V15.6.0.

10 Giordani, M., Mezzavilla, M., and Zorzi, M. (2016). Initial access in 5G mmWave cellular networks. *IEEE Communications Magazine* 54 (11): 40–47.

11 3GPP. (2016). Study on LTE-based V2X services. TR 36.885, V14.0.0.

12 3GPP. (2018). Study on evaluation methodology of new vehicle-to-everything (V2X) use cases for LTE and NR (Release 15). TR 37.885 V15.2.0.

13 M., Fallgren et al., "Multicast and Broadcast Enablers for High-Performing Cellular V2X Systems," in *IEEE Transactions on Broadcasting* 65 (2): 454–463.

14 Roger, S., Martin-Sacristan, D., Garcia-Roger, D. et al. (2019). Low-latency layer-2-based multicast scheme for localized V2X communications. *IEEE Transactions on Intelligent Transportation Systems* 20 (8): 2962-2975.

15 Nekovee, M., von Wrycza, P., Fresia, M. et al. (2015). Millimeter-wave based mobile radio access network for fifth generation integrated communications (mmMAGIC). In: *Proceedings of European Conference on Networks and Communications (EuCNC).*

16 Li, Y., Luo, J., Castaneda, M., Stirling-Gallacher, R., Xu, W., and Caire, G. (2018). On the beamformed broadcasting for millimeter wave cell discovery: performance analysis and design insight. *IEEE Transactions on Wireless Communications* 17 (11): 7620–7634.

17 Li, Y., Luo, J., Castaneda, M. et al. (2017). Analysis of broadcast signaling for millimeter wave cell discovery. In: *Proceedings of IEEE 86th Vehicular Technology Conference (VTC Fall).*

18 Ozdemir, M. K., and Arslan, H. (2007). Channel estimation for wireless OFDM systems. *IEEE Communications Surveys and Tutorials* 9 (2): 18–48.

19 Siomina, I., Furuskär, A., and Fodor, G. (2009). A mathematical framework for statistical QoS and capacity studies in OFDM networks. In: *Proceedings of IEEE 20th Int'l Symposium on Personal, Indoor and Mobile Radio Communications (PIMRC).*

20 Sesia, S., Toufik, I., and Baker, M. (2011). *LTE – The UMTS Long Term Evolution: From Theory to Practice*, 2e. Wiley.

21 Fodor, G., Pap, L., and Telek, M. (2019). Recent advances in acquiring channel state information in cellular MIMO systems. *Infocommunications Journal* 11 (3): 2–12.

22 Gesbert, D., Kountouris, M., Heath Jr. R. W., Chae, C.-B., and Salzer, T. (2007). Shifting the MIMO paradigm: from single-user to multiuser communications. *IEEE Signal Processing Magazine* 24 (5): 36–46.

23 Marzetta, T. (2010). Noncooperative cellular wireless with unlimited numbers of base station antennas. *IEEE Transactions on Wireless Communications* 9 (11): 3590–3600.

24 Médard, M. (2000). The effect upon channel capacity in wireless communications of perfect and imperfect knowledge of the channel. *IEEE Transactions on Information Theory* 46 (3): 933–946.

25 Hassibi, B., and Hochwald, B. M. (2003). How much training is needed in multiple-antenna wireless links? *IEEE Transactions on Information Theory* 49 (4): 951–963.

26 Kim, T., and Andrews, J. G. (2005). Optimal pilot-to-data power ratio for MIMO-OFDM. In: *Proceedings of IEEE Global Communications Conference.*

27 Kim, T. and Andrews, J. G. (2006). Balancing pilot and data power for adaptive MIMO-OFDM systems. In: *Proceedings of IEEE Global Communications Conference.*

28 Marzetta, T. (2006). How much training is needed for multiuser MIMO? In: *Proceedings of IEEE Asilomar Conference on Signals, Systems and Computers (ACSSC).*

29 Laiyemo, A. O. (2018). *High speed moving networks in future wireless systems.* PhD dissertation, Oulu University, Finland.

30 Liang, L., Peng, H., Li, G. Y., and Shen, X. S. (2017). Vehicular communications: a physical layer perspective. *IEEE Transactions on Vehicular Technology* 66 (12): 10647–10659.

31 Fodor, G., Marco, P. D., and Telek, M. (2015). Performance analysis of block and comb type channel estimation for massive MIMO systems. In: *Proceedings of 1st Int'l Conference on 5G for Ubiquitous Connectivity.*

32 Mehana, A. H., and Nosratina, A. (2012). Diversity of MMSE MIMO receivers. *IEEE Transactions on Information Theory* 58 (11): 6788–6805.

33 Eraslan, E., Daneshrad, B. and Lou, C.-Y. (2013). Performance indicator for MIMO MMSE receivers in the presence of channel estimation error. *IEEE Wireless Communications Letters* 2 (2): 211-214.

34 Guo, K., Guo, Y., Fodor, G., and Ascheid, G. (2014). Uplink power control with MMSE receiver in multi-cell MU-Massive-MIMO systems. In: *Proceedings of IEEE Int'l Conference on Communications (ICC)*: 5184–5190.

35 Abrardo, A., Fodor, G., Telek, M., and Moretti, M. (2019). MMSE receiver design and SINR calculation in MU-MIMO systems with imperfect CSI. *IEEE Wireless Communications Letters* 8 (1): 269–272.

36 Moghadam, N. N., Shokri-Ghadikolaei, H., Fodor, G. et al. (2017). Pilot precoding and combining in multiuser MIMO networks. In: *Proceedings of IEEE Int'l Conference on Acoustics, Speech and Signal Processing (ICASSP).*

37 Zhao, P., Fodor, G., Dan, G., and Telek, M. (2018). A game theoretic approach to setting the pilot power ratio in multi-user MIMO systems. *IEEE Transactions on Communications* 66 (3): 999–1012.

38 Zhao, P., Fodor, G., Dan, G., and Telek, M. (2019). A game theoretic approach to uplink pilot and data power control in multicell multiuser MIMO systems. *IEEE Transactionson Vehicular Technology* 68 (9): 8707–8720.

39 Fodor, G., Marco, P. D., and Telek, M. (2015). On minimizing the MSE in the presence of channel state information errors. *IEEE Communications Letters* 19 (9): 1604–1607.

40 Fodor, G., Marco, P. D., and Telek, M. (2016). On the impact of antenna correlation and CSI errors on the pilot-to-data power ratio. *IEEE Transactions on Communications* 64 (6): 2622–2633.

41 Petrovic, D., Rave, W., and Fettweis, G. (2007). Effects of phase noise on OFDM systems withand without PLL: characterization and compensation. *IEEE Transactions on Communications* 55 (8): 1607–1616.

42 Molisch, A. F. (2005). *Wireless Communications*. Wiley.

43 Dahlman, E., Parkvall, S., and Sköld, J. (2018). *5G NR: The Next Generation Wireless Access Technology*. Academic Press.

44 Manolakis, K., Xu, W., and Caire, G. (2017). Synchronization signal design and hierarchical detection for the D2D sidelink. In: *IEEE 51st Asilomar Conference on Signals, Systems and Computers*.

45 Hisham, A., Yuan, D., Ström, E.G., and Brännström, F. (2020). Adjacent channel interference aware joint scheduling and power control for V2V broadcast communication. Accepted for publication, *IEEE Trans. on Intelligent Transport Systems*.

46 Peng, B., Hu, C., Peng, T. et al. (2013). A resource allocation scheme for D2D multicast with QoS protection in OFDMA-based systems. In: *Proceedings of IEEE 24th Int'l Symposiumon Personal, Indoor, and Mobile Radio Comm. (PIMRC)*.

47 Soret, B., Mogensen, P., Pedersen, K.I., and Aguayo-Torres, M.C. (2014). Fundamental tradeoffs among reliability, latency and throughput in cellular networks. In: *Proceedings of IEEE Global Communications Conference*.

48 Popovski, P., Nielsen, J. J., Stefanovic, C., de Carvalho, E., Ström, E. G., et al. (2018). Wireless access for ultra-reliable low-latency communication: principles and building blocks. *IEEE Network* 32 (2): 16–23

49 Johansson, N.A., Wang, Y.E., Eriksson, E., and Hessler, M. (2015). Radio access for ultra-reliable and low-latency 5G communications. In: *IEEE Int'l. Conference on Communications Workshops (ICC)*.

50 Aygun, B., Mandelli, S., Wild, T. et al. (2017). Side-link assisted hybrid automatic repeat request for ultra-reliable low latency communications. In: *21st Int'l. ITG Workshop on Smart Antennas*.

51 Laneman, J.N., and Wornell, G. W. (2003). Distributed space-time-coded protocols for exploiting cooperative diversity in wireless networks. *IEEE Transactions on Information Theory* 49 (10): 2415–2425.

52 Yiu, S., Schober, R., and Lampe, L. (2006). Distributed space-time block coding. *IEEE Transactionson Communications* 54 (7): 1195–1206.

53 Dammann, A., and Kaiser, S. (2001). Performance of low complex antenna diversity techniques for mobile OFDM systems. *In: 3rd Int'l. Workshop on Multi-Carrier Spread-Spectrum & Related Topics*.

54 Schellmann, M., and Soni, T. (2019). Ultra-reliable V2X communication: On the value of user cooperation in the sidelink. In: *IEEE European Conference on Networks and Communications (EuCNC).*

55 5GCAR Project Deliverable 3.1. (2018). Intermediate 5G V2X Radio. Available at https://5gcar.eu/deliverables/

56 Manuzzato, E., Tamminen, J., Turunen, M. et al. (2016). Digitally-controlled electrical balance duplexer for transmitter-receiver isolation in full-duplex radio. In: *22th European Wireless Conference.*

57 Han, X., Huo, B., Shao, Y. et al. (2017). RF self-interference cancellation for full-duplex communication with microwave photonic technique. *In: Int'l. Topical Meeting on Microwave Photonics (MWP).*

58 Campolo, C., Molinaro, A., Berthet, A. O., and Vinel, A. (2017). Full-duplex radios for vehicular communications. *IEEE Communications Magazine* 55 (6): 182–189.

6

Network Enhancements

Toktam Mahmoodi[1], Ricard Vilalta[2], Apostolos Kousaridas[3], Massimo Condoluci[4], and Panagiotis Spapis[3]

[1] *King's College London, UK*
[2] *Centre Tecnològic de Telecomunicacions de Catalunya, Spain*
[3] *Huawei German Research Center, Germany*
[4] *Ericsson Research, Sweden*

It is widely accepted that the fifth generation (5G) cellular network will benefit the automotive sector, one of the most prominent verticals. Connected automated driving (CAD) and connected road user (CRU) services in automotive cover a wide range of use cases, which introduce stringent requirements such as ultra-low latency communications, ultra-high data rates, and timely service delivery. It is clear that vehicle-to-everything (V2X) communications are impacting 5G requirements and 5G architectural design, in terms of enhancements of system architecture, security, and privacy. The focus of this chapter is on cellular V2X (C-V2X) while some of the technical solutions have a broader applicability. Hence, this chapter focuses on necessary network enhancements.

We present an overview of the work performed in the domain of V2X network systems and architecture, including five areas of architectural enhancements that shape the fundamentals of flexible and agile V2X communication networks:

1. The application of a network-slicing paradigm for V2X, which relies on a set of network slices belonging to the types already standardized: enhanced mobile broadband (eMBB), ultra-reliable low-latency communication (URLLC), and massive machine type communication (mMTC).
2. Relationship of the proposed vehicular-slicing architecture with current network function virtualization (NFV) and software-defined network (SDN) developments.
3. Introduction of a cloudified architecture, which provides flexibility for radio access network (RAN) design in terms of how RAN functionalities can be placed in different network domains, splitting the radio and baseband functionalities between central cloud and distributed entities.
4. Enhancements of network procedures through the formation of local end-to-end (E2E) radio data paths over the air interface between base station and user equipment (Uu interface), proposed to enable the fast and guaranteed transmission of localized data

* With contributions from Laurent Gallo.

Cellular V2X for Connected Automated Driving, First Edition.
Edited by Mikael Fallgren, Markus Dillinger, Toktam Mahmoodi, and Tommy Svensson.

traffic among the involved devices, satisfying their quality of service (QoS) requirements and the features of V2X communication.
5. Addressing a multi-operator environment for V2X communications, since the end devices cannot be assumed to operate under a single operator.

These five areas of architectural enhancements are presented in Section 6.1–6.5. Finally, the chapter is summarized in Section 6.6.

6.1 Network Slicing

The set of use cases under consideration to be supported by 5G is vastly wider than those of previous generations, resulting in significant heterogeneity in the communication needs and traffic patterns to be supported by cellular networking. From the perspective of traffic features, these use cases can be split into three groups: eMBB, URLLC, and mMTC. These groups reflect the different needs of high bitrates, high reliability with low latency, enhanced coverage, and massive numbers of connections.

It becomes challenging for a monolithic general-purpose network to fulfill these heterogeneous requirements. In this picture, network slicing represents a novel architectural paradigm allowing a communication system to be built that, even relying on the same physical infrastructure, is composed of a collection of slices running in parallel and independently from each other. Each slice can be considered an end-to-end logical network that provides specific network capabilities and network characteristics designed to address the needs of a well-defined set of services. Network slicing is under discussion in several forums such as the 3rd Generation Partnership Project (3GPP), European Telecommunications Standards Institute (ETSI), International Telecommunication Union (ITU), and next generation mobile networks (NGMN), and an overview of the work on this topic can be found in [1].

In the remainder of this section, the different aspects of network slicing within 3GPP networks will be briefly discussed, and more detailed discussions of the features of network slicing tailored to V2X scenarios will be provided.

6.1.1 Network Slicing and 3GPP

Considering the domain of a mobile network operator (MNO), a network slice consists of entities including the user equipment (UE), the RAN, and the control plane and user plane of the core network - i.e. up to a point that the user-plane function (UPF) interfaces with the data network. Network slices natively enable the support of features such as using QoS differentiation for specific applications. In particular, within one network slice, different applications can be assigned to specific bearers (in the RAN) and flows (in the core network), which are treated with different priority levels This allows traffic differentiation among applications associated with the same slice.

Overall, network slicing within the 3GPP domain can be split into slicing for RAN and slicing for the core network. Network slicing for RAN imposes several challenges, mainly related to the fact that the spectrum is a limited resource. As a consequence, careful spectrum management is provisioned to avoid inefficiencies due, for instance, to excessive fragmentation of the spectrum among different slices [2]. Three main options for RAN slicing can be considered depending on how deeply the slicing concept is extended into the radio protocol stack:

- Full traffic isolation between slices, where transmission functionality is shared between all the network slices, while all other functions are slice-specific (i.e. each slice has its own independent radio stack starting above the physical layer). Although this allows the highest degree of traffic isolation between slices, it may result in poor flexibility in terms of RAN resources.
- Intermediate degree of isolation and customization achieved by having physical (PHY) layer and medium access control (MAC) layers in common to all slices, while the radio link control (RLC) layers and above are slice-specific.
- RAN sharing through advanced QoS treatment. In this case, slicing is enforced in the core network, and different slices use a shared RAN where advanced QoS mechanisms are used to achieve differentiation of traffic treatment at the RAN level. Despite the obvious advantages in terms of multiplexing gain, this approach is the most challenging for ensuring traffic isolation.

From a 3GPP perspective, slicing in a RAN is assumed to be achieved through adequate radio resource management, such as proper scheduling mechanisms at the MAC layer. These aspects of radio resource management, such as scheduling, are not standardized in 3GPP but are left to implementation and configuration. To guarantee adequate configurability of the radio interface, to meet various requirements, several enhancements are available for new generation RAN (NG-RAN), to fulfill strict requirements and also guarantee that configurations can be tuned to achieve adequate degrees of RAN slicing. These enhancements include:

- *Flexible numerology*, where different radio configurations can be used simultaneously to support numerologies for URLLC, eMBB, and mMTC within the same band
- *Mini-slot*, for fast, flexible scheduling (e.g. for URLLC); and downlink preemption for urgent URLLC transmissions, which are allowed to preempt resources of services with lower priority
- *Grant-free uplink* for reducing scheduling overhead: e.g. for URLLC

From a core network perspective, network slicing can be achieved by using enablers such as SDN and NFV [3], which are discussed in more detail in the following sections of this chapter. To fully exploit the benefits of softwarization and virtualization,[1] functionalities in the core network of 5G systems are designed with a paradigm switch with respect to 4G systems from network nodes to network functions (NFs). This design approach removes the anchor to the implementation and deployment of specific network nodes with dedicated hardware running custom firmware, and instead allows exploiting software functions that can run on virtualized environments. This brings benefits not only in terms of flexibility

1 Although virtualization is an important technical enabler for network slicing, some functions that are critical in terms of hard real-time requirements might also be deployed on dedicated hardware (at least, in early deployment phases).

for deployment, instantiation, and reconfiguration, but also in terms of the possibility of using ad hoc functionalities tailored for specific services that complement particular network functions. These enhancements can of course be exploited to achieve network slicing within the core network. Consider the following possibilities:

1. Only necessary network functions (NFs) will be instantiated within a slice, or only needed functionalities or software components of an NF (e.g. some NFs or some functionalities of an NF may not be needed for a particular slice). Each NF can have an amount of computational, storage, and networking resources adequate to meet the service level agreement (SLA) of the services associated with it. This would allow, for instance, a higher degree of redundancy for NFs belonging to slices with requirements for high reliability and availability.
2. NFs can be chained in the most convenient way, taking service requirements into consideration. NFs can be deployed in or moved to the most appropriate locations, considering the requirements of the services associated with them. For instance, using central clouds, edge clouds, or intermediate data centers/aggregation points allows deploying NFs according to different latency requirements, where granularity can be different for user- and control-plane NFs (e.g. control plane NFs may be located in central clouds, while user plane NFs may be located in edge clouds).

6.1.2 Network Slicing and V2X

Automotive applications embrace a large set of services. Given that these services impose a specific set of requirements, the assumption for V2X scenarios is that vehicles will be served by a composition of slices belonging to the standard types already defined by 3GPP (i.e. eMBB, URLLC, and mMTC). This is because, from a requirement perspective and considering the flexibility provided by the definition of slicing, it is not necessary to define new slice types. A possible example of a network slicing configuration for V2X communication is shown in Figure 6.1, along with the possible tenants of the different slices, taking into consideration that V2X services may be offered by various providers. The concept of multi-tenancy is notably leveraged for V2X scenarios, wherein the tenant is the subject (company, vertical, or service provider) that may benefit from the services supported by the relevant slice(s).

The communication infrastructure in Figure 6.1 is as follows. For RAN, the three possible options for slicing that were briefly introduced in Section 6.1.1 are shown. The core network is split into an edge cloud and a central cloud to reflect the geographical locations where network functions can be deployed. In particular, the edge cloud is in the proximity of UEs, while the central cloud is located in a more remote location where NFs common to all slices are located in addition to slice-specific NFs.

Three tenants are shown in Figure 6.1: (i) the MNO, referred to as the "operator" in the figure; (ii) the road operator; and (iii) a generic original equipment manufacturer (OEM). All slices[2] are deployed on the same physical infrastructure belonging to and being managed by the MNO. In the case study presented in Figure 6.1, the MNO manages the mobile broadband slice, which supports services such as infotainment.

2 For explicative purposes, in Figure 6.1, different slices are shown associated with different services. In practice, services might be gathered within one slice, i.e. a slice is not dedicated to one service.

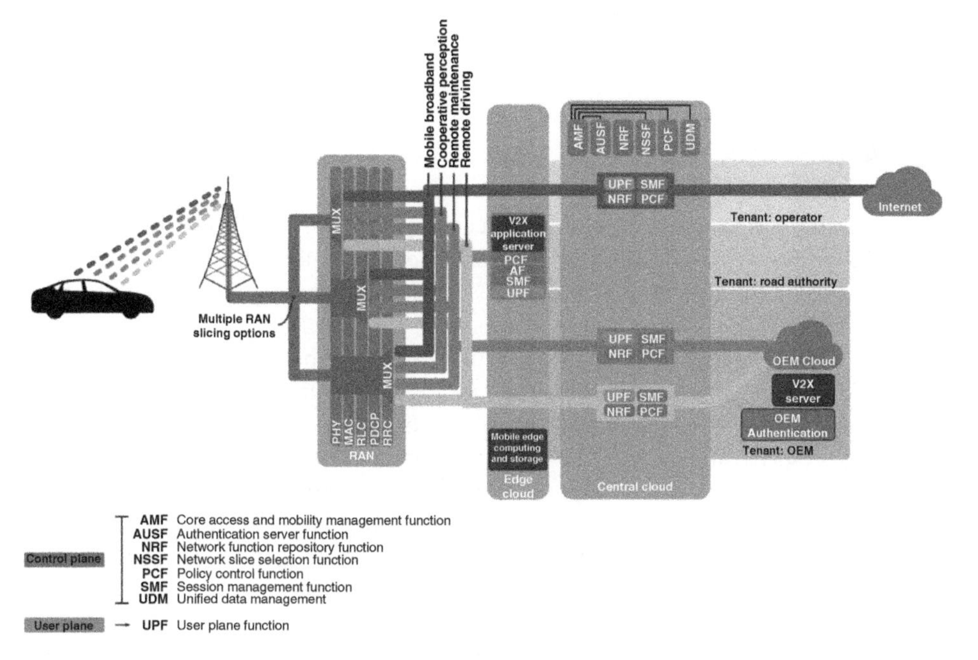

Figure 6.1 Possible example of network-slicing configuration for V2X communication.

The road operator is interested in cooperative intelligent transportation system (C-ITS) services, such as hazard warnings and in-vehicle signage. But in the future, road operators might be interested in offering more advanced services such as cooperative perception. In this case, messages are transmitted and received by vehicles to spread and acquire instantaneous information about the presence and behavior of other vehicles or road users in their proximity. For cooperative perception, information may then be considered time- and location-sensitive, which may require low latency and high reliability during transmission/delivery of messages. In our example shown in Figure 6.1, it is assumed that NFs are mostly allocated in the edge cloud, as close as possible to the final users. An alternative deployment might consider that sufficient transport network resources toward the central cloud are allocated for the slice, thus supporting the service by allocating NFs at the central cloud. In another example, the road authority might offer cooperative maneuver services, such as the lane-merge assistance from 5GCAR [4]. In this case, the service may benefit from an edge-cloud deployment of NFs.

Several services can be offered by the OEM. In Figure 6.1, two examples with different service-level requirements are shown: remote maintenance and remote (teleoperated) driving. For remote maintenance, an eMBB slice could be used to retrieve data from the on-board sensor to forward plan the maintenance of a large vehicular fleet. The remote driving service has more stringent requirements. Low latency, high data rates, and high reliability are required in the uplink to provide real-time video flow and/or instantaneous sensor data to remote drivers. Low latency and high reliability are required in the downlink to deliver driving commands to the vehicle. In Figure 6.1, NFs for these services are located at the operator's central cloud, but various purposes and degrees of redundancy can be used to guarantee meeting the different requirements of these two services. In both cases, however, different deployments can be considered, e.g. the OEM hosting the V2X application server on the company's premises.

6.2 Role of SDN and NFV in V2X

Network slicing was introduced in the previous section. To provide the flexibility requirements of the network slices, SDN and NFV technologies are needed. This section gives a brief introduction to these technologies and later describes the role of SDN and NFV in network slicing.

SDN is a consolidated network architecture paradigm that provides network programmability by separating the control plane logic from the data plane forwarding infrastructure. This decoupling provides novel benefits to network operators, such as saving on capital expenditures (CAPEX) by replacing dedicated hardware network equipment with software-driven network elements and operation expenditures (OPEX) through the introduction of faster new services. SDN has been accompanied by new open standard interfaces, which allows programming the forwarding logic in physical devices from remote, centralized entities generally defined as SDN controllers.

NFV displaces traditional network functions (based on dedicated hardware middleboxes) into software appliances that can be dynamically deployed, scaled, and migrated in and between commercial off the shelf (COTS) servers. This initiative benefits network operators by decreasing their OPEX and CAPEX. The NFV ETSI Industry Specification Group (ISG) is responsible for defining the necessary architecture and interfaces to fulfill this vision represented in [5], whose main architectural building blocks are as follows:

- *Virtualized network functions (VNFs) and element managers (EMs)*: VNFs are the logical entities representing the network functions being performed over a non-virtualized network. EM performs management over one or several VNFs.
- *NFV infrastructure (NFV-I)*: Hardware resources, including computing, storage, and network to provide the infrastructure for VNF.
- *NFV management and orchestration (MANO)*: Responsible for the allocation of requested network services (defined as an interconnected set of VNF) on top of a dedicated NFV-I. It is composed of (i) the virtualized infrastructure manager (VIM) that provides the functionalities for controlling the interaction of a VNF with infrastructure resources; (ii) VNF managers, responsible for the lifecycle management of each VNF; and (iii) the NFV orchestrator (NFV-O), responsible for orchestration and management to provide network services on NFV-I.
- *Operations support systems and business support systems (OSS/BSS)*: Represent the traditional network operator systems in charge of operations, administration, maintenance, and network monetization. The network service deployment is triggered by OSS/BSS.

ETSI NFV analyzed the network slice data model to provide a unified data model including 3GPP network slices and ETSI NFV network services. The ETSI report, published in [6], says that a network slice subnet can be an NFV network service. Moreover, 3GPP network functions can be described as VNF and physical network functions (PNFs). With regard to the NFV MANO framework, [6] proposes a new element in OSS/BSS to handle the network slice lifecycle. This element is connected through NFV standard interfaces to the NFV-O and EMs.

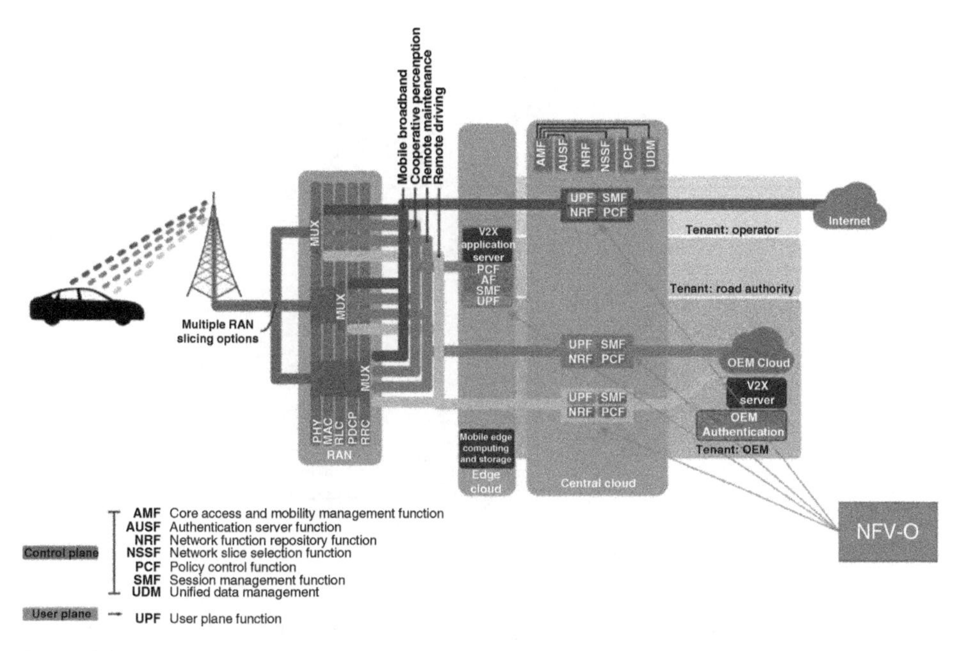

Figure 6.2 NFV and network slicing in a proposed V2X architecture.

The analysis described in [6] allows us to focus on how V2X network slices are deployed from the NFV orchestration perspective. Figure 6.2 shows the location of the NFV-O in the previously presented network slicing scenario. Note that a single NFV-O might be used, as it can handle multiple slice deployments as requested by OSS/BSS. Related topics are detailed later, such as VNF function placement (edge or core locations), integrating of multi-access edge computing (MEC), and mobility support for edge computing functions.

E2E connectivity is of extreme importance in network slicing. This is where SDN becomes essential. Figure 6.3 shows the SDN controller and how it controls network elements that interconnect network domains. While VNFs are deployed on top of NFV-I points of presence (NFV-I-PoP), network elements need to interconnect these NFV-I-PoP and provide network overlays for necessary traffic flows that are characterized to provide QoS.

Some of the challenges that SDN faces in V2X networks are the following:

- Necessary extensions for network traffic flow identification in deployed SDN switches and SDN controllers. For example, general packet radio services (GPRS) tunneling protocol (GTP) headers need to be identified.
- Introduction of network service header that can act as a QoS flow identifier [7].
- Extension of SDN controller hierarchy/peer for multi-domain provisioning.
- Mobility extensions for flow migration.

Finally, Figure 6.4 shows our proposed integrated NFV/SDN architecture for V2X network, with deployed orchestrator, NFV-O and manager, i.e. virtual network function manager (VNFM). The objective is to deploy network slices, which are tailored for specific QoS needs and use cases such as (i) mobile broadband, (ii) cooperative perception, (iii) remote maintenance, and (iv) remote driving.

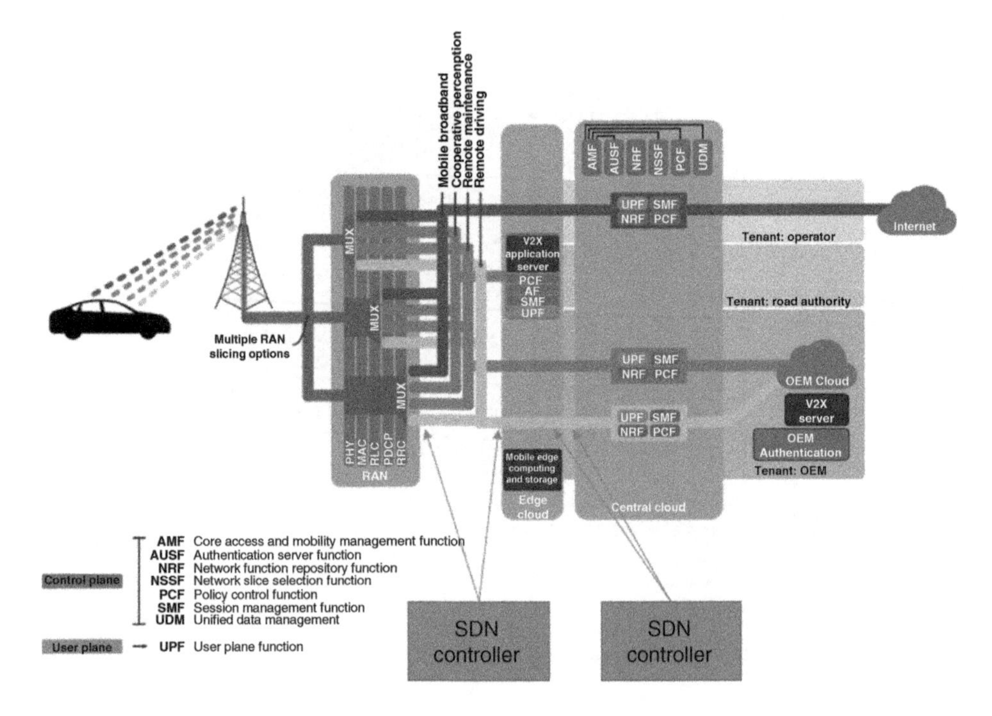

Figure 6.3 SDN in V2X architecture.

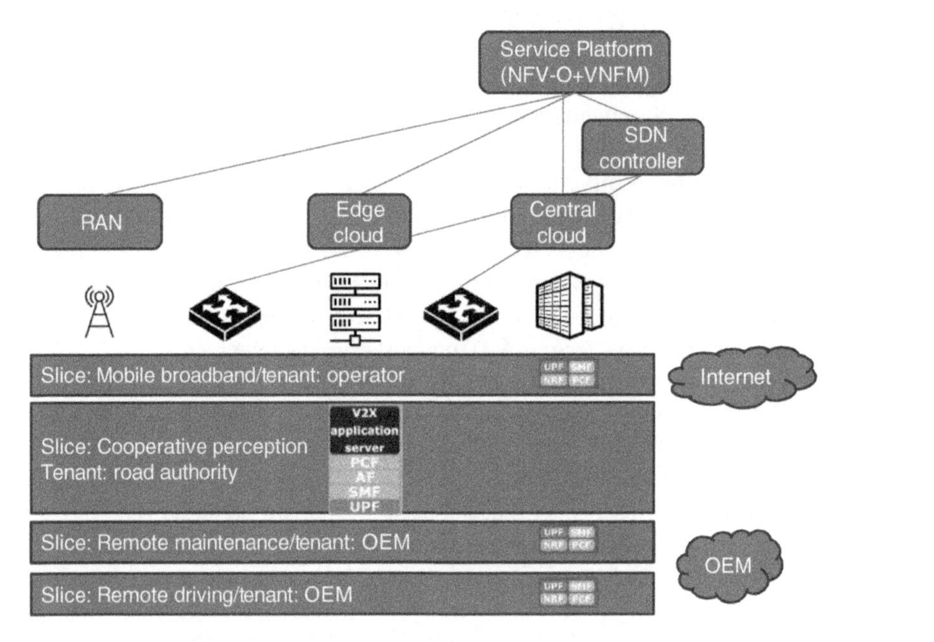

Figure 6.4 NFV/SDN integrated architecture.

NFV-O, as part of a global service platform, is responsible for the deployment of located network services over multiple NFV-I, which may be located at the network edge or core. Moreover, NFV-O is responsible for establishing the necessary traffic flows that conform to the network slice and interconnect the deployed network services.

6.3 Cloudified Architecture

A flexible network architecture is envisioned as one of the properties of 5G V2X networks, to enable seamless integrated connectivity for multi radio access technology (RAT) and multi-link operation where ultra-low latency and ultra-high reliability are supported for critical automotive communications. Such flexibility can be foreseen in the software-based network control [8], placement of network functionalities [9], and design of RAN [10], which then can be realized through network slicing, as detailed earlier in this chapter.

Flexibility in the RAN design varies in terms of how RAN functionalities are placed in the fog, i.e. splitting radio and baseband functionalities between central cloud and distributed entities. The emergence of an open RAN (ORAN) alliance and further study of how to offer such flexibility and possibilities for the fronthaul link [11] demonstrates that the industry needs to bring such flexibility into action.

3GPP has introduced eight options for splitting RAN functionalities [12]. Among these options, we examine latency and jitter with the split between packet data convergence protocol (PDCP) and RLC, between MAC and PHY, and intra-PHY on an experimental platform (further explained in [10] and [13]). A summary of pros and cons of each splitting point is detailed in Table 6.1. The study is performed with traffic models of three classes of service

Table 6.1 Pros and cons of different functionality splits.

	Pros	**Cons**
PDCP-RLC	- Hybrid automatic repeat request (HARQ) is in radio units (RUs), enabling fast retransmission - Fronthaul network can handle traffic from different bearers with different priorities. - Low overhead on the fronthaul.	- Only radio resource control (RRC) and packet data convergence protocol (PDCP) are centralized. - Fronthaul traffic grows with user-plane (UP) / control-plane (CP) traffic load for each bearer.
MAC-PHY	- Layer-3 and Layer-2 are centralized. - Multiplexes multiple bearers into one transmission block.	- HARQ in the central unit may be challenging to meet HARQ time requirements. - Overhead on the fronthaul depends on PDCP, RLC, and MAC headers and the size of resource blocks.
Intra-PHY	- Architecture is closed to full centralization. - Fronthaul load does not depend on the number of bearers and/or UEs.	- Fronthaul load increases with the bandwidth, number of sectors, and antennas. - Higher latency as the RU must receive all packets related to resource elements before starting the inverse fast Fourier transform (IFFT).

Source: [10].

in 5G, including URLLC, communicating CAD or CRU messages; mMTC traffic, representing vehicles' sensors messages; and eMBB traffic, representing infotainment traffic in the vehicle [14]. The ultimate aim of this study is to show which split performs best for which application classes in V2X, assuming a split can be achieved more dynamically and as part of network slicing as discussed in the previous sections.

6.4 Local End-to-End Path

In many V2X use cases (e.g. cooperative maneuvers, sensor information sharing, video sharing, intra-platoon communication), data traffic that is exchanged among vehicles, i.e. vehicle-to-vehicle (V2V), has localized significance. This means communicating vehicles that participate in the same use case are located in the same geographical region, and there is no need to access a remote server – e.g. V2X application server, intelligent transportation system (ITS) cloud server – although multiple transmission modes (unicast, broadcast, multicast) might be required. For localized V2X communications, either the cellular network (the Uu) interface or the shortrange cellular communication (the PC5) interface could be used, depending on the radio conditions and the environment where the V2V use case takes place. Specifically, the New Radio-Uu (NR-Uu) interface could provide guaranteed QoS (i.e. high reliability, low latency), in the case of, e.g. no line-of-sight among communicating vehicles, poor PC5 radio conditions, or high PC5 interference due to vehicles' high density.

Existing cellular solutions based on the Uu interface may need updates to more efficiently support the challenging performance requirements of localized CAD and CRU, including the need for fast and guaranteed transmission of localized data.

This study targets enhancements of network procedures through the formation of local E2E radio data paths over the Uu interface to enable the fast and guaranteed transmission of localized data traffic among involved devices, satisfying their QoS requirements and the features of the V2X communications, as initially presented in [15]. *End-to-end* means the (user plane) radio data paths are established among involved communicating end devices (i.e. vehicles), while *local* means paths are established via base stations (BSs). Instead of using solutions such as local UPFs that have a local UP end-point rather than reaching the UPF within the core, in this solution, core network nodes do not participate in user plane transmissions since data traffic is localized and handled directly among BSs. Local E2E paths via the BS can support various communication modes (unicast, multicast, broadcast) without the need to interact with other entities such as a multimedia broadcast multicast service (MBMS). Although in the proposed solution, data traffic does not pass through core network nodes, online and offline charging methods could be used, specifically with the support of the 5G system architecture's charging function (CHF). For instance, a session management function (SMF) used to establish local E2E paths could also trigger the appropriate event and support the collection of required charging information [16].

Localized communication through the Uu interface requires the introduction of a data routing/forward function at the BS that transmits data packets, e.g. among vehicles, in a fast and guaranteed way without traversing any core network entity (i.e. user plane). This routing table in the BS maps and connects the uplink (UL) and downlink (DL) radio bearers of different UEs to form local radio paths and, consequently, achieve faster forwarding of

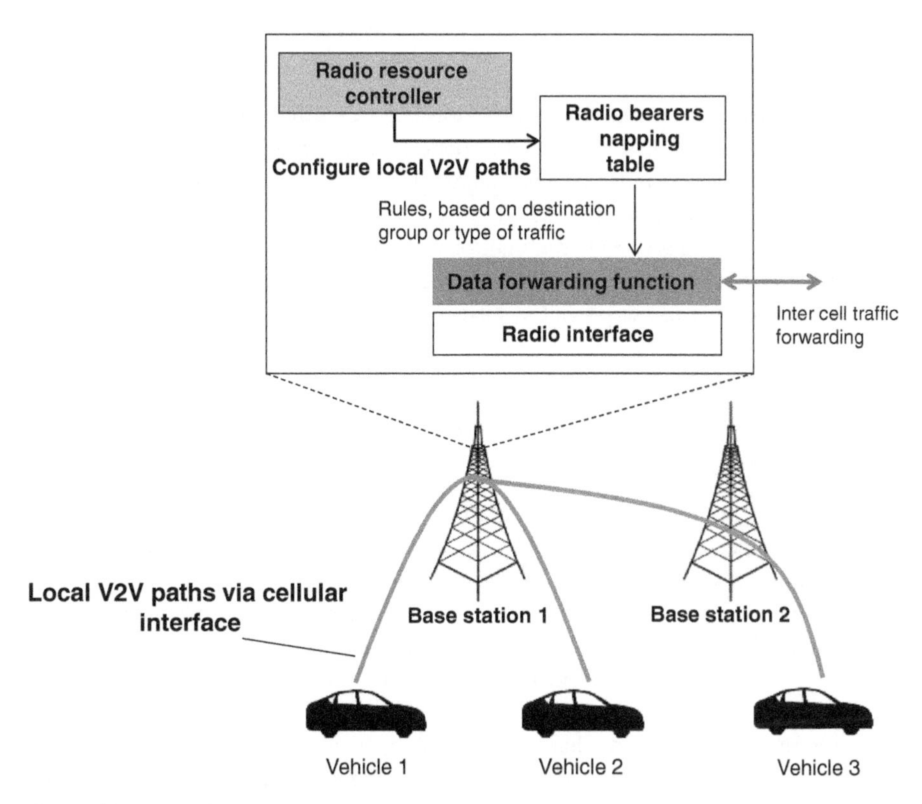

Figure 6.5 Fast V2V paths via a cellular interface.

localized V2X traffic (user plane latency reduction). Based on the type of traffic, the BS routing table forwards a data packet to one or more UEs in the same or neighboring cells (e.g. multicast, unicast transmissions). If UEs are attached to different cells, a possible solution that requires further investigation would use the Xn[3] interface for path establishment and data packet exchange. Figure 6.5 provides an overview of the involved entities and interfaces.

A UE requests the establishment (or update) of local cellular V2V paths using radio resource control (RRC), non-access stratum (NAS) protocols for localized V2X traffic and to transmit/receive data packets over a local E2E path. The type of service and the identifiers of other involved UEs in the corresponding V2V service are information that the initiating UE should provide and are used to establish paths as well as configure routing tables. RRC and NAS protocols need an extension to support the establishment, update, and release of local cellular V2V paths between UEs and BSs as well as to update and configure the routing table used for forwarding localized data traffic. Based on these RRC or NAS messages, core access and mobility management functions (AMF) and SMF functions can control the establishment, modification, and release of this new type of link (i.e. local cellular V2V paths) as well as update and configure the routing tables introduced at the BSs to form V2V paths for localized V2X traffic over the Uu interface.

3 Xn interface is a newly specified interface in 3GPP between RAN nodes in standalone operation.

6.5 Multi-Operator Support

Since the early stages of V2X communication, it has been identified that V2X communication needs to be de facto addressed for a multi-operator environment since the end devices cannot be assumed to operate under a single operator [17]. UEs crossing country borders are a similar problem that requires dealing with multiple operators; the European Automotive Telecom Alliance (EATA) has already highlighted this issue [18].

The 5G Automotive Association (5GAA) has come up with use cases related to vehicular networking [19]. Multiple use cases (UCs) have a global scope, implying that the UE is expected to

- Cross borders and thus switch operators
- Be present in areas where multiple operators offer coverage

Several of the UCs have stringent latency constraints [20], such as:

- Cross-traffic left-turn assist (high automation), where autonomous cars exchange planned, future trajectories with each other. Based on those, more accurate estimations of possible collisions are possible. The latency requirement is 10 ms.
- High definition (HD) sensor sharing, where a vehicle is expected to use its own sensors (e.g. HD camera, lidar) and sensor information from other vehicles to perceive its environment (e.g. come up with a 3D model of the world around it) and safely perform an automated driving lane change. The latency requirement is 10 ms.
- See-through for pass maneuver, where a driver of a vehicle that signals an intention to pass a remote vehicle (RV) uses a video stream (from the RV) showing the view in front of the RV. The latency requirement is 50 ms.
- Vulnerable road user (VRU) discovery, where a vehicle is informed about the presence of a vulnerable road user. The latency requirement is 100 ms (recommended 20 ms).

Solutions available in the literature can be based on using the already available Uu interface and existing data paths, where the data for two devices attached to different operators must cross all domains (i.e. access, backhaul, and core) of the two operators, incurring a delay that is not acceptable for most cases of V2X communication. On the other hand, using the sidelink (i.e. PC5) interface requires coordination among the operators to use spectrum resources for ensuring proper reception of the information – without necessarily ensuring high reliability if the communication takes place in an uncoordinated manner (Mode 4, PC5). Other solutions rely on schemes that facilitate using both interfaces but either require RAN sharing among operators or force devices to roam from one operator to another. The first approach, which is based on RAN sharing, requires agreements in all the coverage areas to ensure proper operation of the network; the other, based on force roaming, introduces unacceptable delays for the transition from one operator to the other because of the required attachment procedure. When edge computing is considered, the previous problem is even more complicated.

Multi-operator V2X communication using PC5 may happen using:

(1) Mode 1 and 3, by applying either shared carriers and time-division multiplexing (TDM) schemes, or separate carriers that do not face resource collision between different public land mobile networks (PLMNs) but requires carrier switching (with interruption and delay) or multiple receiver (Rx) chains (which increases the cost).

(2) Mode 2 and 4, by applying either shared carriers or separate ones, requiring carrier switching (with interruptions and delay) or multiple Rx chains (costly). Every time the spectrum configuration needs to be updated, the operators have to communicate the change to the UEs, directly or indirectly.
(3) These cases should also include PC5 in unlicensed or licensed spectrum (which would require agreements among MNOs for a shared carrier). The configuration should be static or update every time a reconfiguration occurs. Also, reconfigurations and carrier switching come with delays that may hinder meeting some requirements. And in the case of a shared carrier in licensed spectrum, agreements between operators are required.
(4) Multi-operator V2X communication using Uu may take place using typical Uu communication through normal UPF functions (i.e. home routing) or local breakout schemes to communicate with vehicles of the other operator. The latter approach requires implementations with local breakout dependent on the topology (e.g. BSs in the highway should have access), thus increasing complexity. Finally, edge cloud factors should be considered; for example, in the case of HD maps, vehicles associated with different edge clouds owned by different operators should communicate. This requires MECs to share the same schemes with local breakout solutions, described earlier, to reduce the time for sharing this information; or use alternative approaches (e.g. shared MEC functions). Additionally, multicast/groupcast requires duplication of messages in the two (or more) operators' spectrum.

Challenges related to increased delays or limited reliability can be solved if the multi-operator situation is reorganized to a single operator based on agreements among operators for a regional split. Splitting the overall area into regions that only one operator is responsible for simplifies the multi-operator environment and enables efficient V2X communication. Additionally, new ways to charge and to split costs and investments fairly or proportionally among the involved operators could be envisioned. PC5 communication is efficiently handled by one operator without requiring complex coordination among multiple operators. Also, edge cloud solutions do not require further enhancements since a single operator offers the service (Figure 6.6).

In general, such an approach would require the implementation of handover between the operators or would require the UE to disconnect from one operator and reconnect to another. The former solution is described in detail in [21] but is not generally implemented among operators. The second approach would lead to unacceptable issues.

To minimize transition time from one operator to another when crossing the boundaries of an area, devices may be registered in advance with all available operators. For example, in the case of a vehicle, while performing a typical attachment to its home public land mobile network (HPLMN), the subscriber server of the HPLMN can trigger attachment (i.e. registration of the vehicle) to all other available operators through the core network. When attached to all the networks, a device is considered "connected" to only one of them and in an "idle" state in the others. Which operator a UE should be connected to is dictated by the network by direct indication from the operator, via pre-configuration, or through tracking area updates (TAUs).

One key requirement of this solution is enabling vehicles (i.e. UEs) to transit smoothly from one operator's area to another. Thus, we may define transition areas. For a UE to be

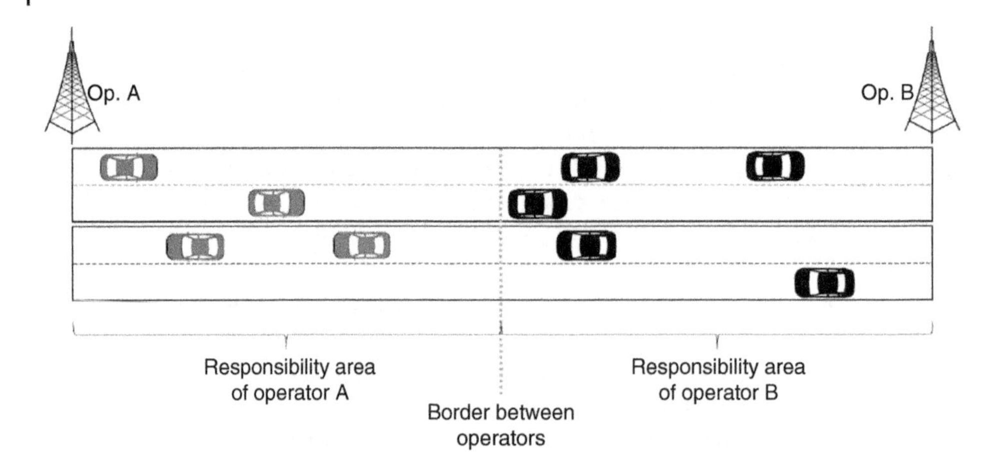

Figure 6.6 Regional split of a highway between operator A (Op. A) and operator B (Op. B).

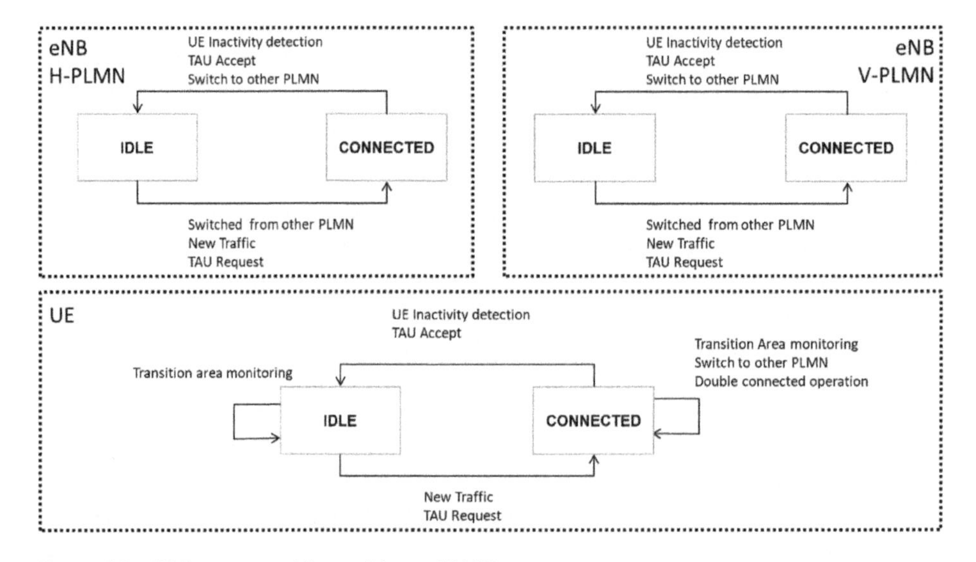

Figure 6.7 RRC state transitions with two PLMNs.

able to listen to messages (e.g. emergency notification messages), three options are possible (shown in Figure 6.7):

The UE is CONNECTED to both operators, listens to both operators' signaling channels, and can be scheduled to receive from and transmit to both of them. In this case, the UE has two independent reception chains, which increases the cost.

The UE is CONNECTED to one operator and IDLE with the other but listens to the downlink signaling channel with a higher frequency.

The UE is IDLE with both operators but listens to the downlink service requests of both operators with a higher frequency compared to standard operation.

If the UE participates in active communication, it should not switch operators directly because (i) doing so interrupts communication and (ii) it cannot be ensured that other UEs (i.e. vehicles) participating in the collaborative maneuver will change operators at the same time. Thus, coordination is required among the vehicles that participate in active communication before switching to a new operator.

6.6 Summary

This chapter has provided an overview of potential network enhancements that can facilitate or improve the delivery of CAD and CRU services. The solutions presented here expanded on infrastructure solutions such as slicing, SDN, NFV, and cloud RAN to bring more flexibility and agility to network design, to network procedures that create more optimal data paths, and, finally, to system-level solutions that support multi-operator connections during long-distance commutes. The network enhancements introduced in this chapter could be deployed individually or as a group; in most cases, the potential performance improvement does not depend on the deployment of other solutions.

References

1 5GCAR. (2018). Initial design of 5G V2X system level architecture and security framework. 5GCAR Deliverable 4.1.

2 Jiang, M., Condoluci, M., and Mahmoodi, T. (2017). Network slicing in 5G: An auction-based model. In: *Proceedings of IEEE International Conference on Communications (ICC).*

3 Condoluci, M. and Mahmoodi, T. (2018). Softwarization and virtualization in 5G mobile networks: benefits, trends and challenges. *Computer Networks* 146: 65–84.

4 5GCAR. (2019). The 5GCAR demonstrations. 5GCAR Deliverable 5.2.

5 ETSI. (2017). Network functions virtualisation (NFV). GS NFV 002, architectural framework, V1.1.1.

6 ETSI. (2017). Network Functions Virtualisation (NFV), Release 3. Evolution and ecosystem; report on network slicing support. GR NFV-EVE 012 V3.1.1.

7 Quinn, P., Pignataro, C., and Elzur, U. (2017). Network Service Header (NSH). IETF draft.

8 Mahmoodi, T. and Seetharaman, S. (2014). Traffic jam: Handling the increasing volume of mobile data traffic. *IEEE Vehicular Technology Magazine* 9 (3): 56–62.

9 Vizzareta, P., Condoluci, M., Machuca, C.M. et al. (2017). QoS-driven function placement reducing expenditures in NFV deployments. In: *Proceedings of IEEE International Conference on Communications (ICC).*

10 Mountaser, G., Condoluci, M., Mahmoodi, T. et al. (2017). Cloud-RAN in support of URLLC. In: *Proceedings of IEEE GLOBECOM.*

11 O-RAN Fronthaul Working Group. (2019). Control, user and synchronization plane specification. Technical Specification ORAN-WG4.

12 3GPP. (2016). Study on new radio access technology: Radio access architecture and interfaces. R3-161687, Draft TR 38.801.

13 Mountaser, G., Rosas, M.L., Mahmoodi, T., and Dohler, M. (2017). On the feasibility of MAC and PHY split in cloud RAN. In: *Proceedings of IEEE Wireless Communications and Networking Conference (WCNC).*

14 5G PPP. (2016). 5G PPP use cases and performance evaluation models.

15 Kousaridas, A. and Zhou, C. (2018). Local end-to-end paths for low latency vehicular communication. In: *Proceedings of IEEE 87th Vehicular Technology Conference (VTC Spring).*

16 3GPP. (2018). Telecommunication management, charging management, 5G data connectivity domain charging. TS 32.255, Release 15.

17 3GPP. (2016). Study on LTE-based V2X services. TS 36.885, Release 14.

18 Jonnaert, E. (2017). EATA - European Automotive and Telecom Alliance. Committee of the Regions – CORAI group.

19 5GAA. (2019). C-V2X use cases: Methodology, examples and service level requirements.

20 5GCAR. (2019). 5GCAR scenarios, use cases, requirements and KPIs. 5GCAR Deliverable 2.1.

21 3GPP. (2019). Procedures for the 5G System (5GS). TS23.502.

7

Enhancements to Support V2X Application Adaptations

Massimo Condoluci[1], YunXi Li[1], Laurent Mussot[2], Apostolos Kousaridas[3], Maliheh Mahlouji[4], and Toktam Mahmoodi[4]

[1] *Ericsson Research, Sweden*
[2] *Orange Labs Networks, France*
[3] *Huawei German Research Center, Germany*
[4] *King's College London, UK*

The vehicular ecosystem is composed of a plethora of applications with a wide range of use cases, including applications for safety, vehicle operations management, convenience, autonomous driving, platooning, traffic efficiency and environmental friendliness, society, and community [1]. This is reflected in a heterogeneous set of requirements, where some use cases require very high data rates or timely service delivery, ultra-low communication latencies, high reliability, or a combination of these. To support this heterogeneity, radio and core network enhancements have been introduced to mobile networks to support tight requirements, quality of service (QoS) frameworks have been developed to support traffic management, and network slicing has been designed to support the simultaneous handling of different verticals [1]. However, in addition to the complexity of handling traffic with many different demands, vehicular applications are expected to run connected automated driving (CAD) and connected road user (CRU) services in multifaceted scenarios. This complexity is due to the fact that vehicles communicate among themselves, either directly via a short-range (i.e. sidelink a.k.a. PC5) or long-range (i.e. Uu) link or via a combination of the two, as well as with road infrastructure, road users, edge and remote clouds, etc. As a consequence, the application-design perspective becomes more involved, requiring vehicle-to-everything (V2X) applications – i.e. CAD and CRU services – to be able to adapt to various deployment or network conditions without impacting service availability. For instance, a CAD/CRU service may run with a different configuration depending on the area: for instance, because of different network capabilities in the areas where the service is running. Another example is a CAD/CRU service that adapts its configuration while running to react to network changes while a vehicle is moving, which may involve changing service settings (e.g. change to a different set of active sensors, change of transmission bitrate) as well as changing the network technologies in use (e.g. switching between short- and long-range or using them simultaneously).

While designing mechanisms to support the adaptation of V2X applications is up to vehicle manufactures, an important consideration is how to enhance network capabilities to

Cellular V2X for Connected Automated Driving, First Edition.
Edited by Mikael Fallgren, Markus Dillinger, Toktam Mahmoodi, and Tommy Svensson.

enable and support application adaptation. This chapter discusses two primary facets of supporting an enhanced application adaptation. The first is network enhancements: in particular, enhancing the interaction between network and application. The aim is to enhance network knowledge about V2X service needs as well as enhance application knowledge about network capabilities that target V2X services. Second, the chapter discusses how to apply these enhanced interactions to V2X use cases. It focuses on the application design of options for achieving joint use of short- and long-range (sidelink or PC5 and Uu, respectively) links to help applications adapt to varying network conditions and deployment environments. In this case, solutions can be adopted at either the application level or network layers. Note that the focus of this chapter is on cellular V2X (C-V2X), while some of the technical solutions have a broader applicability.

The remainder of this chapter is structured as follows. Section 7.1 provides an overview of 3GPP mechanisms for supporting application adaptation and discusses some approaches to application adaptation presented in the literature. Section 7.2 focuses on enhancements to application and network interaction, and Section 7.3 presents and analyzes several options for joint use of short- and long-range links. Final remarks are provided in Section 7.4.

7.1 Background

The design of fifth generation (5G) systems by the 3rd Generation Partnership Project (3GPP) includes features that enhance networks by supporting more effective application adaptation. One of the new features is related to the definition of a network function: the application function (AF), which allows an application server to interact with the 3GPP core network [2]. The interaction happens via a network exposure function (NEF), which is in charge of translating network information before exposing it to relevant AFs. With this framework, the 5G system can expose capabilities and events to AFs, then enhancing the capabilities of the application to adapt to network capabilities as well as influence network behavior. For instance, as discussed in [3], the AF can request that a data session to user equipment (UE) be set up with a specific QoS (e.g. low latency or jitter) and priority handling; the AF may also change the QoS by providing a different QoS reference parameter while the session is ongoing. If an AF session can adjust to different QoS parameter combinations, the AF may provide alternative service requirements containing one or more QoS reference parameters in a prioritized order (which indicates a preference for the QoS requirements with which the service can operate). The first QoS reference parameter in the alternative service requirements reflects the QoS requirement of the highest priority for the AF session and contains the same QoS reference parameter provided with the service information. This parameter provided by the AF is used by the network when generating notifications about QoS fulfillment. To accomplish this, 3GPP in [2] defined the notification control procedure, which is one of the QoS parameters associated with a QoS flow and which, if marked, indicates that notifications are requested from the new-generation radio access network (NG-RAN) when the guaranteed flow bit rate (GFBR) can no longer (or can again) be guaranteed for a QoS flow during the lifetime of the QoS flow. Notification control can be used for a guaranteed bit rate (GBR) QoS flow if the application traffic can adapt to the change in the QoS (e.g. if the AF can trigger rate adaptation). If, for a given GBR QoS flow, notification control is enabled and the NG-RAN determines that the GFBR can no longer be guaranteed, NG-RAN sends a notification and may provide values for the GFBR, the packet delay budget (PDB), and the packet error rate (PER): parameters it can guarantee. If the NG-RAN has received alternative QoS profile(s) for this QoS Flow and supports the alternative QoS profile handling, the NG-RAN further checks whether the values of the GFBR, PDB, and PER parameters that the NG-RAN can guarantee match any of the alternative QoS profile(s); if there is a match, the NG-RAN indicates the reference to the alternative QoS profile(s) together with the notification to the core network. This notification can be exposed by NFs to the AF.

3GPP defines in [4] the capability for the AF to influence traffic routing, where an AF may request to influence network routing decisions for user-plane traffic of protocol data unit (PDU) sessions. Another feature defined by 3GPP is the procedure for future background data-transfer negotiation. The AF can negotiate with the network regarding transfer policies for future background data transfers: i.e. a desired time window for the background data transfer, a reference to a charging rate for the time window, network area information, and optionally a maximum aggregated bitrate.

These enhancements allow easier interaction between the network and applications compared to previous systems. Nevertheless, the interaction focuses primarily on QoS-related features, which, although undoubtedly important, represent only one aspect of possible application adaptations for V2X services.

Another essential aspect of supporting application adaptation is adjusting to network capabilities and their variations while a CAD/CRU service is running. In V2X scenarios, moving vehicles may experience different channel conditions, network loads, etc., that may impact achievable performance. Application adaptation has primarily been studied with a focus on rate adaptation, which can be achieved in several ways. The first approach relies on using end-to-end rate-adaptation mechanisms, e.g. [5, 6]. These mechanisms are based on measurements performed on both the sender and receiver sides. In particular, end-to-end adaptation is performed on the transmitter side, and explicit feedback is transmitted by the receiver. Also, 3GPP has focused on rate adaptation with network support [7]. Focusing on continuous media delivery, 3GPP defined the progressive download and dynamic adaptive streaming over hypertext transfer protocol (3GP-DASH) framework. Two components are defined within this framework: DASH clients, which can be on both the back-end and device sides; and DASH-aware network elements (DANEs) on the network side. DASH clients and DANEs exchange messages to improve the efficiency of streaming sessions. These messages include information about real-time operational characteristics of networks, servers, proxies, and caches as well as DASH client performance and status. Although rate adaptation is an important part of enabling applications to react to changes in network performance, it may be too limited for V2X use cases. Additional aspects of application adaptation may also be relevant for V2X application adaptation.

Scheduling is another area of research for supporting adaptation for ongoing CAD/CRU services. For V2X scenarios, the focus is primarily on designing enhanced algorithms to meet the target service requirements in mobility scenarios. A detailed list of research on scheduling for V2X scenarios can be found in [8]. The assumption in most of this literature is that data to be delivered to vehicles will be used immediately after reception. In some cases, e.g. high definition (HD) map dissemination, data to be delivered to a vehicle is available before the vehicle approaches the target area. This means that, potentially, the data transfer can be "relaxed" to avoid network congestion. A solution developed for "low-priority" data transfer is based on the idea of adapting the file-delivery window to reduce congestion. In this domain, low extra delay background transport (LEDBAT) [9] represents one of the established solutions introduced to provide "less than best effort" data transfer for applications without strict constraints on data-delivery deadlines or throughput. Approaches like LEDBAT are effective in guaranteeing low-priority data transfer without significantly impacting network load, potentially enable a cost reduction for data transfers, and can be successfully applied to use cases such as software updates. Nevertheless, their use in case of service requirements with relaxed transmission windows but with spatial/time deadlines may be challenging. This is of primary importance for some V2X services (e.g. HD map updates) where data content should be delivered before the vehicle enters the geographical area the information refers to.

Another aspect of application adaptation is related to enabling joint use of short- and long-range links, thus allowing an application to adapt to the most suitable interface (or to simultaneous use) according to e.g. specific service requirements or to specific conditions. Within the communication protocol stack, several options are available for implementing this joint use. These options are described in Section 7.3, along with the associated pros and cons.

7.2 Enhanced Application-Network Interaction for Handling V2X Use Cases

As discussed earlier, the interaction between network and application has been primarily conceived in terms of supporting enhanced rate-adaptation capabilities and other QoS-related aspects. The availability of information about features other than rate adaptation may enable novel possibilities, supporting not only application adaptation but also network adaptation. An enhanced interaction between application and network that extends the legacy QoS-related interaction is a key enhancement to support V2X scenarios, especially in use cases with stringent requirements or with particular constraints in addition to QoS ones. The remainder of this section focuses on possible approaches for enhancing application-network interaction and discusses further improvements based on an extended exchange of information between network and application.

7.2.1 C-V2X Connectivity Negotiation

One aspect to consider when enhancing interaction between application and the core functions of a mobile network is that information exchange usually focuses on QoS-related information, which is essential but may be of limited scope in V2X scenarios. Other parameters may also become relevant to negotiate optimal C-V2X connectivity. For example, spatial/time information associated with e.g. a CAD/CRU service is very relevant, given the mobility involved in V2X use cases (e.g. it is important that information refers to specific areas or time intervals). Examples of such information are the location where a CAD/CRU service is running and how long it is expected to run, as well as the receiver's current location, speed, intended trajectory, etc. This enhanced data can be taken into consideration by mobile network for optimizing C-V2X connectivity. On the application side,

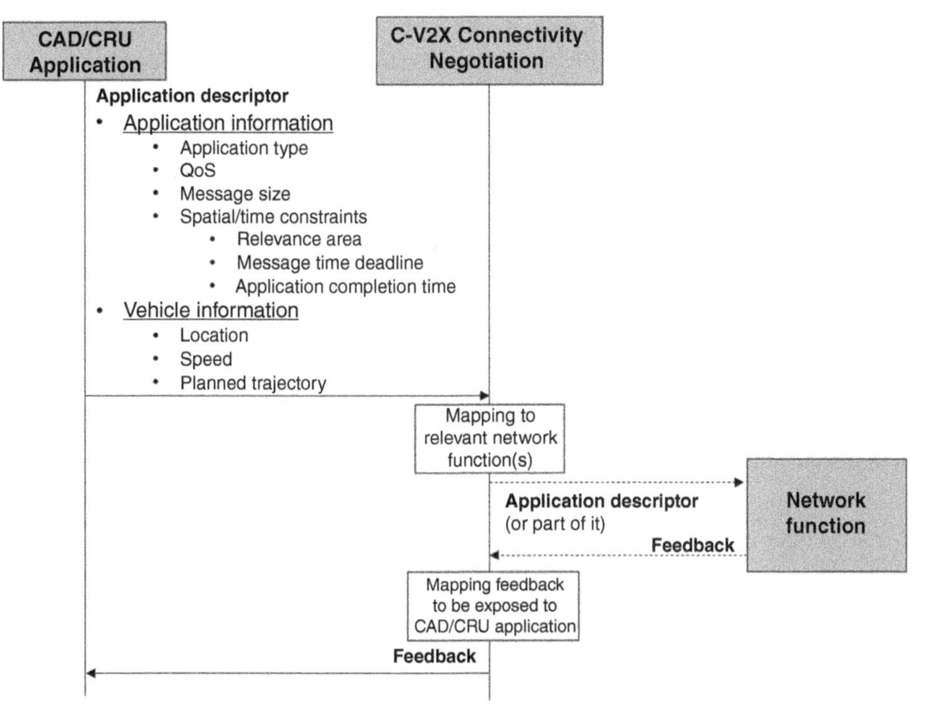

Figure 7.1 Flow diagram and information exchanged for C-V2X connectivity negotiation.

beneficial information includes expected achievable performance in a particular area/time or network capabilities in transferring a given message by a specified deadline. Such information can then be used, e.g. by a CAD service to adapt the vehicle's driving status (such as speed, route, level of automation) as a result of its increased network capabilities awareness.

C-V2X connectivity negotiation enables such interactions enhancing the current QoS-based negotiation procedures. In addition, C-V2X connectivity negotiation facilitates the introduction of V2X-specific features within the network by gathering service-specific information and providing it to other network functions, which can then use it for service-specific enhancements.

Figure 7.1 shows a flow diagram and information exchanged for C-V2X connectivity negotiation. When mapped to a 5G system, a CAD/CRU service can be a CAD/CRU application connected to an AF to interact with a 5G core network. C-V2X connectivity negotiation can be considered as an enhanced network functionality for existing NFs such as a NEF or policy control function (PCF). The CAD/CRU application provides a descriptor containing two categories of information:

- Application information:
  - Application type (e.g. video, voice, remote control).
  - QoS information. This may be in the form of service requirements that are then mapped to a QoS profile. It may indicate the requested QoS from the application; or it can be a request asking to receive information about the anticipated QoS achievable over the mobile network.

  - Message size. This may be relevant for file-transfer (such as HD map acquisition and software updates) and can indicate the overall or partial size of the message to be transferred.
  - Spatial constraints. A relevance area can provide a spatial deadline for a message, i.e. the message should be transferred before the vehicle leaves/enters the relevance area. This information may also tell where the CAD/CRU application is being used (for use cases with confined areas, such as lane merges).
  - Time constraints. These include e.g. message-transfer deadlines and length of application duration (how long the application is expected to run).
- Vehicle information:
  - Location (e.g. current location of the vehicle) and/or planned trajectory (e.g. waypoints with associated timestamps).
  - Speed (e.g. average speed, or speed may be obtained based on timestamps and waypoints included in the planned trajectory).

The C-V2X connectivity negotiation functionality in the mobile core network processes the information included in the application descriptor to understand whether it can provide feedback directly to the V2X service or if other NFs should be contacted. In this case, the C-V2X connectivity negotiation functionality maps and eventually translates the application descriptor to the relevant NF (or NFs), which process it and then provide feedback to the C-V2X connectivity negotiation functionality, which is finally provided to the CAD/CRU application. The content of the feedback may vary depending on the application descriptor. Examples of feedback include (i) achievable performance expected to be supported in a particular location; (ii) whether the message is expected to be transferred by the associated deadline; (iii) whether the desired QoS is expected to be fulfilled for the entire application duration time indicated in the descriptor.

An example of C-V2X connectivity negotiation applied to a V2X use case could be a file transfer to update an HD map. In this case, the descriptor indicates the amount of data to be transferred to a vehicle together with a message-transfer deadline and the vehicle's expected trajectory. The feedback provided by the C-V2X connectivity negotiation may indicate to the application whether the message-transfer deadline can be met and whether to start/pause the delivery to support efficient file delivery.

7.2.2 Use-Case-Aware Multi-RAT Multi-Link Connectivity

To manage traffic flows and guarantee QoS for various V2X use cases, one possible solution is deploying multiple radio access technology (multi-RAT) multi-link connectivity. Enhancement of application-network interactions seems to be a prerequisite for enabling multi-RAT multi-link connectivity; however, this solution would provide much more flexibility and efficiency to support diverse V2X use cases.

Multi-RAT multi-link connectivity is an extended version of the dual connectivity solution defined in earlier standards, such as 3GPP Release 12. *Dual connectivity* refers to the simultaneous use of radio resources provided by two eNBs connected via an X2 interface. However, multi-RAT multi-link connectivity allows simultaneous use of radio resources of two or more eNBs by one or more RATs, which allows smarter use of resources to enable demanding V2X use cases.

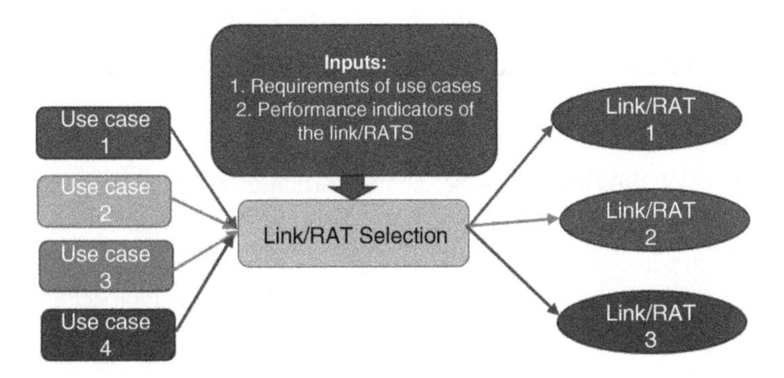

Figure 7.2 Multi-link/RAT selection problem.

The link/RAT-selection mechanism can be designed as a cell-association problem, which leads to a static solution; or it can be designed as a use-case-aware selection, which leads to a dynamic solution. Because of high mobility and the dynamic nature of V2X traffic, as well as the diverse use cases with different QoS requirements, the use-case-aware selection approach seems to be a more reasonable idea. The problem that needs to be solved dynamically is mapping traffic to the best links/RATs (Figure 7.2).

Since V2X use cases and their QoS requirements vary, the combination of RATs and/or links that guarantees the requirements of those use cases may also be diverse (Figure 7.2). For instance, while the class of cooperative safety use cases is highly delay-sensitive, and persistent coverage is crucial in cooperative manoeuvers, cooperative navigation mostly requires larger bandwidth from the network.

The simultaneous use of all available links/RATs provides new multiplexing and diversity sources (extra degrees of freedom) and mitigates the fading effect of wireless channels. The link/RAT multiplexing gain is achieved by using different links/RATs for concurrent transmission of packets over various available resource blocks to improve the data rate. For example, the link/RAT multiplexing could be suitable for use cases such as cooperative navigation. On the other hand, diversity gain is achieved either by selecting the best link/RAT or by transmitting the same data over different available links/RATs, which improves reliability.

Finding the best combination of available links/RATs is challenging, given vehicles' mobility and highly dynamic network congestion. The link/RAT selection is determined based on many conditions, the most important of which are satisfying QoS and improving spectrum use. However, RAT-selection approaches can vary, such as network-driven, policy-based, device-driven, etc. Here we briefly describe those approaches:

Network driven: When a session is established, or during an ongoing session, the network decides which link/RAT to use.

- o Pros: Efficient use of links/RATs, taking into account the dynamic behavior of traffic and the network
- o Cons: Overhead/delay in session establishment

Policy-based: During an ongoing session, the network designs policies and sends them to UEs, and then the UE uses them to control the traffic.

- Pros: Lower overhead/delay compared to the network-driven approach. Dynamic traffic behavior is reflected and handled trough thresholds with a network-assisted approach.
- Cons: Lower efficiency since policies cannot directly take into consideration the instantaneous status of the network.

User-driven: UEs decide which link/RAT to use based on information received from the network (e.g. status of available links/RATs).

- Pros: Lower latency/overhead compared to the network-driven approach. Limited dynamic traffic behavior compared to the policy-based and network-driven approaches.
- Cons: Lower efficiency since UEs cannot directly take into consideration the instantaneous status of the network.

7.2.3 Location-Aware Scheduling

Location-aware scheduling is a functionality that could be used to support the transfer of messages with relaxed transfer windows but still with spatial/time deadlines for completing the transfer. The idea is that a mobile network considers various pieces of information associated with a data transfer:

Message information, e.g. message size.

Spatial information, e.g. relevance area, indicating whether the message transfer should be completed before the vehicle enters (or leaves) the relevance area (this can also be provided in the form of a time deadline).

Vehicle information: location or speed (or planned trajectory, e.g. in the form of waypoints with associated timestamps).

This information is used by location-aware scheduling to optimize message delivery (e.g. dynamic QoS treatment), under the constraint that message delivery can be completed before the vehicle reaches the geographical or time deadline. Given these inputs, location-aware scheduling provides the following as output:

Scheduling information, such as transfer planning, etc.

QoS requirements, such as a QoS profile, etc. This information is tailored to offer a low-priority transfer within the deadline. The outputs can be either used by the mobile network or provided to the application. In the first case, network entities such as RAN or user-plane nodes can use information about transfer planning to indicate the expected traffic load and information about associated QoS requirements to enforce the correct QoS treatment for data transfers. For instance, the priority of data transfers for vehicles that are closer to the relevance area can be increased to guarantee timely transfer completions. If location-aware scheduling delivers information about transfer planning to the application, then the application knows the best time to trigger the message transfer.

In the remainder of this section, we introduce an example to further explore the location-aware scheduling functionality. This example (see Figure 7.3) involves an HD map update use case where different messages, i.e. layers 1–2 and layer 3, should be delivered to a certain vehicle. These messages have associated sizes and areas of relevance and should be received before entering those relevance areas. Figure 7.3 shows the joint use of C-V2X connectivity negotiation and location-aware scheduling. The CAD/CRU application provides C-V2X connectivity negotiation functionality with an application descriptor, which includes message information (relevance area and message size) for two messages (layer

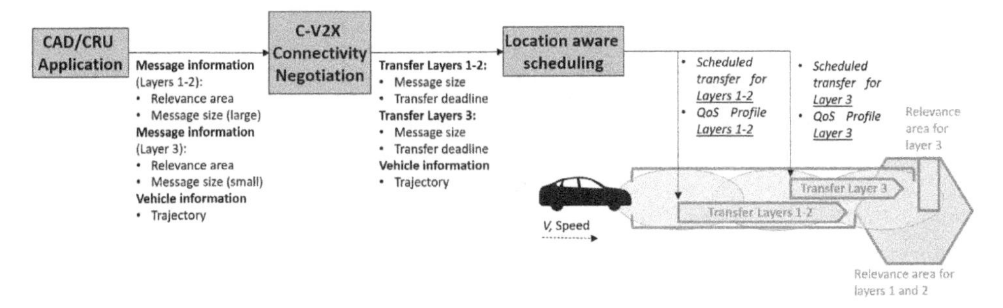

Figure 7.3 Example of joint use of C-V2X connectivity negotiation and location-aware scheduling applied to an HD map update.

1, 2 and layer 3) to be transferred, along with vehicle information. The C-V2X connectivity negotiation functionality processes the received application descriptor and computes the transfer deadline for each message to be transferred. The information about the messages to be transferred (size and deadline) together with the vehicle information are provided by the C-V2X connectivity negotiation functionality to the location-aware scheduling. The location-aware scheduling then takes into account the information received to run the schedule planning for the application. In this example, the location-aware scheduling sees that the sizes of layers 1 and 2 are larger than that of layer 3 and also that, from a geographical perspective, the area of relevance for layers 1 and 2 is closer to the vehicle than that for layer 3. Based on this information, the location-aware scheduling schedules the transfer of layers 1 and 2 in a time slot before the transfer of layer 3. From a network perspective, the load can be kept low because the message transfers are spread into two different time windows. From an application perspective, the application is guaranteed that information is received before the vehicle approaches the relevant geographical area.

7.3 Redundant Scheduler for Sidelink and Uu

One of the main challenges raised by some V2X use case requirements and safety-critical automotive applications is achieving high communication reliability (up to 99.999%) while fulfilling a low-latency budget on the order of few milliseconds. To increase reliability, a common approach is to apply retransmissions whenever the receiver cannot retrieve a message successfully. Various techniques could be applied at the radio link level to improve the communication link. However, reliability will eventually decrease under adverse radio conditions, leading inevitably to the case where the receiver is out of coverage of the corresponding radio technology.

Generally speaking, transmitting the same message over a different path is one way to overcome the limitations of a single radio link at any given time. C-V2X, by definition, offers two possible paths: over the PC5 interface for sidelink and direct communications and through the cellular network infrastructure over the Uu radio interface. Those interfaces are linked to different radio frequency bands: long term evolution (LTE) Band 47 (ITS 5.9 GHz band) for sidelink communications, and several LTE bands or new radio (NR) band options for V2X communications over the Uu interface matching the 3GPP standards. This

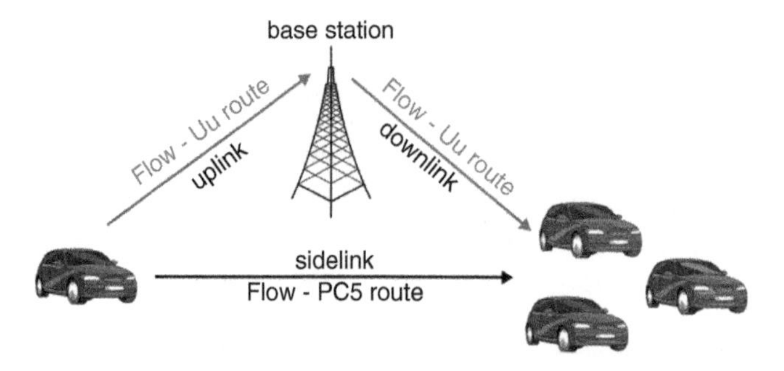

Figure 7.4 Multicast V2V connectivity over PC5 and Uu routes concurrently.

heterogeneity could lead to various situations in terms of radio and channel load conditions for both interfaces, as well as different propagation signal behaviors and communication paths. Designing a multipath protocol or, more specifically, a redundant scheduler using both interfaces concurrently is the primary idea investigated in this chapter. Previous studies have clearly shown that better reliability can be achieved when Uu and PC5 links are used simultaneously [10]. As illustrated in Figure 7.4, data flows are duplicated and transmitted on both interfaces with an example of a vehicle-to-vehicle (V2V) connectivity in a multicast and broadcast fashion.

In the following subsections several options envisioned to design a redundant scheduler are described, depending primarily on the target protocol layer of the OSI model or following the European Telecommunications Standards Institute (ETSI) intelligent transportation systems (ITS) architecture example. Note that other implementation options could be considered.

7.3.1 Application or Facilities Layer

A redundant scheduler can be implemented at the application layer or facilities layer when the ETSI ITS architecture [11] is used. A similar approach should be applied when other ITS standards are examined, such as e.g. Institute of Electrical and Electronics Engineers (IEEE) wireless access in vehicular environments (WAVE).

This option should be the easiest to implement and, by design, limits redundancy to a specific subset of applications. Furthermore, the duplication is RAT-agnostic, meaning flows can be routed over any available radio interface. However, it requires all UEs involved in the communication to implement the ETSI ITS stack the same way or use the same application, and the receiving UEs must be capable of receiving on both interfaces. Being located in a higher layer of the stack, this solution enables no interaction with the lowest layers, and this might limit the possibility of further optimizations. For example, change of routing due to a RAT failure would be required to be handled at a higher layer of the stack with potential consequences in terms of reactiveness (e.g. delay to discover failure).

An example of an implementation following the ETSI ITS architecture is shown in Figure 7.5.

A message duplicated at the facilities layer, such as a decentralized environmental notification message (DENM) [12], needs to use two different stacks, eventually relying on different transport and network protocols.

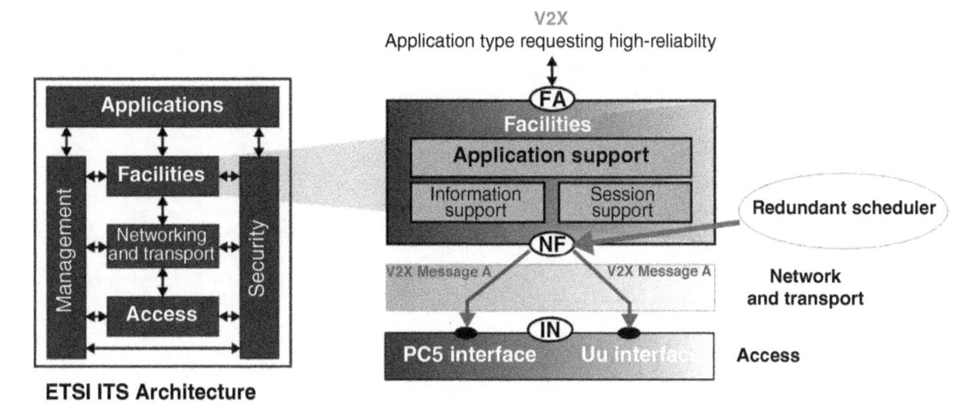

Figure 7.5 Redundant scheduler applied at the facilities layer within the ETSI ITS architecture.

For instance, some road safety and traffic efficiency ITS applications could be designed to use non-Internet Protocol (non-IP) protocols and rely primarily on short-range communication technologies. For example, within the ETSI ITS architecture, a typical use of basic safety applications is to encapsulate V2X messages with the Bidirectional Transport Protocol (BTP) transport protocol [13], or the Geonetworking Network Protocol [14]; or to apply a non-IP type header with the matching V2X message family corresponding to the protocol (value set to 3 for ETSI-ITS), as specified in [15, 16]; and then connect with the PC5 interface directly configured to transport non-IP SDU types, as illustrated in Figure 7.6.

Alternatively, the same messages could be carried out over the Uu interface with a different networking and transport stack, following Figure 7.7.

However, C-ITS security and certificate policies are currently based on existing ETSI ITS standards requiring the Geonetworking layer, which requires including the entire stack up to the Geonetworking level (V2X message, BTP, Geonetworking) before carrying the

V2X Message (CAM,DENM,..)
BTP
Geonetworking
Non-IP Header (V2X message family)
PDCP (Non-IP SDU type)
RLC
MAC
PHY

Figure 7.6 Typical ETSI ITS protocol stack over the PC5 interface.

V2X Message
(CAM,DENM,..)
UDP/TCP
IP
PDCP
RLC
MAC
PHY

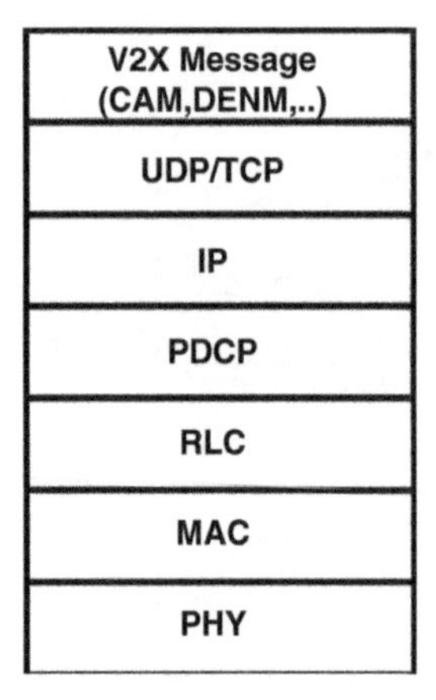

Figure 7.7 Typical ETSI ITS protocol stack over the Uu interface.

V2X Message
(CAM,DENM,..)
BTP
Geonetworking
UDP
IP
PDCP
RLC
MAC
PHY
V2X Message
(CAM,DENM,..)
BTP
Geonetworking
TCP
IP
PDCP
RLC
MAC
PHY
V2X Message
(CAM,DENM,..)
BTP
Geonetworking
MQTT
TCP
IP
PDCP
RLC
MAC
PHY

Figure 7.8 Example of the ETSI ITS protocol stack supporting the Uu-CSP over the Uu interface.

payload over a UDP or TCP flow, depending on the implementation choice. This is also pointed out and pushed by 5GAA, which specifies a communication system profile for C-ITS implementation using the long-range cellular Uu interface (referred to as Uu-CSP); see Figure 7.8. The main scope of Uu-CSP is to harmonize the higher-layer profile to ensure the interoperability of C-ITS services with different communication technologies, e.g. short-range and long-range communications.

There is still an ongoing study item in ETSI-ITS Working Group 5 (WG5) [17] evaluating message signing at the facilities layer to overcome this limitation. Generality speaking, if message duplication is managed at higher layers, how to deal efficiently with all possible different protocol stacks must be considered.

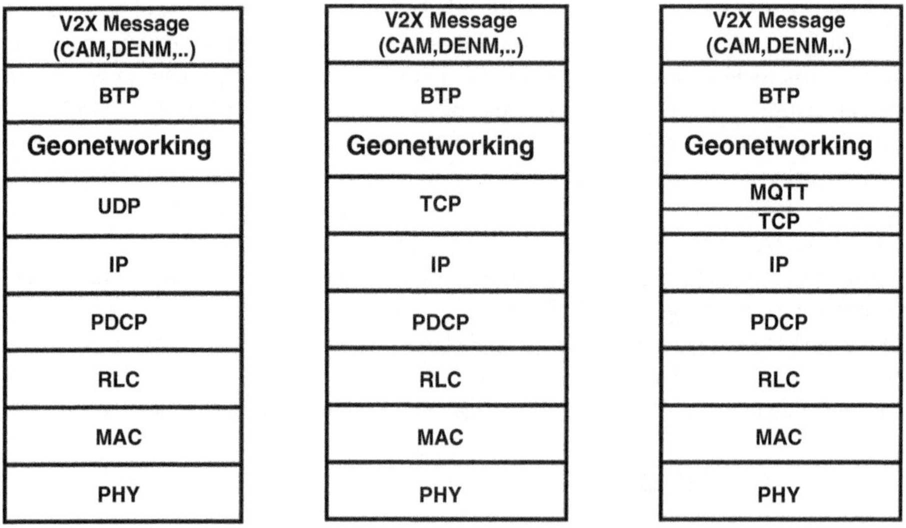

7.3.2 Transport Level

Another approach for delivering the same flows over Uu and PC5 interfaces relies on an appropriate multipath protocol with a redundant scheduling policy at the transport layer.

When looking at Transmission Control Protocol (TCP)/IP models, multipath protocols with eventual redundancy capabilities have already been proposed, such as multipath TCP [18] and multipath extensions for quick UDP internet connections (QUIC) based on User Datagram Protocol (UDP) [19]. For this reason, unlike the facilities layer implementation, other non-vehicular applications running on the same UE can benefit from the scheduler if needed. Figure 7.9 illustrates such an implementation option.

On the flip side, multipath transport protocols are currently only designed for IP traffic, effectively excluding non-IP flows, which are commonly used in ITS based on short-range communications. Figure 7.10 represents the case of an ETSI ITS architecture when a redundant scheduler is applied at the UDP/TCP level and therefore applies only to IP flows.

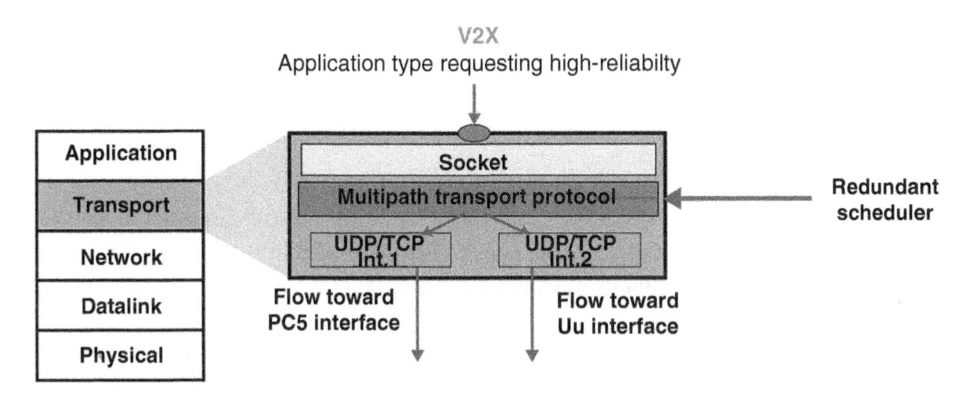

Figure 7.9 Multipath Transport protocol over the Uu and PC5 interfaces in IP mode.

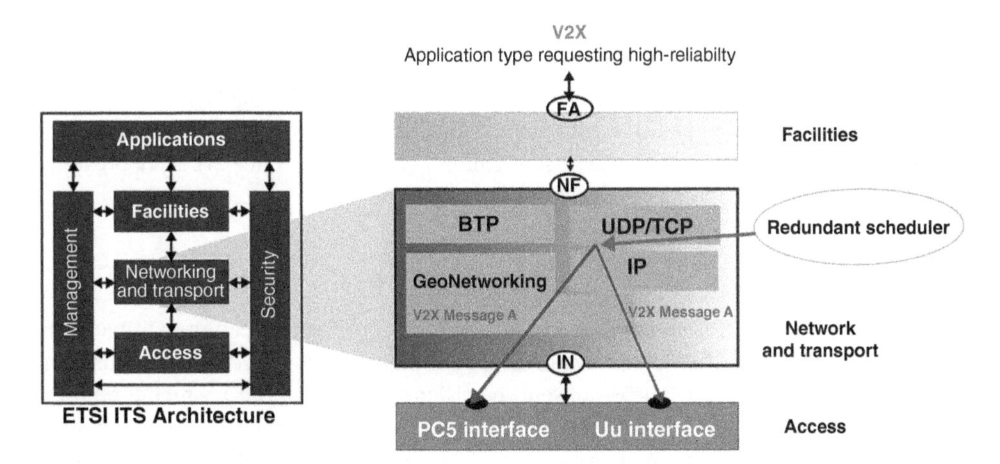

Figure 7.10 Redundant scheduler applied at the UDP/TCP level within the ETSI ITS architecture.

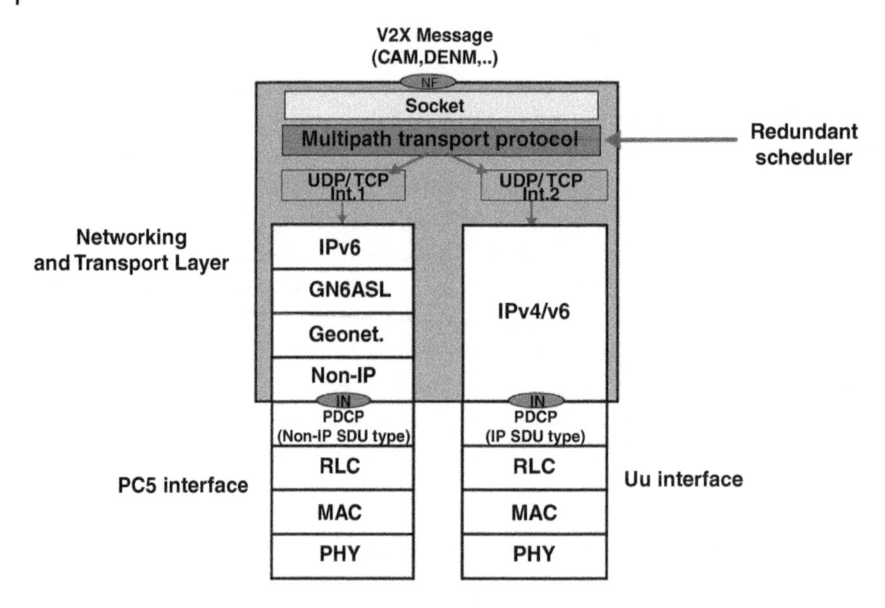

Figure 7.11 Details of redundant scheduler applied at the UDP/TCP level within the ETSI ITS architecture.

As shown in the previous subsection, it would be technically feasible to include non-IP flows as well to encapsulate the V2X message/BTP/Geonetworking stack into a UDP or TCP packet, for instance, but doing so would require extra overhead and lead to a lack of efficiency.

Furthermore, different options for interfacing with the PC5 interface may arise in further amendments of ITS standards. These might require configuring the PC5 interface to work directly in IP mode; or they might extend the ETSI ITS standards [13] to introduce other types of transport like IPv6 packets over Geonetworking (already specified in [20]), so the V2X message family specified in [16] applies to IPv6 packets in addition to the current BTP over Geonetworking defined in [20] subpart 1, referred to as Geonetworking to IPv6 adaptation sub-layer (GN6ASL). This option is summarized in Figure 7.11.

Finally, similarly to the facilities layer implementation, due to its location in the higher layers of the protocol stack, redundant scheduling policy at the transport layer prevents optimized use of duplication when links are saturated.

7.3.3 RRC Level

Last but not least, the RRC sublayer represents another convenient location to implement a redundant scheduler, due to the availability of already-standardized components. For instance, RRC is defined for Release 14 semi-persistent scheduling (SPS), which is defined for both the Uu and PC5 interfaces. Furthermore, Mode 3 sidelink already enables cross-carrier scheduling, which represents a solid base to define an extension designed to support Uu + PC5 mode.

However, it is difficult to estimate the impact in terms of modifications needed to the RRC sublayer to implement the features of a redundant scheduler. Furthermore, a redundant scheduler at RRC sub-layer requires the eNB or gNB to be tightly interconnected, to enable efficient duplicate usage of both interfaces for a vehicle-to-infrastructure (V2I) or vehicle-to-network-to-infrastructure connectivity.

7.4 Summary

This chapter discussed enhancements for application adaptation of V2X services. The chapter provided an overview of features introduced by 3GPP in 5G networks that allow applications to influence network decisions, with a focus on QoS-related negotiation. We also summarized some of the proposals in the literature for supporting application adaptation with a focus on rate adaptation and other performance-oriented parameters. The chapter provided some insights about network enhancements that could be adopted to enhance the interaction between network and application to enhance network knowledge about V2X service needs and, on the other hand, to enhance application knowledge about network capabilities to target V2X services. Toward this end, we presented examples of how to apply such enhanced interactions to V2X use cases. Finally, the chapter reviewed several design options for joint use of short- and long-range links to empower the capabilities of application adaptation given varying network conditions and deployment environments.

An empowered exchange of information among applications and mobile networks is expected to enable more advanced mechanisms of application adaptation, focusing not only on performance adaptation but also on vehicle adaptation (speed, trajectory, etc.). Enhancing this exchange of information will make CAD/CRU applications aware of expected network performance and capabilities at a particular time or in a specific area and let applications use this information to negotiate C-V2X connectivity. At the same time, the mobile network can optimize resource usage to optimally serve applications under any traffic and load conditions [21]. As an example, an empowered exchange of information between application and mobile network enables fully multi-link multi-RAT connectivity, e.g. by enabling the mobile network to selectively steer different traffic flows to the most adequate links based on application needs and network conditions. In addition, enhanced reliability can be provided by using multiple links as well as increased data rates.

References

1 5GAA. (2019). C-V2X use cases: Methodology, examples and service level requirements. White paper.

2 3GPP. (2017). Group services and system aspects; system architecture for the 5G system. TS 23.501, Stage 2, Release 15, v15.0.0

3 3GPP. (2019). Policy and charging control framework for the 5G system (5GS). TS 24.501, Stage 2, Release 16, v16.2.0.

4 3GPP. (2019). Group services and system aspects; procedures for the 5G system (5GS). TS 23.502, Stage 2, Release 16, v16.0.2.

5 Alvestrand, H., Lundin, H., and Holmer, S. (2012). A Google congestion control algorithm for real-time communication on the world wide web. Tech. Rep. 3, IETF.

6 Johansson, I. (2014). Self-clocked rate adaptation for conversational video in LTE. ACM SIGCOMM Capacity Sharing Workshop.
7 3GPP. (2017). Dynamic adaptive streaming over HTTP (3GP-DASH). TS 26.274, Release 15.
8 Sami, M., Noordin, N.K., Khabazian, M. et al. (2016). A Survey and Taxonomy on Medium Access Control Strategies for Cooperative Communication in Wireless Networks: Research Issues and Challenges. *IEEE Communications Surveys & Tutorials* 18 (4): 2493–2521.
9 Ros, D. and Welzl, M. (2013). Assessing LEDBAT's delay impact. *IEEE Communications Letters* 17 (5): 1044–1047.
10 Lianghai, J., Weinand, A., Han, B., and Schotten, H. (2018). Multi-RATs support to improve V2X communication. In: *Proceedings of IEEE Wireless Communications and Networking Conference (WCNC).*
11 ETSI. (2010). Intelligent Transport Systems (ITS), communications architecture. EN 302 665, v1.1.1.
12 ETSI. (2014). Intelligent transport systems (ITS); vehicular communications; basic set of applications; part 3: specifications of decentralized environmental notification basic service. EN 302 637-3, v1.2.2.
13 ETSI. (2017). Intelligent transport systems (ITS); vehicular communications; geonetworking; part 5: transport protocols; sub-part 1: basic transport protocol. EN 302 636-5-1, v2.1.1.
14 ETSI. (2017). Intelligent transport systems (ITS); vehicular communications; geonetworking; part 4: geographical addressing and forwarding for point-to-point and point-to-multipoint communications; sub-part 1: media-independent functionality. EN 302 636–4-1, v1.3.1.
15 3GPP. (2017). Universal Mobile Telecommunications System (UMTS); LTE; proximity-services (ProSe) user equipment (UE) to ProSe function protocol aspects. TS 24.334, Stage 3, Release 14.
16 3GPP. (2017). LTE; user equipment (UE) to V2X control function; protocol aspects. TS 24.386, Stage 3, Release 14.
17 ETSI. (2018). Intelligent transport systems (ITS); security; pre-standardization study on ITS facility layer security for C-ITS communication using cellular Uu interface. ETSI ITS WG5 WI on TR 103 630.
18 IETF. (2013). TCP extensions for multipath operation with multiple addresses. RFC 6824.
19 De Coninck, Q. and Bonaventure, O. (2017). Multipath QUIC: Design and evaluation. In: *Proceedings of 13th International Conference on emerging Networking EXperiments and Technologies (CoNEXT).*
20 ETSI. (2013). Intelligent transport systems (ITS); vehicular communications; geonetworking; part 6: internet integration; sub-part 1: transmission of IPv6 packets over geonetworking protocols. EN 302 636-6-1, v1.2.1.
21 Ye, Q., Li, J., Qu, K. et al. (2018). End-to-end quality of service in 5G networks: examining the effectiveness of a network slicing framework. *IEEE Vehicular Technology Magazine* 13 (2): 65–74.

8

Radio-Based Positioning and Video-Based Positioning

Kai Cordes[1], Hellward Broszio[1], Henk Wymeersch[2], Stephan Saur[3], Fuxi Wen[2], Nil Garcia[2], and Hyowon Kim[4]

[1] *VISCODA, Germany*
[2] *Chalmers University of Technology, Sweden*
[3] *Nokia Bell Labs, Germany*
[4] *Hanyang University, South Korea*

For a decade, non-satellite positioning has attracted the interest of researchers and industries as one of the key enablers for location-based services. In this regard, a number of solutions, as well as commercial products, have been developed for a multitude of application scenarios in home and office, warehouse, industrial, and hospital environments. For 5G Communication Automotive Research and innovation (5GCAR), the primary field of interest is positioning for vehicular-to-everything (V2X) applications. Ultimately, the 3rd Generation Partnership Project (3GPP) standardization group also initiated work items on this topic to empower fifth generation (5G) technology for a reliable and accurate positioning system that can be used anywhere and anytime. In scenarios involving unconnected road users, positioning is done using additional sensors such as video cameras.

This chapter provides an overview of state-of-art radio- and image-based positioning tailored to the localization of connected autonomous vehicles. More specifically, we address the technology and performance requirements of legacy solutions along with the details of time-based, angular-based, and video-based positioning. Some of the proposed methods closely follow the long term evolution (LTE) and the new radio (NR) standards; others show more innovative solutions that can achieve a high level of accuracy. Finally, the chapter concludes with a comprehensive set of simulations highlighting the pros and cons of each solution.

Different generations of mobile communication standards have provided location estimation capabilities through measurements such as cell-identity document (cell-ID) in second generation (2G) and third generation (3G), and time-difference-of-arrival in 3G and fourth generation (4G) relying on dedicated positioning reference signals [1]. 4G provides high availability but, due to the limited bandwidth, lacks resolution. This renders mobile localization only suitable for specific applications, such as emergency call localization and, in combination with global positioning system (GPS), outdoor suburban localization. Large bandwidths have been extensively explored in ultra-wide bandwidth

* With contributions from Silvio Mandelli, Marouan Mizmizi, and Anastasios Kakkavas.

Cellular V2X for Connected Automated Driving, First Edition.
Edited by Mikael Fallgren, Markus Dillinger, Toktam Mahmoodi, and Tommy Svensson.

(UWB) localization, which has remained a niche application due to the heavy reliance on dedicated infrastructure (and thus limited availability) [2]. With the availability of higher carrier frequencies, there are new opportunities via three interrelated arguments [3]: (i) limited diffraction and lack of object penetration at high frequencies make the channel sparse, which reduces the need for resolution in the first place; (ii) more antenna elements can be used within a given area, which leads to higher angular resolution; and (iii) larger available bandwidth leads to better delay resolution. Combined, these three compound effects lead to lower localization errors. Similarly, the resolved multipath can be used to generate a partial environment map, though it is generally of lower accuracy. Experimental results have been reported in [4] for 802.11 standards at 60 GHz and in [5] for frequencies above 200 GHz, with additional experimental setups listed in [6].

Video-based object detection, localization, and tracking have been a broad research field for the last 40 years. Recently, the computer vision community has made significant advances, enabling new applications for automated driving. Camera-based systems are available at a small cost and usually require minimal installation effort. *Computer vision* and *machine learning* algorithms provide efficient, accurate solutions for object detection, classification, and reconstruction. Many cameras are already installed to monitor motorways and city regions for various applications, such as estimating speed, counting vehicles, recognizing license plates, and traffic analysis. The localization of vehicles in monocular video is an active research field [7–15].

In the 5GCAR project, the targeted use case for video-based vehicle positioning is *Lane Merge Coordination* (Figure 8.1). The goal is the sharing and coordination of driving trajectories among a group of vehicles to improve traffic safety and efficiency [16]. For this purpose, the accurate localization of vehicles is crucial. Vehicle localization is required for a large observation area and, thus, for large distances from the camera. In [17], it is shown that for a lane change at 100 km/h, at least a 100 m range is required. In an automated cooperative maneuver, trajectory recommendations are computed for the

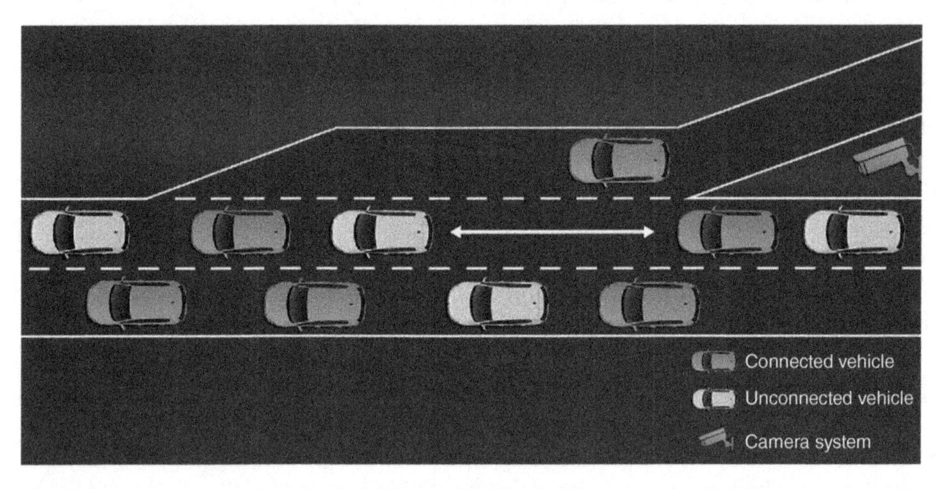

Figure 8.1 In the *Lane Merge Coordination* use case, trajectory recommendations are computed to enable the automated merge of an incoming vehicle. Connected vehicles on the main lane cooperatively open a gap for the incoming vehicle while considering unconnected vehicles. The accurate external localization of unconnected vehicles is crucial.

participating, connected vehicles. Camera-based localization is required for unconnected, i.e. non-communicating, vehicles that are not equipped with self-localizing technology. They must be considered during lane merge coordination for the computation of trajectory recommendations for the connected cars. Cooperation among all participating vehicles enables the execution of joint maneuvers for automated driving. As explained in Chapter 2 (cf. Table 2.2 and Table 2.3), accurate vehicle positioning is a key performance indicator (KPI) of primary importance. The challenge for video-based positioning is to design an accurate real-time system for the large target area. This can be achieved with minimal installation effort such as a small number of cameras installed at given positions (masts) next to lane intersection and an easily reusable calibration procedure.

Radio-based positioning is presented in Section 8.1 of this chapter, and video-based positioning is presented in Section 8.2. Finally, conclusions are presented in Section 8.3.

8.1 Radio-Based Positioning

In this section, we describe the main requirements for radio positioning and how Release 16 addresses these requirements, and showcase the potential of moving to higher carrier frequencies, as investigated in 5GCAR. Finally, we list several challenges for radio positioning.

8.1.1 Use Cases and Requirements

As we enter the era of autonomous driving, precise positioning of road users becomes an increasingly important topic. At the same time, we have to deal with multiple facets in many regards. First and foremost, the required positioning accuracy (mean distance between true and estimated position) depends on the use case. Additionally, the precision of positioning describes how a series of measurements are distributed around their mean value. These use case requirements are often combined as a percentage of measurements for which the positioning error must be smaller than a certain threshold. A second important source of diversity is the environment. A dense urban scenario probably has significantly more road users per area unit than a rural environment. On a highway, the speed is higher, but all road users move in the same direction on given lanes, which in the end simplifies position estimation. Also, the use cases of interest depend on the environment. Automated parking and vulnerable road user (VRU) protection may be applied in a city center, but not on the highway. No positioning method fits all use cases, performance requirements, and scenarios; rather, we will see a mix of different methods.

In the scope of the introduction of the fifth generation of mobile radio (5G), 3GPP working groups have identified and characterized various use cases in the domain of V2X communication for which knowledge of road user positions is required [18, 19]. The following are some of the most prominent use cases.

- *Traffic monitoring, management, and control*: Such a service is supervised by a local authority in order to optimize traffic flow and provide related recommendations to road users. For that purpose, the absolute two-dimensional position of the road user needs to be determined with sufficient accuracy to identify the lane (1 m lateral to the driving direction and 3 m longitudinal) [18].

- *Road user charging*: This use case tracks the road user to levy a charge based on road infrastructure usage. The required accuracy is the same as for traffic monitoring [18].
- *Platooning*: This use case coordinates the behavior of a group of vehicles. The vehicles move similar to a train, with virtual strings attaching them. Reliable control of the relative distance and orientation between the vehicles is a crucial requirement, whereas absolute position is less important. The required relative positioning accuracy is 10 cm lateral to the driving direction and 50 cm longitudinal [19].
- *Advanced driving*: This use case family summarizes all aspects of automated driving, such as cooperative collision avoidance and lane merging. It has similar requirements to platooning [19].

It is important to note that these use case studies do not presume radio access technology (RAT) based positioning methods only. They can be complemented by any other means (radar, video, sensor sharing) to achieve the ultimate performance target. However, the 5G system is expected to provide communication channels, measurement procedures, and the network architecture framework. In addition to positioning accuracy, latency, update rate, data rate, availability, reliability, and the number of supported road users are important performance metrics to enable these use cases. This chapter focuses on positioning accuracy.

8.1.2 Radio-Based Positioning in New Radio Release 16

3GPP recently concluded a work item on radio-based positioning in Release 16 [20]. The specification includes a reference signal design, measurements, and procedures. In short, the following positioning methods are supported:

- Downlink time difference of arrival (DL-TDOA)
- Uplink time difference of arrival (UL-TDOA)
- Downlink angle of departure (DL-AOD)
- Uplink angle of arrival (UL-AOA)
- Multi-cell round trip time (MC-RTT)

Here, we discuss the two primary features of Release 16 positioning: downlink positioning reference signal (DL PRS) design, and the definition of PRS resource sets.

Downlink positioning reference signal (DL PRS) [21]: A new reference signal is specified in the Release 16 work item. There are two main enhancements over LTE. The first addresses the allocation of radio resources. NR allows more flexibility for the frequency comb size and number of PRS symbols. Whereas in LTE, the comb size (number of subcarriers defining the PRS resource grid which is equivalent to the number of mutually orthogonal PRS resources) was fixed to 6, in NR, configurations with 2, 4, 6, and 12 subcarriers are possible. Figure 8.2 illustrates the DL PRS resource grids for comb-2, comb-4, and comb-6. The number of PRS symbols in the time direction can be selected accordingly so that an equivalent comb-1 pattern can be generated. Equivalent comb-1 means PRS transmission covers all subcarriers even if distributed over several subsequent orthogonal frequency division multiplexing (OFDM) symbols. In a dense urban environment, it is advantageous to have many different orthogonal PRS resources – in this case, comb-12 – to reduce the interference level. The drawback is that the measurement of a full equivalent comb-1 pattern takes 12 symbols. In a rural environment with fewer network nodes, one would choose a small comb size – in this case, 2 or 4 – since there is less need for interference management. Now the

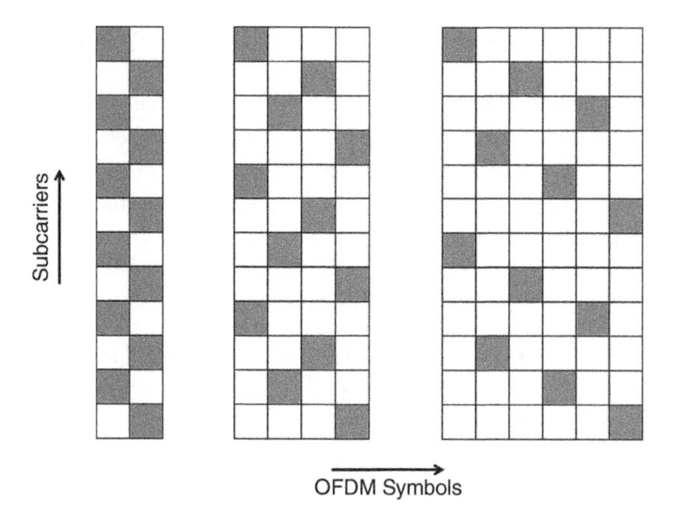

Figure 8.2 NR DL PRS resource allocation with comb-2, comb-4, and comb-6 patterns.

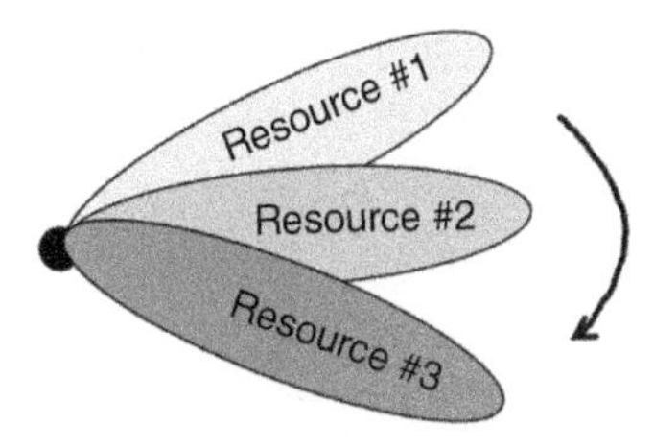

Figure 8.3 PRS resource set with three PRS resources characterized by their beam directions.

equivalent comb-1 pattern is complete after two and four symbols, respectively. However, due to the reduced density of network nodes, PRS coverage can become a problem. Therefore, the number of PRS symbols can be configured independently from the comb size, and repetitions of PRS are possible. As an example, a configuration with 12-symbol duration and comb size 4 can be exploited for three repetitions of the PRS to enhance the coverage.

The second major novelty of NR PRS is the definition of a *PRS resource set* comprising PRS resources with the same configuration. The idea is that the transmission of a single PRS resource corresponds to a beam direction, while a PRS resource set covers the entire area of interest by subsequently transmitting the PRS in different directions. This mechanism is referred to as PRS *beam sweeping*. Figure 8.3 shows an example of a PRS resource set with three PRS resources. Hence, beyond time and frequency, interference can also be controlled in the spatial domain by coordinating PRS resource sets from several transmission points. In the uplink direction, the sounding reference signal (SRS) is used for positioning. There are similar possibilities to configure the resource grid as for DL PRS.

Extended measurements [22]: The full set of measurements to support Release 16 positioning methods in a 5G system are the following:

- *DL relative signal time difference (DL RSTD)*: Time difference of arrival from a pair of network nodes at the road user.

- *UL relative time of arrival (UL-RTOA)*: Time of arrival of the SRS at the network node relative to a reference time.
- *UL angle of arrival (UL AOA)*: Indication of the direction from where the SRS is received.
- *UE receiver (Rx) – transmitter (Tx) time difference and gNB Rx – Tx time difference*: These are required for determining the RTT between the road user and several network nodes.
- *DL PRS reference signal received power (DL PRS-RSRP)*: Measurement at the road user, indicating the reliability of the corresponding downlink measurements.
- *UL SRS reference signal received power (UL SRS-RSRP)*: Measurement at the network node, indicating the reliability of the corresponding uplink measurements.

These measurements (or a subset of them) are provided through a positioning protocol to the location management function (LMF) in the network. The LMF is responsible for parameter configuration, e.g. the comb size of DL PRS, distribution of required assistance data to road users, collecting measurement reports, determining the position of the road users, and providing the results to applications.

3GPP will study the performance potential of Release 16 positioning methods in the scope of Release 17 and will specify further enhancements if needed to fulfill the performance requirements. Today, highly accurate relative positioning between two road users, which is needed e.g. for platooning or overtaking, cannot be realized with Release 16 methods. The reason is that sidelink-assisted positioning, i.e. measurement from direct signal transmission from one road user to the other, is not specified. Hence, 5GCAR has developed solutions beyond Release 16, which we present next.

8.1.3 Radio-Based Positioning Beyond Release 16

The goal of the localization system is to track the state, based on measurements, given a certain mobility model. The state contains the location of the vehicle (including longitude, latitude, and height) as well as the locations of landmarks in the environment (e.g. buildings and objects that yield strong radio reflections). Measurements include the conventional time (difference) of arrival (T(D)OA), as well as angle of arrival (AOA) if the receiver has multiple antennas, and angle of departure (AOD) if the transmitter has multiple antennas. These measurements depend on the state via simple but highly nonlinear relations. The quality of the measurements (i.e. in terms of the resolution and the noise variance) depends on the received signal. This means the quality can be adapted by proper signal design. Finally, there is the dynamic model as well as on-board sensors. Radio localization systems differ in their architectures and signal flow (e.g. based on uplink or downlink or sidelink transmission) but generally loop through five distinct stages: state prediction, signal optimization, transmission, measurement, and state update. In this section, we go deeper into three of these stages: signal optimization, the measurement process, and the state update. But first, we describe the salient properties of the channel, which provides opportunities to extract highly informative high-resolution measurements.

8.1.3.1 The mmWave Channel

Before developing a localization method, we first describe the mmWave channel, which in turn determines the measurement model. Due to the much shorter wavelength at higher carrier frequencies, multiple input multiple output (MIMO) systems consisting of many

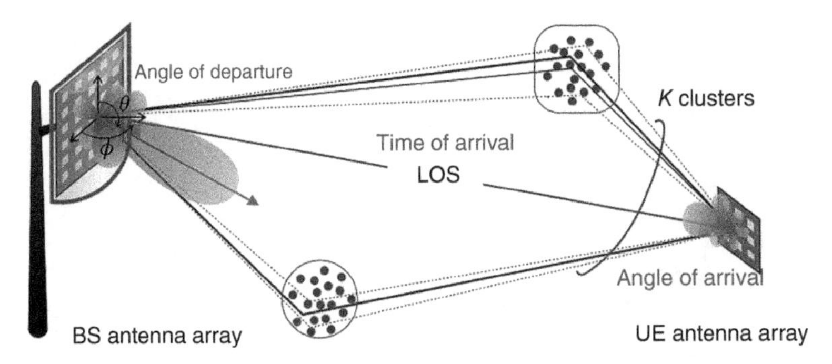

Figure 8.4 The 5G mmWave channel is characterized by few propagation paths, each possibly corresponding to a cluster of rays. Each ray provides information about the AOA, AOD, and TOA.

antennas can be compacted into much smaller sizes, enabling the use of large arrays at base stations (BSs). Due to the large array aperture, a BS with many antennas can estimate the angles of multiple paths very accurately. In fact, the mmWave channel can be considered parsimonious in the sense that only a few multipath components carry non-negligible energy. mmWave propagation is characterized by limited scattering, no diffraction and shadowing, and the existence of only a few propagation paths. Each path is thus a function of the propagation environment and characterized by channel gains, angles of arrival, angles of departure, and delays, as illustrated in Figure 8.4. Propagation paths are of a deterministic specular nature when the surface on which waveforms impinge is sufficiently smooth, or of a stochastic diffuse/scattering nature when the surface is relatively rough, or a combination of both. Hence, in general, each path except the line-of-sight (LOS) path comprises a cluster of paths with similar angles and delays. When the paths within a cluster are not resolvable in either angles or delays, they lead to fluctuations in the received power. This is the model typically assumed in the communication literature [23]. On the other hand, when intra-cluster paths are resolvable, they may carry useful information about the environment.

8.1.3.2 Signal Design

In downlink positioning, a BS emits a series of M pilots, each precoded differently. The user equipment (UE) receives such known pilots (defined at the protocol level) to estimate the AOA, AOD, time of arrival (TOA), and signal strength of the strongest paths. In this section, we investigate the optimal precoders leading to the best AOA and AOD estimation accuracy.

Whether initial access or tracking, the most common procedure for estimating the mmWave channel consists of sweeping the channel with beams at the transmitter and/or receiver [24–26]. By detecting the time at which the received power is the largest, the correct pair of beams can be identified. While intuitive, such beams may not necessarily yield the best achievable estimation accuracy of the AOA or AOD. To the best of our knowledge, the literature in mmWave precoding has not addressed the fundamental limits in terms of AOA and AOD estimation when performing optimal precoding at the transmitter.

We seek to find the best transmit precoders for estimating the AOD and AOA for a single-path channel, assuming that the AOD and AOA are known to lie within specific

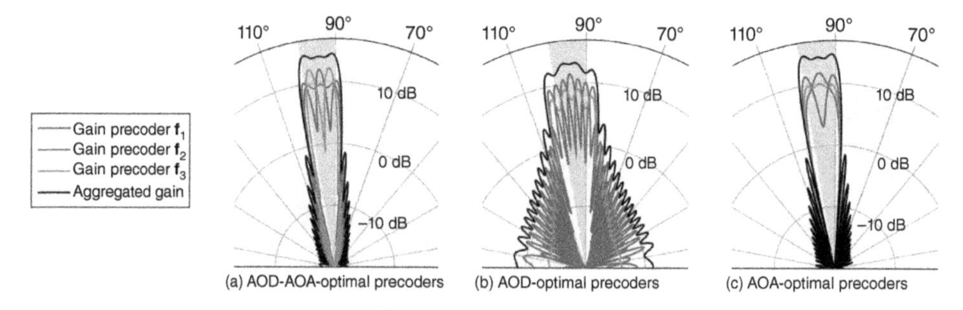

Figure 8.5 Optimal precoders' beampatterns. Number of pilots is $M = 3$.

ranges of angles [27]. Such a priori knowledge on the AOD/AOA channel state information (CSI) is available from initial access or by tracking the AOD/AOA. To abstract the analysis from specific estimators, we use the Cramér-Rao bound (CRB) as a proxy metric for the variance of the AOD and AOA, valid for high SNRs. Determining the AOD yields the direction from the BS to the UE, and determining the AOA provides the UE's orientation with respect to the BS. Thus, for naming purposes and without loss of generality, we refer to the CRB on the AOD as the direction error bound (DEB) and the CRB on the AOA as the orientation error bound (OEB). The optimal precoders are the result of minimizing the worst-case DEB or OEB within the prior ranges. The problem is re-expressed as a convex optimization problem, and because convex problems can be solved accurately, we obtain the optimal precoders and not just an approximation.

Figure 8.5 plots the optimal precoders for three different criteria: (a) optimal precoders when optimizing the AOA and AOD, (b) optimal precoders when optimizing the AOD only, and (c) optimal precoders when optimizing the AOA only. The uncertainty range at the BS is 10°. The AOD-optimal precoders show a large number of ripples. The intuitive explanation is that the UE identifies the AOD by observing the changes in received signal strength and phase changes for the M training sequences. Thus, to increase the AOD estimation accuracy, the precoders' gain and phase for closely spaced angles must be as different as possible. The AOA-optimal precoders maximize the array gain toward the UE. Finally, the AOD-AOA-optimal precoders offer a trade-off between the AOA-optimal and AOD-optimal precoders in terms of gain and "ripples."

The proposed precoders are especially interesting because they are optimal for channel estimation, although they are complex to generate. Complexity can be reduced by storing a precomputed codebook of such precoders. In the case of arrays of antennas with analog frontends, the precoders are constrained to unit-magnitude, and only their phases can be modified. Since the proposed precoders were unconstrained, they can only be applied to digital arrays. For analog arrays, additional constraints must be added to the optimization.

8.1.3.3 The Measurement Process

A cluster can be characterized in multiple ways. Traditionally, a statistical model has been considered, whereby a cluster is modeled via a mean and a spread in both the angle and delay domains [28]. Given this model, second-order estimation methods exist that can accurately estimate the mean and spread of a cluster without pilot data [29]. There is, however, a lack of first-order methods for quickly estimating channel parameters and their spread.

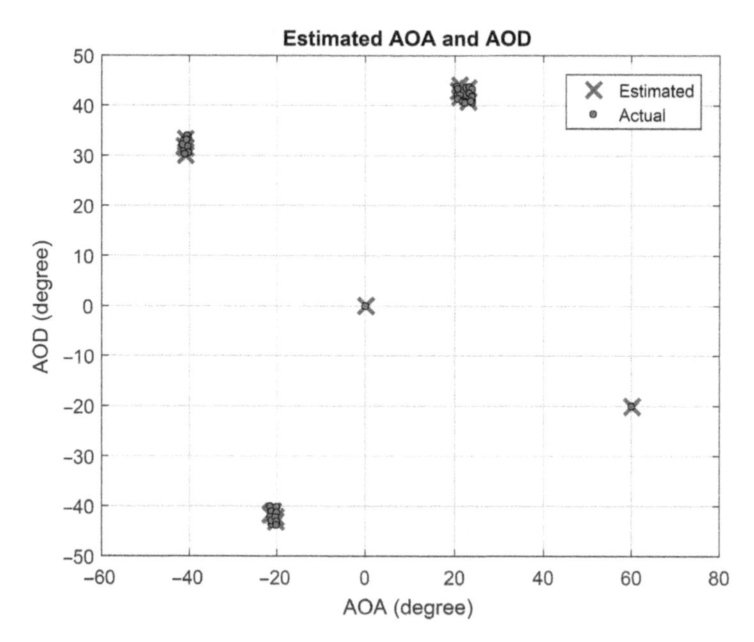

Figure 8.6 For a configuration with 32 antennas at transmitter and receiver, we show the AOA and AOD of the paths in the channel (dots) and those estimated by a tensor-based method. This method can determine the primary clusters.

5G mmWave localization has either considered only the LOS path or treated multipath as purely specular [30]. Standard 5G mmWave channel estimation is based either on compressive sensing approaches [23], which express the sparsity in an appropriate domain; or on tensor decompositions, where the dominant higher-order singular values can be related to the dominant signal paths [31]. These methods do not account for the intra-cluster spread of angles or delay. Tensor-based methods can be used to estimate 5G mmWave channels to recover the angles and delays of individual paths in each non-line-of-sight (NLOS) cluster [32]. These angles and delays are then used for localization and mapping, even when clusters have no specular component, as shown in Figure 8.6.

8.1.3.4 Localization, Mapping, and Tracking

Thanks to the benefits of 5G for positioning [33], the UE can obtain accurate channel estimation parameters such as the TOA, AOA, and AOD by the downlink signal transmitted from the BS. Despite using those accurate channel estimation parameters as the measurements, it is not trivial to estimate the UE position due to the following challenges [34]: the UE heading is not perfectly known, even if the inertial sensor is mounted on the UE; imperfect clock synchronization between the UE and BS causes a clock bias. It is thus required to estimate the UE position, as well as the heading and clock bias. In addition, it is possible to build up a map of the objects by relating the measurements to the objects in the propagation environment. Then the UE uses the measurements that correspond to the objects. However, there are standard challenges in such simultaneous localization and mapping (SLAM), including missed detections, false alarms due to clutter, and uncertainty of data association (i.e. lack of origin-related tags on 5G measurements).

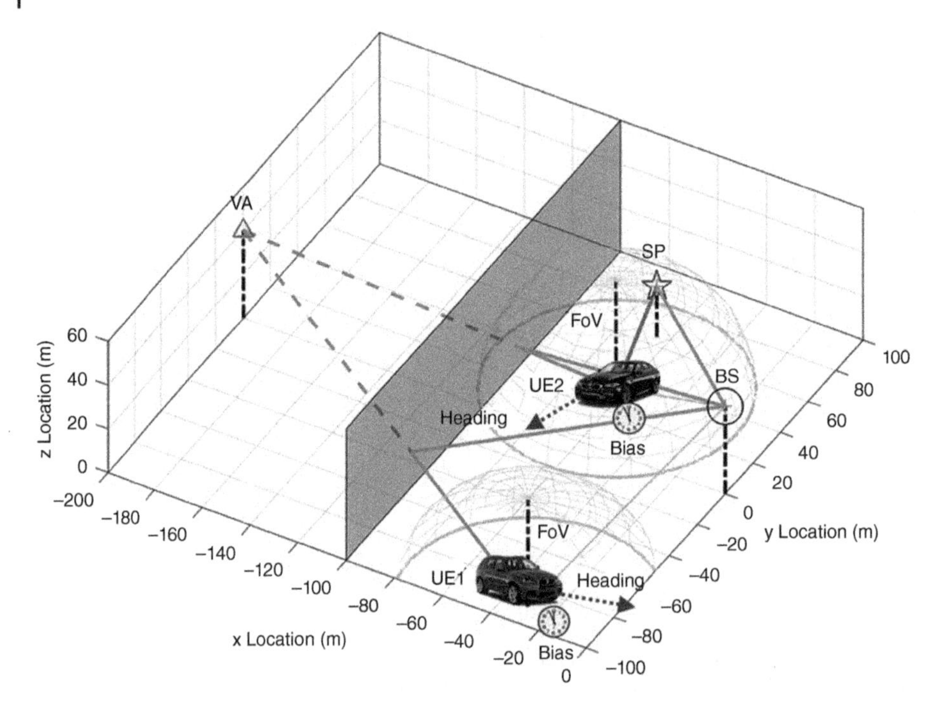

Figure 8.7 Scenario with the propagation environment (BS, VA, and SP) and two vehicles (position, heading, and clock bias). The half-sphere indicates the FoV for the SP.

The problem of 5G localization, mapping, and tracking can be tackled with different methods: by using geometric relations, the state of the user and map can be estimated by Bayesian and non-Bayesian approaches in [35] or through optimization [36]. However, challenges such as missed detections, false alarms, and uncertainty of data association are not considered. In [34], the message passing (MP) approach is used, and a SLAM method is proposed by executing belief propagation on a factor graph. Explicit data association is required in [34, 37], but uncertainty of the data association is not considered. The joint probabilistic data association scheme is proposed with the MP approach in [37]. In addition, random-finite-set approach for the SdAM problem has been developed [38, 39]. However, these methods have not yet been applied to a 5G scenario, as shown in Figure 8.7. In the 5G scenario, the propagation environment is characterized by reflection surfaces, such as walls, and scattering points (SPs), such as the small obstacles. Reflection surfaces are parameterized by fixed virtual anchors (VAs). These different types of objects generate multipath measurements. In [40], a method for 5G localization, mapping, and tracking is introduced. The method consists of a multi-model probability hypotheses density filter (PHD) and map fusion routine. In the multi-model PHD filter, the UE estimates its state and the position of objects. The objects represent a statistical map as viewed by the UE, and the map is integrated into the BS. Then the UE reuses the integrated map information to improve SLAM performance.

To demonstrate the efficiency of the introduced method, the results are obtained in a two-vehicle scenario, where all propagation paths are used with the estimated propagation

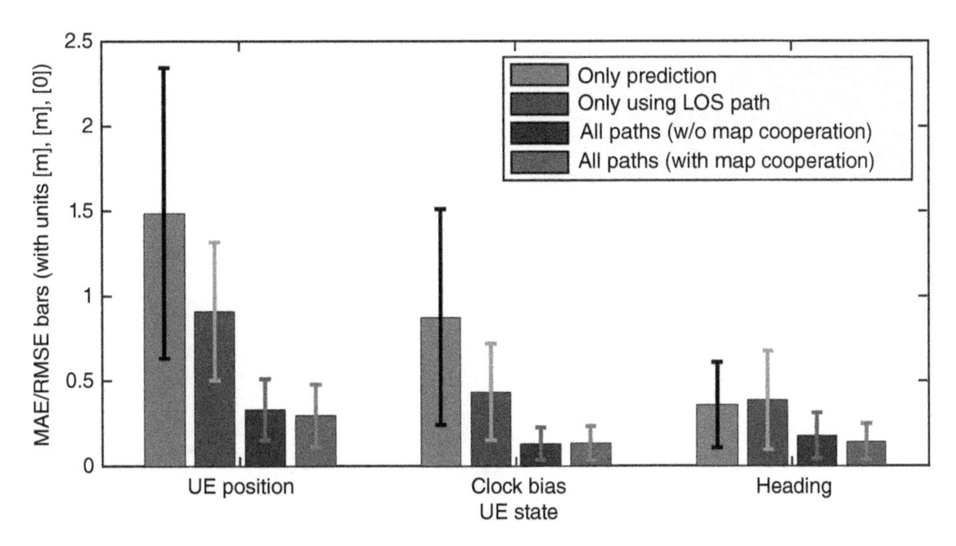

Figure 8.8 MAE and RMSE bars of UE state estimates (location, clock bias, and heading).

environment consisting of one BS, four VAs, and four SPs. The VA has no field of view (FoV), but the SP has a limited FoV. The performance of UE state estimation is evaluated by the mean absolute error (MAE) and root mean square error (RMSE) bar. For mapping performance, the generalized optimal sub-pattern assignment (GOSPA) distance is computed. Figure 8.8 shows the performance of UE state estimation for four cases, as follows: (i) executing prediction only with the UE dynamics; (ii) performing only UE estimation with the LOS measurement; (iii) executing the multi-model PHD filter; and (iv) executing the multi-model PHD filter and map fusion routine. It is clearly shown that there are performance gains when the propagation environment is reused. Figure 8.9 shows the mapping performance for the SP. It is demonstrated that when the map information is cooperatively used, performance is improved.

As an extension of the 5G scenario, following vehicular networks could be considered: for instance, (i) regarding other UEs as moving SPs; (ii) multi-cell with high-intensity clutter and a high rate of misdetections; and (iii) sparsely executing SLAM with the collected 5G measurements (e.g. multi-scanning), rather than at every time step.

8.1.4 Technology Component Complementation

5GCAR has spent much effort developing complementary methods that can be applied instead or in combination with the framework presented in Section 8.1.3. *Trajectory and collision prediction* are tailored to the VRU protection application. In the near future, we will see a mix of self-driving and human-controlled vehicles in dense urban environments with pedestrians, cyclists, and scooters crossing streets. The possibility of using the 5G mobile radio network for collision avoidance is an attractive new application. First, we determine the probability density function (pdf) of the positions of the road users at a certain point in time in the future, typically in the range between 0.5 and 3 seconds, in the following referred to as *time window*. To do so, we apply a motion model, e.g. the assumption that

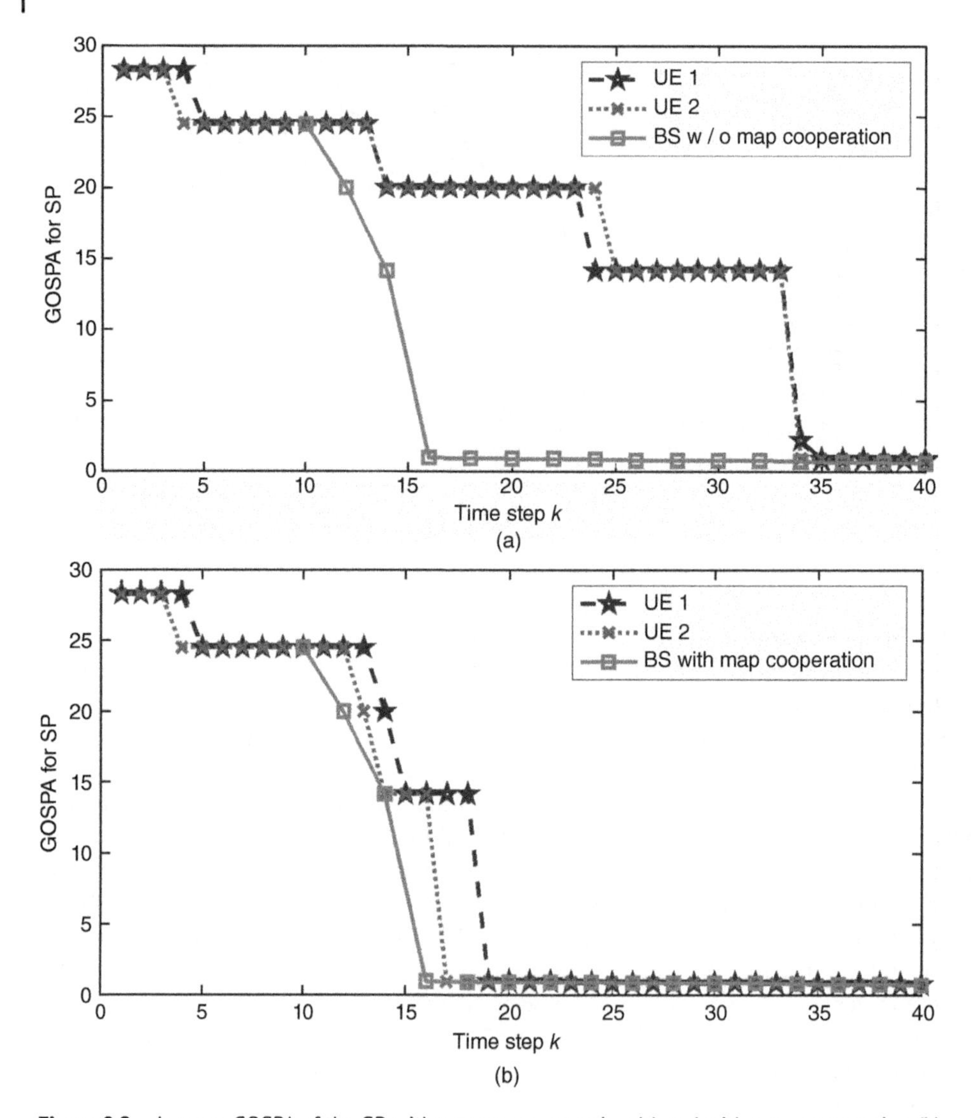

Figure 8.9 Average GOSPA of the SP without map cooperation (a) and with map cooperation (b).

the road user will move with constant velocity along the lane [41]. The shape of the pdf depends at a minimum on the size of the road user (its projection on the ground), its speed, and the time window. Obviously, the larger the time window is, the more indistinct the pdf becomes. The probability of a collision between road users is calculated from the two pdfs. Intuitively, in the area where significant sectors of the pdfs overlap, we observe a high collision probability, as illustrated in Figure 8.10. The application sends an alert message when the collision probability exceeds a threshold. Simulation results have shown that a good compromise is a time window around two seconds with an alarm threshold of 90%. False alarms can then be almost eliminated at the cost of a very short reaction time [42].

Highly accurate *relative positioning* between two vehicles is sufficient for many V2X-related applications, e.g. overtaking and platooning. 5GCAR has studied performance

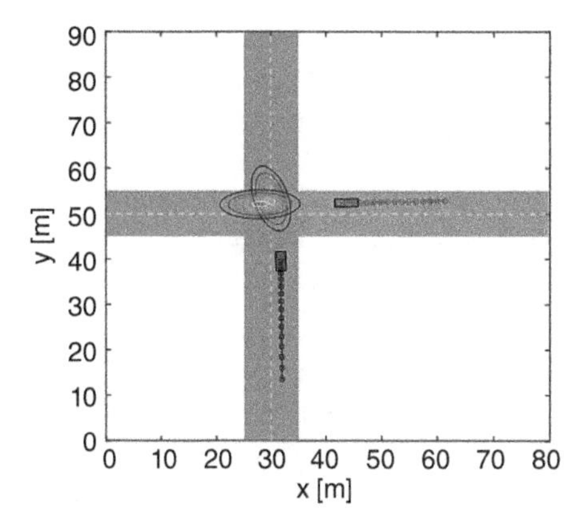

Figure 8.10 Illustration of collision probability as the superposition of pdfs of the future position of two road users. Source: Reproduced with permission from Dr. Marouan Mizmizi.

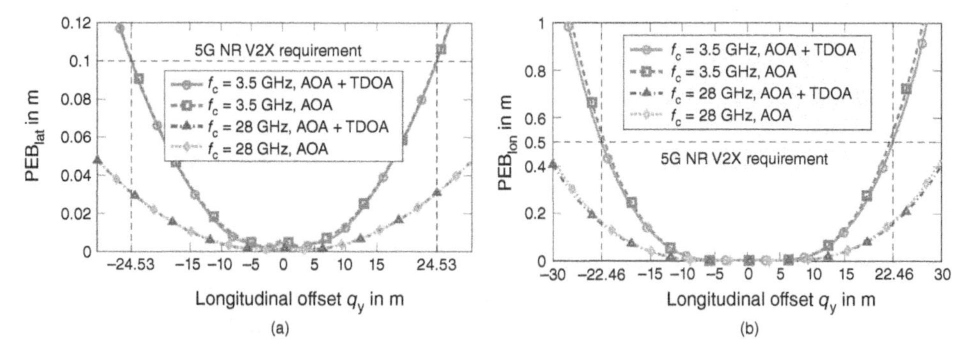

Figure 8.11 Lateral and longitudinal position error bound for the overtaking scenario. Source: Reproduced with permission from Anastasios Kakkavas.

bounds for multi-array V2V relative positioning (sidelink transmission at 3.5 and 28 GHz). The primary achievement was understanding the significance of angular measurements compared to delay measurements with their potential impact on reference signal design in Release 17 and beyond. The 28 GHz configuration can provide better positioning accuracy pertaining mainly to its higher angular resolution, as a result of the larger number of antennas that can be packed into the same physical area. In Figure 8.11, we plot the lateral and longitudinal positioning errors PEB_{lat} and PEB_{lon} as functions of the longitudinal offset q_y between a vehicle and an overtaking vehicle, when both AOA and time difference of arrival (TDOA) or only AOA measurements are used. In the example, each vehicle has three arrays with at least one LOS link to the other vehicle. We found that pure angular measurements are sufficient for the overtaking scenario. However, for the platooning use case, the combination with TDOA measurements provides a significant additional gain [43].

8.1.5 Limitations of Radio-Based Positioning

While radio-based positioning is an attractive and cost-efficient solution due to the expected widespread availability of 5G network infrastructure, technical challenges limit achievable performance. In the following, we discuss the most important aspects. In this context, 5GCAR has intensively studied alternative and complementary concepts, as presented in Sections 8.1.3 and 8.1.4. These can be the foundation for possible improvements in future Releases (17 and beyond) of the 5G specification.

Conventional radio-based positioning requires good propagation conditions between the road user and a sufficiently large number of network nodes. The number of hearable nodes may be limited due to large distances – too low signal-to-interference-plus-noise ratio (SINR). Particularly in the uplink, due to the lower transmit power, this can become an issue. As an example, to determine the two-dimensional position of a road user, in theory, at least two TDOA measurements are needed: i.e. at least three network nodes must be involved in the procedure. The expected accuracy increases with a higher number of available measurements given that the location management function (LMF) can determine and select the most reliable measurements. Not only the number of hearable network nodes but also the geometric relation among them has a significant impact on positioning accuracy. For angle measurement, it is obvious that an error of one degree translates to a growing position error with increasing distance between transmitter and receiver. However, a fixed time measurement error can cause variant position errors, depending on the geometric relation. This effect is referred to as geometric dilution of precision (GDOP).

A requirement that is particular to time measurements is the need for exact synchronization among all transmitting (downlink) or receiving network nodes (uplink). Note that an unknown clock offset of 1 ns corresponds to a distance error of 30 cm; it is not trivial to achieve this level of synchronization accuracy, especially in a macro-cellular network, as is assumed for V2X. A closely related challenge is the need for exact knowledge of the location of the antenna panels of the network nodes. Using GPS for network node calibration is insufficient when the target position accuracy of a road user is less than 1 m.

Another challenge is the mobile radio channel between transmitter and receiver. The received signal is affected by reflection, diffraction, and scattering of radio waves. Typically, the impulse response is characterized by a series of peaks, each representing one propagation path. Direct LOS may even be blocked by obstacles. The dominant path may then not be the first arriving path, and time and angle measurements may be misleading. This challenge can be solved with the availability of mmWave spectrum and higher bandwidths (see Section 8.1.3.1).

Finally, the limited availability of pilot signals used for time measurements – in 3GPP, referred to as PRS – can be a problem in dense urban environments. 3GPP has specified concepts in Release 16 to overcome this problem. In the downlink, network nodes close to each other use disjoint radio resources for PRS. Orthogonality can be achieved in time (through a muting pattern) and frequency (using different groups of subcarriers; see Figure 8.2). However, each network node shares its PRS resources with other network nodes, with overlapping coverage areas eventually leading to interference. PRS interference can be mitigated by using pseudo-noise sequences and spatial separation with beamforming (see Figure 8.3). PRS interference also limits the achievable positioning accuracy.

8.1.6 Summary

Radio-based positioning is an attractive method because the existing infrastructure for mobile communications can be reused. In Release 16, 3GPP specified a set of methods that still have to prove their ability to fulfill the performance requirements for V2X use cases. These methods rely on a combination of AOA, AOD, TOA, and TDOA measurements to provide high accuracy in the lateral and longitudinal directions. Such measurements also allow mapping of landmarks in the environments (reflecting surfaces and scatter points). Channel-estimation methods can obtain these measurements from received waveforms and provide them as input for localization and SLAM algorithms. We also highlighted several challenges and limitations of radio-based positioning, such as synchronization requirements and the impact of interference from different propagation paths.

The error metrics are root mean square error for position and generalized optimal sub-pattern assignment for landmarks. 5GCAR has developed complementary solutions beyond Release 16, which indicate that the required accuracy of less than 1 m for absolute positioning and less than 10 cm for relative positioning can be achieved.

8.2 Video-Based Positioning

This section evaluates the achievable accuracy of video-based vehicle positioning. Therefore, the accuracy achieved in the 5GCAR *Lane Merge Coordination* use case, using a set of monocular cameras for vehicle positioning, is compared to the accuracy achieved in literature.

Current approaches for vehicle localization from monocular video incorporate machine learning techniques to estimate an object's two-dimensional (2D) [7, 8] or three-dimensional (3D) bounding box [10–15]. Since extensively trained 2D models are available, most 3D models are based on 2D position proposals and extend their localization to 3D in a second step [11–15]. Most 3D localization approaches using machine learning are trained and evaluated on the publicly available KITTI data set [9]. This data set provides ground truth (i.e. manually annotated) locations of vehicles for a camera positioned on the roof of the ego-vehicle.

For evaluation, the average localization precision (ALP) is used [12, 14, 15], which measures the distance between the localized object and the ground truth location provided by the benchmark. The location of an object is deemed correct if the 3D distance to the ground-truth location is smaller than a threshold t_{ALP}. The ratio of correct matches determines the ALP. Current approaches show an ALP between 34% and 71% for $t_{ALP} = 1\ m$ and an ALP between 54% and 90% for $t_{ALP} = 2\ m$ [12, 14, 15], depending on the difficulty of the observed scene. The cameras used in the KITTI benchmark are mounted on top of a moving car. Thus, compared to the data used in the 5GCAR project, the viewpoint in KITTI is more challenging due to the smaller height of the camera. On the other hand, the KITTI benchmark only includes vehicle distances of up to 70 m, while the distances to objects in the 5GCAR scenario are much larger.

8.2.1 Vehicle Positioning System Setup

For the targeted lane merge use case, vehicles on the road are localized using a roadside camera system. The camera system consists of monocular video cameras with small overlapping regions in their views. The union of these views observes the relevant scene content, which is required for the lane merge.

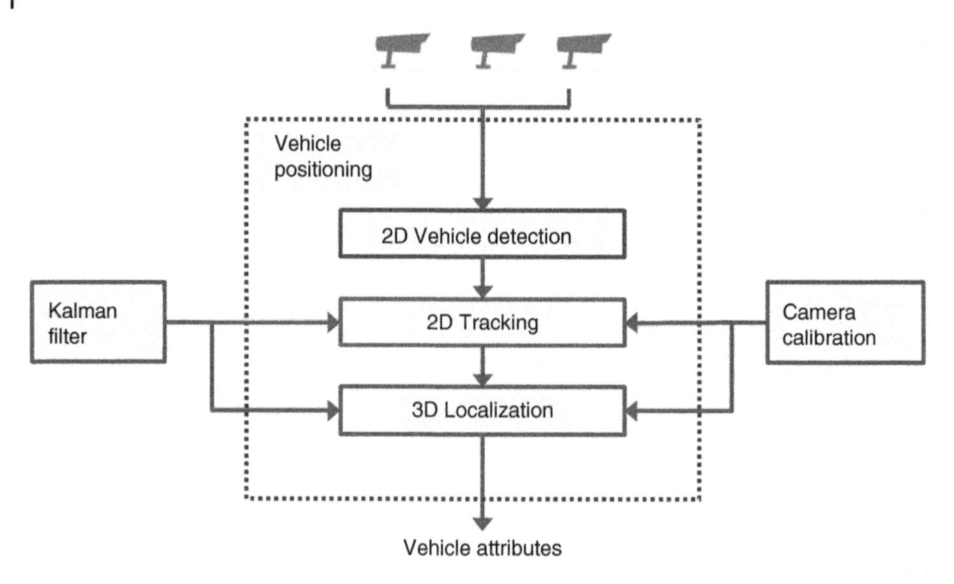

Figure 8.12 Workflow of video-based vehicle positioning. The camera system computes vehicle attributes such as localization, heading, and speed based on 2D vehicle detections in an image, camera calibration, and Kalman filtering.

The processing pipeline is shown in Figure 8.12. Vehicle localization consists of detection, tracking, and the final 3D localization step, as proposed in [44]. The object detector computes image locations of vehicles. The vehicle tracking and 3D localization incorporate the previously determined camera calibration. For tracking, the prediction of the object location in the next image is used. For filtering in the temporal domain, a Kalman filter is employed. For each synchronized input image set from the cameras, vehicle attributes are computed, leading to the KPI *update rate* (cf. Section 8.1.1). *Latency* is determined from the time between image capture and the finalization of processing (Figure 8.12). Capturing and processing require one frame each, leading to a latency of 67 milliseconds (camera frame rate: 30 Hz). Vehicle attributes include the vehicle position, heading, speed, and acceleration and a timestamp representing the point in time of the image capture. The timestamp is required for the data fusion assigning positions from the cameras to locations provided by the connected cars, i.e. cars with self-localization technology. Since the latency of the system is known, the correct image capture time can be used for the assignment.

The observed area includes main and entering lanes within a range of up to 180 m. To cover the required visual range in the camera images, three cameras are used. They are mounted on two masts at 10 m height. The distance between the masts is 24 m. Two cameras attached to the first mast observe the main lane. Their objective is to localize vehicles at small (camera 1 with small focal length $f = 8\ mm$) as well as large distances (camera 2 with large focal length $f = 25\ mm$). The third camera (camera 3 with $f = 8\ mm$) at the second mast observes the entering lane. The camera views are shown in Figure 8.13. A top view computed from the calibrated cameras is visualized in Figure 8.14.

Figure 8.13 Synchronized camera views of the three cameras with $f = 8$ *mm* (left and right) and $f = 25$ *mm* (center). Feature tracks on the background are used to compensate for mast movement, while feature tracks on the vehicles provide stable tracking. The camera's fields of view for these images are illustrated in Figure 8.14 (right).

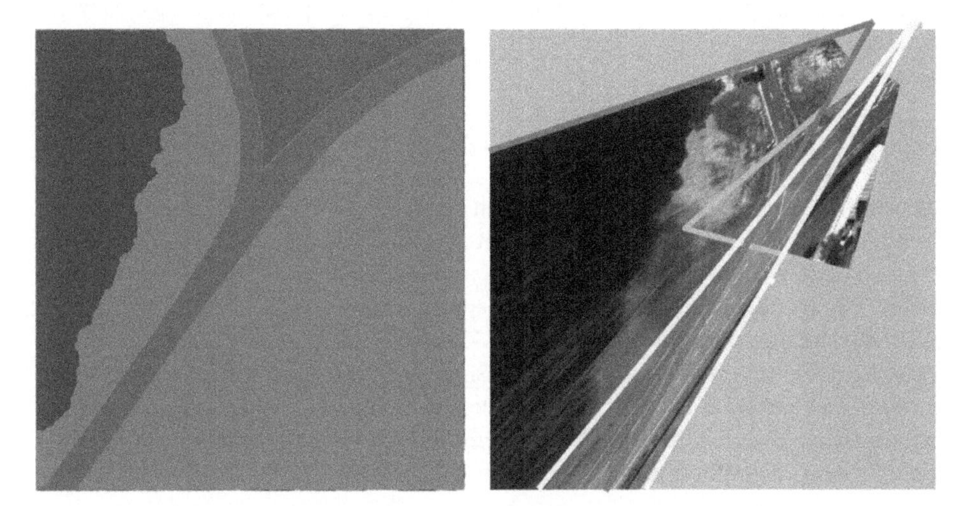

Figure 8.14 The lane merge target region with two main lanes and one acceleration lane. The left image shows a top view, while the right image is computed from the three cameras installed at two masts using the proposed multi-camera calibration. The medium, narrow, and wide camera views in the right image correspond to the left, middle, and right image in Figure 8.13.

8.2.2 Multi-Camera Calibration

For the mapping between a localized point $p \in \mathbb{R}^2$ in the camera plane and the corresponding three-dimensional points $P \in \mathbb{R}^3$ in the global coordinate system, accurate camera parameters are required. These parameters include extrinsic as well as intrinsic parameters. The extrinsic parameters determine the rotation and translation of the camera coordinate system with respect to the world coordinate system. The intrinsic parameters, such as the focal length and the dimensions of a pixel, describe the camera. The camera calibration procedure determines these parameters; they have a significant influence on the accuracy of the system.

For the 5GCAR project, the calibration is done using landmarks on the road, measured with GPS real time kinematics (RTK), as well as measured distances between the cameras

[45]. The positions of the landmarks are selected such that their projections are easily localized in each of the camera views: for example, edges of street markings and the base point of a street sign. These positions are annotated manually in the images. GPS coordinates are transformed into the local coordinate system [46]. The calibration procedure minimizes reprojection error, i.e. the sum of distances between reprojections $p_{j,k}$ of 3D points P_j using the camera $A_k \in \mathbb{R}^{3\times4}$, $p_{j,k} = A_k P_j$, and positions $\hat{p}_{j,k}$ (annotated landmarks) in image k:

$$\varepsilon_{RMSE} = \sum_{j=1}^{J_k} d(\hat{p}_{j,k}, p_{j,k})^2. \tag{8.1}$$

Minimizing the reprojection error ε_{RMSE} provides the camera projection matrices A_k, which are used to map 3D points to image points. To map image points to 3D points, the depth of the point is required. For a point on the road, the depth is computed by intersecting the corresponding line of sight with the ground plane. The top view showing the images projected to the ground plane on the right in Figure 8.14 visualizes the computed camera parameters.

8.2.3 Vehicle Detection

Although deep learning–based approaches can estimate a 3D model of an object in the detection step, most of these approaches use mature 2D object detectors [7, 8] for hypothesis generation of object locations. Due to the lack of large databases, 3D model approaches are trained and evaluated only on the KITTI benchmark [9]. For the lane merge use case, significantly different viewpoints are required. Thus, a combination of 2D detection, 2D tracking, and 3D localization is followed [44]. Since the *YOLOv3-608* detector [8] provides accurate 2D detections due to its extensively learned network, this approach is used for vehicle detection. It is real-time capable and provides detections in the form of bounding boxes. The example results are shown in Figure 8.15.

(a)

(b)

Figure 8.15 Examples of vehicle detection with their detection probabilities for the cameras observing the main lanes (camera 1 and camera 2).

8.2.4 Vehicle Tracking

For robust tracking of vehicles, detected feature points in vehicle regions are used [47]. The localization and speed in the previous frame are incorporated to predict a feature point in the next image, which stabilizes tracking. Feature-based object tracking provides robust results, especially in the case of occlusions and missing detections. Examples of feature-tracking trajectories are shown in Figure 8.16. The feature points on the background are used to estimate background motion to compensate for mast movement due to wind.

Figure 8.16 Vehicle tracking based on feature point trajectories for each detected object in camera 1 (a) and camera 2 (b).

8.2.5 Vehicle Localization

The 3D localization of a vehicle in the global coordinate system uses 2D-3D correspondence following from the camera calibration. Due to the relatively high viewpoints of the cameras (10 m), the center of gravity of the 2D bounding box approximates the center of gravity of the 3D bounding box. Thus, a 3D localization is obtained by projecting the center of the 2D bounding box onto a plane with height h_r. The projection plane with a height > 0 can be interpreted as the compensation for the error of the center of gravity approximation [44]. The experiments shown in Section 8.2.6 used $h_r = 0.7\ m$. This method cannot take different vehicle shapes into account.

The results from each of the three processed cameras are merged into the final tracking result, providing one localization for each vehicle in the entire scene. Object merging between the cameras is performed using the overlapping regions in the camera views. Vehicles from different views are associated with their minimal 3D distance in the global coordinate system. As mentioned previously, the localization of a vehicle is determined by projecting the 2D image location into a reference plane of height h_r in the global coordinate system.

8.2.6 Accuracy Evaluation

In the 5GCAR demonstration, five cars are involved in a lane merge. Three of these cars are equipped with self-localization technology (our experiments used the *Emlid Reach M+ RTK module*). Their on-board units (OBUs) combine global navigation satellite system (GNSS) RTK data (localization, heading) with vehicle measurements (speed, acceleration, yaw rate). The GNSS RTK unit receives correction signals from an RTK server and uses a GNSS RTK antenna positioned on the roof of the vehicle. The antenna is installed above the vehicle's center of gravity. The update rate of the GNSS RTK module is 5 Hz. The GNSS RTK accuracy seems satisfactory in most cases. However, some visible inaccuracies occurred during the experiment. The module also computes the heading of the vehicle based on GNSS RTK positions. Measurements of speed, acceleration, and yaw rate are determined from the in-vehicle system and synchronized with data from the OBU. The vehicle data is sent to the server via a cellular network. Synchronization with the video data is performed using the network time protocol (NTP) and UNIX timestamps. One of the connected vehicles (5008) is in the acceleration lane. The other two connected vehicles (V60, DS7) are supposed to open the gap for the incoming vehicle (5008). Two vehicles are unconnected: for the lane merge use case, they have to be detected and localized by the cameras. To accurately evaluate the video-based localization, the three connected cars are considered; for the entire lane merge scenario, up to five vehicles are localized by the system.

Vehicle localization example results for the lane merge scenario are shown in Figure 8.17, displaying the x- and y-position (in meters) with respect to the time t (in seconds). The connected vehicles are V60, 5008, and DS7; 5008 is the vehicle in the acceleration lane. The video-based positions of Veh 2, Veh 3, and Veh 4 match the GNSS RTK positions, e.g. the Veh 2 positions correspond to the DS7 GNSS RTK positions. Thus, Veh 1 and Veh 5 correspond to the unconnected vehicles in this example.

For the detailed analysis of the localization error, the resulting vehicle positions from video and GNSS RTK are matched, and their distances d(5008), d(DS7), and d(V60) are computed. These results are shown in Figure 8.18, assuming the GNSS RTK localization as ground truth. Positive values indicate that the estimated distance of the vehicle is too large; negative values show an underestimated vehicle distance. Since localization via projection does not consider the vehicle shape (especially the height), the ranges of d differ from one car model to another. However, the maximum distance between video-based and GNSS RTK localization is approximately 2 m for the showcased lane merge sequence. The distance increases slightly with the distance between the vehicle and the camera d_{VC}.

The evaluation of a larger data set [48], i.e. 20 sequences from three different recording days, is shown in Figure 8.19. As explained in Section 8.2, the ALP [12, 14, 15] is reported, counting the number of correct vehicle positions, i.e. localization with an error less than $t_{ALP} = 1\ m$ and $t_{ALP} = 2\ m$, respectively. The ratio of correct matches determines the ALP, cf. Figure 8.19 (a). The absolute numbers of detected vehicles (approximately 250 detections per vehicle and sequence) are shown in Figure 8.19 (b). Both measures are visualized with respect to the distance between the vehicle and the camera d_{VC}. Since this distance d_{VC} is valid for only one reference position, this evaluation is limited to two cameras located on the

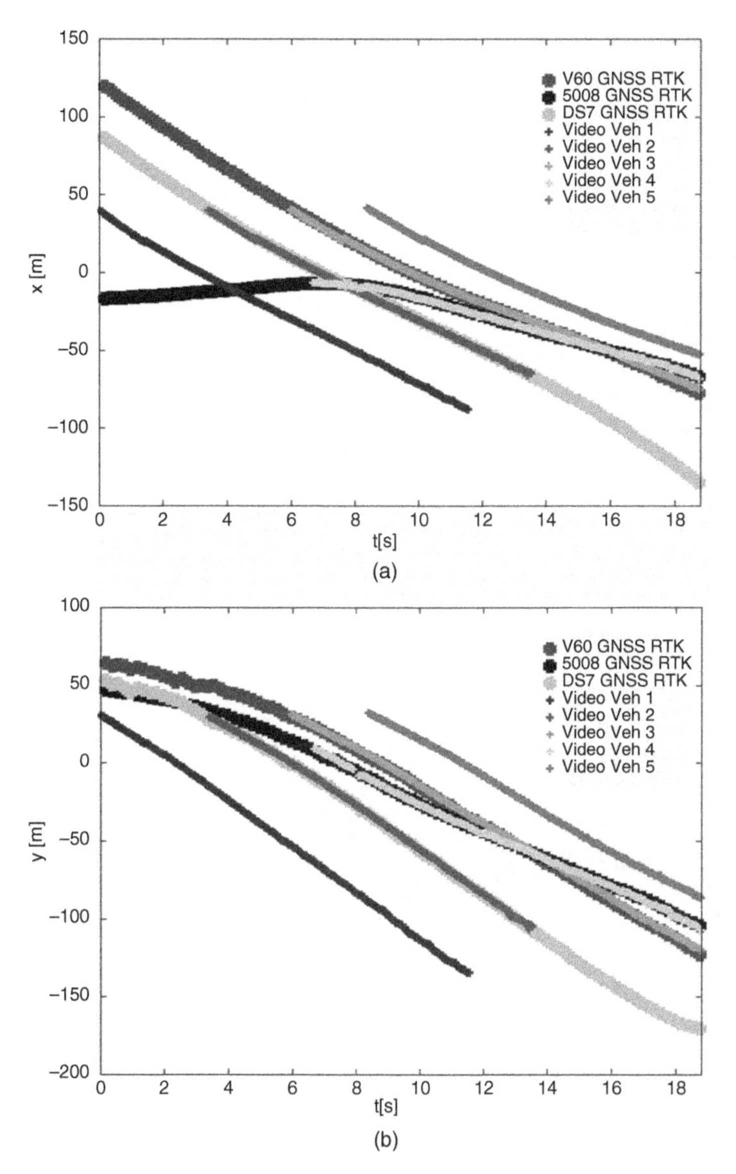

Figure 8.17 Localization (a: x-coordinate, b: y-coordinate) of the vehicles during the lane merge over time. The larger dots denote the GNSS RTK results (three connected vehicles) while the crosses show the video-based positions (five participating vehicles).

same mast (camera 1 and camera 2), using the mast position as a reference. These cameras primarily observe the main lanes (Figure 8.13) and, thus, all participating vehicles.

The highest accuracies are achieved for distances between 90 m and 130 m. The ALP for $t_{ALP} = 2\ m$ remains at a high level, even for large distances (67% for 170 $m < d_{VC} <$ 180 m). For $t_{ALP} = 1\ m$, the localization precision is significantly lower. Due to the multi-camera approach, using a camera with a small focal length for small distances and a large focal

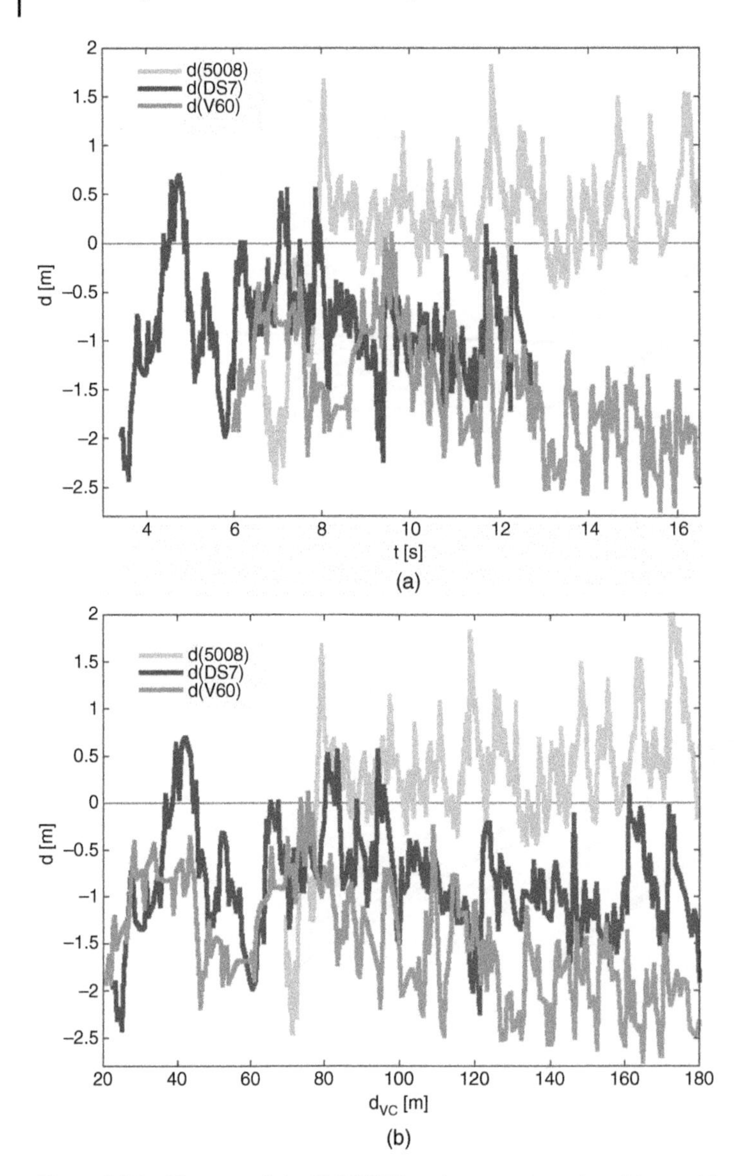

Figure 8.18 Distance of the GNSS RTK and camera-based positions with respect to time t (a) and distance d_{VC} (b) between the vehicle and the camera system. Due to the selected viewpoints for the cameras (Figure 8.13), d_{VC} increases with t.

length for large distances, accuracy does not decrease much with distances larger than 130 m. Compared to the state of the art in vehicle localization (Section 8.1.1), the presented real-time approach provides competitive performance considering the large range of vehicle camera distances of 50 $m < d_{VC} <$ 180 m: while showing inaccuracies for $t_{ALP} = 1\ m$ (ALP: 37.8%), it provides an ALP of 81.6% for $t_{ALP} = 2\ m$. The inaccuracies for $t_{ALP} = 1\ m$ are due to (i) not considering variable vehicle models for the localization (Figure 8.18) and (ii) inaccuracies in the GPS RTK localization, which provides the ground truth for the evaluation.

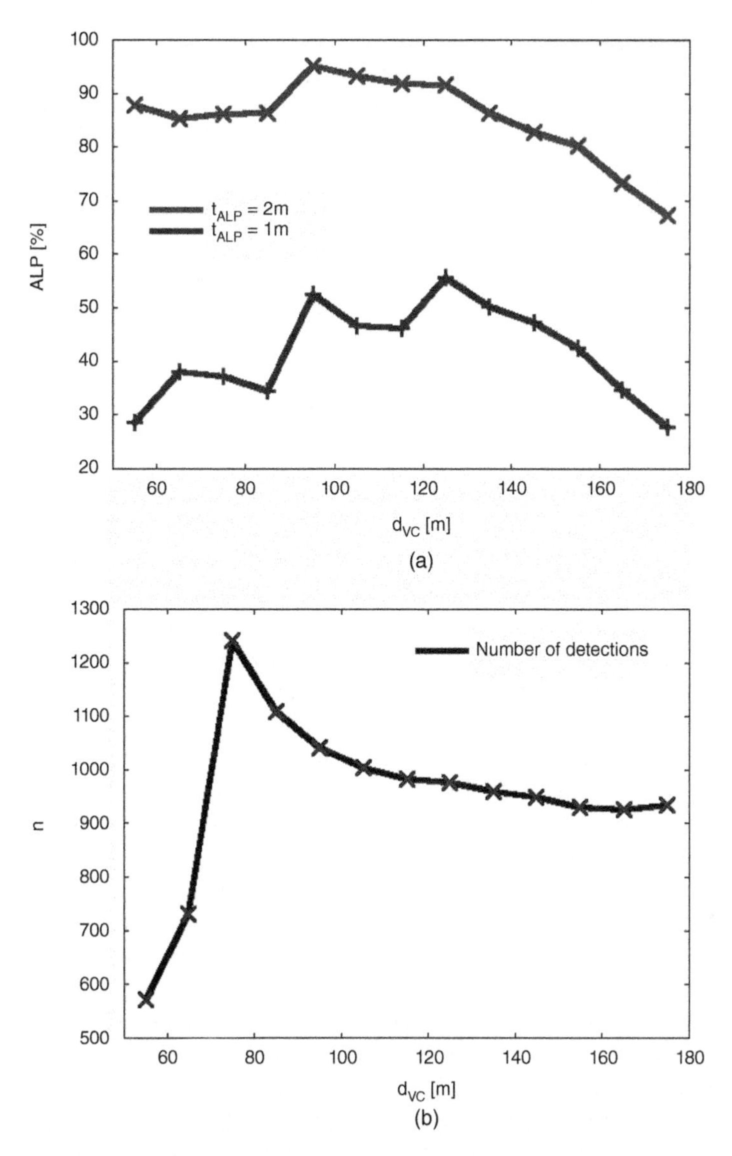

Figure 8.19 Evaluation of the average localization precision (ALP) for 20 sequences and the three connected cars using the GNSS RTK as ground truth localization (a). The absolute number of detected vehicles is shown in (b). Both measures are evaluated piecewise for different distances d_{VC} between vehicle position and the camera system.

8.2.7 Summary

This section evaluated video-based localization of vehicles using monocular video from multiple cameras for a lane merge application. The key features are (i) the combination of multiple cameras with different points of view as well as carefully selected focal lengths to cover the large area required for trajectory planning; and (ii) real-time performance, since trajectory recommendations are computed and distributed online. The localization approach consists of machine learning–based object detection in the 2D image domain and projection of 2D localization onto the 3D global coordinate system.

Vehicle localization is based on 2D detection in each image. Highly accurate camera calibration is employed to map to the 3D coordinate system. The 3D localization approximates the 3D center of gravity of a vehicle with the 2D center of gravity in the image. To compensate for the error, the projection to the 3D global coordinate system uses a reference plane of height h_r.

The localization evaluation uses the ALP as an error measure. The results show competitive performance compared to state-of-the-art vehicle localization while providing real-time processing.

8.3 Conclusions

This chapter presented two approaches for positioning vehicles: 5G radio-based positioning and localization using external sensors (a roadside camera system). 5G radio positioning is an attractive method since the existing infrastructure for mobile communications can be reused. It does not require dedicated devices or signals and can piggyback on the present communication infrastructure. The ability to measure angles in 5G compared to 4G, which only measured delays, provides new opportunities for improving resolution and accuracy, and the ability to use multipath information. 5GCAR has developed complementary solutions to achieve the required accuracy of less than 1 m for absolute positioning and less than 10 cm for relative positioning for optimal conditions. However, several factors may limit performance, such as time-synchronization accuracy and NLOS scenarios.

Video-based positioning requires additional infrastructure (camera installation), but the camera locations are flexible. In the 5GCAR project, a mast with 10 m height is chosen for installation. Thus, NLOS situations can be avoided. Accuracies of less than 1 m are possible, even for large distances between vehicle and camera (evaluated up to 180 m). Performance limits for the localization accuracy of video-based positioning are expected for adverse visibility conditions caused by disadvantageous weather (fog) or lighting (dawn, night). Given good conditions, higher precision and thus greater reliability compared to radio-based positioning are possible. An increase in accuracy, for normal as well as adverse conditions, is expected in the near future since machine learning techniques will provide improved models for 3D object localization. The primary motivation to use video-based vehicle localization in the 5GCAR project is capturing unconnected road users. For this scenario, using additional sensors, such as roadside cameras, is crucial.

Considering the strengths and limitations of both sensor types, the combination of both approaches using sensor fusion should lead to superior results for localization accuracy and reliability.

References

1 Campos, R.S. (2017). Evolution of positioning techniques in cellular networks, from 2G to 4G. *Wireless Communications and Mobile Computing*.

2 Sahinoglu, Z., Gezici, S., and Guvenc, I. (2008). *Ultra-Wideband Positioning Systems*. Cambridge University Press.

3 del Peral-Rosado, J.A., Raulefs, R., López-Salcedo, J.A., and Seco-Granados, G. (2017). Survey of cellular mobile radio localization methods: from 1G to 5G. *IEEE Communication Surveys and Tutorials* 20 (2): 1124–1148.

4 Palacios, J., Casari, P., Assasa, H., and Widmer, J. (2019). LEAP: location estimation and predictive handover with consumer-grade mmWave devices. In: *Proceedings of IEEE Conference on Computer Communications*, 2377–2385.

5 Aladsani, M., Alkhateeb, A., and Trichopoulos, G.C. (2019). Leveraging mmWave imaging and communications for simultaneous localization and mapping. In: *Proceedings of IEEE International Conference on Acoustics, Speech and Signal Processing (ICASSP)*, 4539–4543.

6 del Peral-Rosado, J.A., Seco-Granados, G., Raulefs, R. et al. (2018). Whitepaper on new localization methods for 5G wireless systems and the internet-of-things. Research Gate: Technical Report.

7 Ren, S., He, K., Girshick, R., and Sun, J. (2015). Faster r-cnn: towards real-time object detection with region proposal networks. In: *Advances in Neural Information Processing Systems* (eds. C. Cortes, N.D. Lawrence, D.D. Lee, et al.), 91–99. Curran Associates.

8 Redmon, J. and Farhadi, A. (2017). YOLO9000: better, faster, stronger. In: *Proceedings of IEEE Conference on Computer Vision and Pattern Recognition.*

9 Geiger, A., Lenz, P., and Urtasu, R. (2012). Are we ready for autonomous driving? the KITTI vision benchmark suite. In: *Proceedings of IEEE Conference on Computer Vision and Pattern Recognition.*

10 Chen, X., Kundu, K., Zhang, Z. et al. (2016). Monocular 3d object detection for autonomous driving. In: *Proceedings of the IEEE Conference on Computer Vision and Pattern Recognition*, 2147–2156.

11 Mousavian, A., Anguelov, D., Flynn, J., and Kosecka, J. (2017). 3d bounding box estimation using deep learning and geometry. In: *Proceedings of the IEEE Conference on Computer Vision and Pattern Recognition*, 7074–7082.

12 Chabot, F., Chaouch, M., Rabarisoa, J. et al. (2017). Deep manta: a coarse-to-fine many-task network for joint 2d and 3d vehicle analysis from monocular image. In: *Proceedings of the IEEE Conference on Computer Vision and Pattern Recognition*, 2040–2049.

13 Xu, B. and Chen, Z. (2018). Multi-level fusion based 3d object detection from monocular images. In: *Proceedings of the IEEE Conference on Computer Vision and Pattern Recognition*, 2345–2353.

14 Xiang, Y., Choi, W., Lin, Y., and Savarese, S. (2015). Data-driven 3d voxel patterns for object category recognition. In: *Proceedings of the IEEE Conference on Computer Vision and Pattern Recognition*, 1903–1911.

15 Li, B., Ouyang, W., Sheng, L. et al. (2019). Gs3d: an efficient 3d object detection framework for autonomous driving. In: *Proceedings of the IEEE Conference on Computer Vision and Pattern Recognition*, 1019–1028.

16 Sequeira, L., Szefer, A., Slome, J., and Mahmoodi, T. (2019). A lane merge coordination model for a V2X scenario. In: *Proceedings of IEEE European Conference on Networks and Communications (EuCNC)*, 198–203.

17 Luo, Y., Xiang, Y., Cao, K., and Li, K. (2016). A dynamic automated lane change maneuver based on vehicle-to-vehicle communication. *Transportation Research Part C: Emerging Technologies* 62: 87–102.

18 3GPP. (2018). Study on positioning use cases. TR 22.872, Release 16, v16.1.0.

19 3GPP. (2018). Study on enhancement of 3GPP support for 5G V2X services. TR 22.886, Release 16, v16.2.0.

20 3GPP. (2019). New WID: NR positioning support. RP-190752, Release 16, 3GPP TSG RAN Meeting #83, Shenzhen, China, March.

21 3GPP. (2019). Change request for 3GPP 38.211 (NR; physical channels and modulation). R1-1913661, Release 16, 3GPP TSG-RAN WG1 Meeting #99, Reno, NV, November.

22 3GPP. (2019). Change request for 3GPP 38.215 (NR; physical layer measurements). R1-1913663, Release 16, 3GPP TSG-RAN WG1 Meeting #99, Reno, NV, USA, November.

23 Alkhateeb, A., El Ayach, O., Leus, G., and Heath, R.W. (2014). Channel estimation and hybrid precoding for millimeter wave cellular systems. *IEEE Journal of Selected Topics in Signal Processing* 8 (5): 831–846.

24 Xiao, Z., He, T., Xia, P., and Xia, X.-G. (2016). Hierarchical codebook design for beamforming training in millimeter-wave communication. *IEEE Transactions on Wireless Communications* 15 (5): 3380–3392.

25 Nitsche, T., Cordeiro, C., Flores, A.B. et al. (2014). IEEE 802.11 ad: directional 60 Ghz communication for multi-gigabit-per-second wi-fi. *IEEE Communications Magazine* 52 (12): 132–141.

26 Wang, J., Lan, Z., Sum, C.-S. et al. (2009). In: *Proceedings of IEEE 70th Vehicular Technology Conference Fall*, 1–6.

27 Garcia, N., Wymeersch, H., and Slock, D. (2017). Optimal precoders for tracking the AoD and AoA of a mmWave path. ArXiv.

28 Fleury, B.H. (2000). First-and second-order characterization of direction dispersion and space selectivity in the radio channel. *IEEE Transactions on Information Theory* 46 (6): 2027–2044.

29 Yucek, T. and Arslan, H. (2008). Time dispersion and delay spread estimation for adaptive OFDM systems. *IEEE Transactions on Vehicular Technology* 57 (3): 1715–1722.

30 Shahmansoori, A., Garcia, G.E., Destino, G. et al. (2018). Position and orientation estimation through millimeter-wave MIMO in 5G systems. *IEEE Transactions on Wireless Communications* 17 (3): 1822–1835.

31 Zhou, Z., Fang, J., Yang, L. et al. (2016). Channel estimation for Millimeter-wave multiuser MIMO systems via PARAFAC decomposition. *IEEE Transactions on Wireless Communications* 15 (11): 7501–7516.

32 Wen, F., Kulmer, J., Witrisal, K., and Wymeersch, H. (2019). 5G positioning and mapping with diffuse multipath, ArXiv.

33 Wymeersch, H., Seco-Granados, G., Destino, G. et al. (2017). 5G mmWave positioning for vehicular networks. *IEEE Wireless Communications* 24 (6): 80–86.

34 Wymeersch, H., Garcia, N., Kim, H. et al. (2018). 5G mmWave downlink vehicular positioning. In: *Proceedings of IEEE Global Communications Conference*, 206–212.

35 Yassin, A., Nasser, Y., Al-Dubai, A.Y., and Awad, M. (2018). MOSAIC: simultaneous localization and environment mapping using mmWave without a-priori knowledge. *IEEE Access* 6: 68932–68947.

36 Palacios, J., Casari, P., and Widmer, J. (2017). JADE: zero-knowledge device localization and environment mapping for millimeter wave systems. In: *Proceeding of IEEE International Conference on Computer Communications*, 1–9.

37 Meyer, F., Kropfreiter, T., Williams, J.L. et al. (2018). Message passing algorithms for scalable multitarget tracking. *Proceedings of IEEE* 106 (2): 221–259.
38 Durrant-Whyte, H. and Bailey, T. (2006). Simultaneous localization and mapping: part I. *IEEE Robotics and Automation Magazine* 13 (2): 99–110.
39 Mullane, J., Vo, B.-N., Adams, M.D., and Vo, B.-T. (2011). A random-finite-set approach to Bayesian SLAM. *IEEE Transactions on Robotics* 27 (2): 268–282.
40 Kim, H., Granström, K., Gao, L. et al. (2019). 5G mmWave cooperative positioning and mapping using multi-model PHD. ArXiv.
41 Mizmizi, M., Mandelli, S., Saur, S., and Reggiani, L. (2018). Robust and flexible tracking of vehicles exploiting soft map-matching and data fusion. In: *Proceedings of IEEE 88th Vehicular Technology Conference (VTC Fall).*
42 Saur, S., Mizmizi, M., Otterbach, J. et al. (2020). 5GCAR demonstration: vulnerable road user protection through positioning with synchronized antenna signal processing. In: *Proceedings of 24th International ITG Workshop on Smart Antennas (WSA).*
43 Kakkavas, A., Castañeda Garcia, M.H., Stirling-Gallacher, R.A., and Nossek, J.A. (2018). Multi-array 5G V2V relative positioning: Performance bounds. In: *Proceedings of IEEE Global Communications Conference (GLOBECOM).*
44 Cordes, K., Nolte, N., Meine, N., and Broszio, H. (2019). Accuracy evaluation of camera-based vehicle localization. In: *Proceedings of IEEE Int'l Conference on Connected Vehicles and Expo (ICCVE)*, 1–7.
45 Cordes, K. and Broszio, H. (2019). Constrained multi camera calibration for lane merge observation. In: *Proceedings of Int'l Conference on Computer Vision Theory and Applications (VISAPP)*, 529–536.
46 Vermeille, H. (2002). Direct transformation from geocentric coordinates to geodetic coordinates. *Journal of Geodesy* 76 (8): 451–454.
47 Shi, J. (1994). Good features to track. In: *1994 Proceedings of IEEE Conference on Computer Vision and Pattern Recognition*, 593–600.
48 Cordes, K. and Broszio, H. (2021). Vehicle lane merge visual benchmark. To be published in: *Proceedings of IEEE Int'l Conference on Pattern Recognition (ICPR).*

9

Security and Privacy

Rémi Theillaud[1], Marc Lacoste[2], David Armand[2], and Yvan Rafflé[2]

[1] Marben Products, France
[2] Orange Labs Services, France

Security and privacy are two key requirements for vehicle-to-everything (V2X) communication and applications. Security and privacy for cellular V2X (C-V2X) need to be considered *across multiple operators' domains* (both within a domain and at external domain interfaces) from many different standpoints, taking into account views such as those shown in Figure 9.1:

- *End-to-end view:* User equipment (UE), radio access network, core network, and clouds.
- *Virtualized infrastructure view:* Infrastructure and virtualized resource or function (owned by mobile network operators or other tenants: third parties/virtual network operators) domains.
- *Virtualized network view:* Software-defined network (SDN)/network function virtualization (NFV) management plane, control-plane (CP), and data plane domains.

Some security and privacy threats are not V2X-specific, but more general in the fifth generation (5G) context; thus, they also apply to V2X:

- *Denial of service attacks*, e.g. by jamming radio access networks, overloading the CP with signaling exchanges, or installing malware in devices (e.g. botnets).
- *Threats to anonymity and privacy* of data across radio access networks.
- *SDN/NFV controllers/orchestrators being compromised or abused* (e.g. tenants' impersonation) [1].
- *Poor isolation of network slices* leading to either data leakage or resource exhaustion.

Other security and privacy objectives must be facilitated to legitimate V2X users and applications as a result of malicious activities:

- *Authentication, authorization, and accounting (AAA)* are also critical security functions to verify that any V2X actor, on-board unit (OBU) or road-side unit (RSU) UE, multi-access edge computing (MEC) application, etc., behaves properly (in the current time and location context) and according to the permissions it is granted by V2X authorities. V2X permissions management may call for device enrollment (to global – per region or per country – or delegate authorities), at which time devices are granted permissions:

Cellular V2X for Connected Automated Driving, First Edition.
Edited by Mikael Fallgren, Markus Dillinger, Toktam Mahmoodi, and Tommy Svensson.

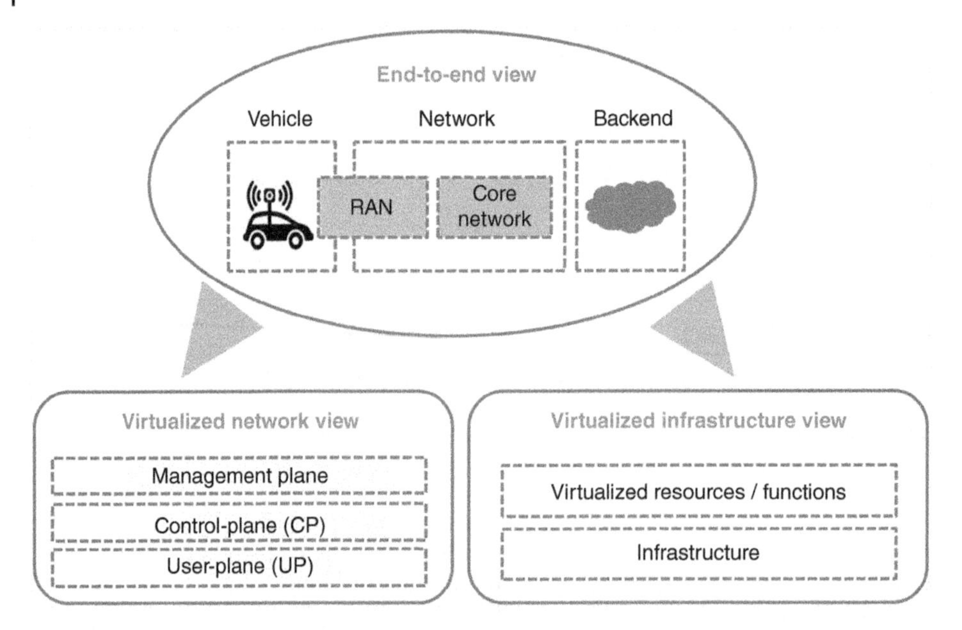

Figure 9.1 Different views of a C-V2X system.

regular vehicles vs. special vehicles (emergency vehicles, road-operator vehicles, police vehicles, public transport/transit vehicles, etc.); roadside units in toll zones, traffic lights, advertising roadworks, etc.

- *V2X messages replay*: Messages may be replayed by an attacker at the same location but at a different time (for instance, messages originally and legitimately sent by an emergency or police vehicle may be replayed later). Messages may also be replayed at a different location, leading to so-called *wormhole attacks*.
- *Liability* in a multi-operator context, or, more generally, in a multi-tenant, multi-service, or multi-country/legislation environment also needs to be considered.

In this chapter, we examine C-V2X security through a more detailed study of different communication formats and interfaces. That includes vehicle-to-network (V2N), in which in a more conventional form, vehicles communicate with the mobile network; vehicle-to-vehicle (V2V), in which vehicles communicate directly with each other; and vehicle-to-infrastructure (V2I), in which vehicles communicate with the wireless-enabled road infrastructure.

In what follows, we look at C-V2X security from the standpoint of V2N communications in Section 9.1 and V2V/V2I interactions in Section 9.2. We also review alternative approaches in Section 9.3 before concluding in Section 9.4.

9.1 V2N Security

In this section, we review overall vehicular security challenges in Section 9.1.1 and then explore isolation challenges in Section 9.1.2. We then present some of the benefits and limitations of software-defined vehicular networking (SDVN) in Section 9.1.3.

9.1.1 Security Challenges

The *connected automated driving* (CAD) ecosystem has a simple three-tier structure: (i) *vehicle*, (ii) *network/edge*, and (iii) *cloud*. The vehicular tier includes cars that communicate with one another. The network/edge tier contains cellular networks and intelligent transport system (ITS) infrastructures. The cloud tier includes clouds (or federation of clouds) forming backend systems.

Increasing levels of *connectivity* and *autonomy* raise acute *security, safety,* and *data-protection* challenges. Vehicles might become attractive targets for hackers, as shown by the rising number of recent attacks [2–4]. In the safety area, a major issue is the *safety-security gap*: the connected vehicle is a cyber-physical system, leading to interaction vulnerabilities between safety and security, whereas threats (malicious) and failures (accidental) are so far handled separately. CAD is also at the heart of a data-shaped ecosystem: vehicle and passenger data are collected, analyzed, and shared with all stakeholders of the CAD ecosystem through multiple channels. Difficult trade-offs must then be made, at rest and in transit, between the integrity of information – which is critical for the safety of vehicles – and the privacy of drivers and passengers [5]. Some root causes for security challenges are shown in Figure 9.2 and may be the source of key business opportunities.

In the vehicle, weak spots are the complexity and vulnerability of currently embedded architectures – notably the proliferation of electric control units (ECUs) and digital process controllers governing safety-critical automotive mechanical systems (engine, battery, tires, steering, etc.), of which some modern vehicles have more than 100 – and the interconnection of a large number of heterogeneous in-vehicle bus hardware architectures. In the network tier, connections should be strongly isolated. Anomalies regarding network isolation should also be detected and mitigated. In the cloud, well-known security challenges from virtualization and network security need to be reconciled with the vehicular dimension, to guarantee end-to-end isolation from vehicle to cloud. Ecosystem security roadblocks also include mobility, interoperability, and responsibility. To meet such challenges, a holistic vision of protection is needed, covering software and hardware, vehicle, network, and cloud tiers; and the full data life cycle.

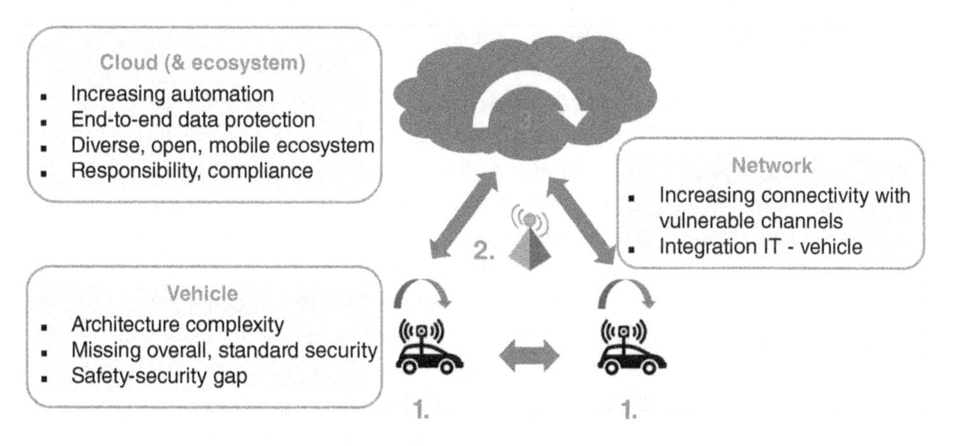

Figure 9.2 Root causes of security challenges.

9.1.2 Isolation Challenges

In the following subsections the isolation challenges, in terms of system isolation and network isolation, are described.

9.1.2.1 System Isolation (Between ECUs)

Safety-critical ECUs in legacy vehicles are connected through a highly vulnerable *in-vehicle bus* - e.g. controller area network (CAN), automotive Ethernet – and to internet-facing systems (e.g. infotainment) via a weakly protected *gateway*. Threats are therefore multiple: to and between ECUs, on the communication bus, and through the gateway. Many network-segmentation solutions have been proposed for securing such in-vehicle architectures, e.g. introducing firewalls or hardening communication protocols, but this remains a major weak spot.

With advances in high-performance computing architectures, ECUs are grouped into *domains*, e.g. advanced driver-assistance systems (ADAS), infotainment, security, radar. The car is then supervised by domain controllers with high computing power – including graphics processing units (GPUs) – that are responsible for isolation within their domain. Smart trade-offs must be made between domains, e.g. security and energy efficiency.

ECUs also become *virtualized* as lightweight execution environments (e.g. virtual machines, containers). Several cloud-isolation challenges then influence the in-vehicle domain. Solutions now mature in other domains (e.g. avionics), such as partitioning hypervisors or open source virtualization (automotive-grade Linux) are key enablers for safe, secure, and reliable communication at the hypervisor level. Trusted execution technologies – e.g. Intel software guards extensions (SGX) enclaves, ARM trust-zone – using hardware-rooted mechanisms also provide a first level of defense for isolation of application-sensitive data despite unsecure operating systems or hypervisors. But those mechanisms remain vulnerable to side-channel attacks against which mitigations need to be put in place at the application, system, and hardware levels [6, 7].

9.1.2.2 Network Isolation (Between Network Slices)

Logical network functions in programmable 5G infrastructures may be grouped in a slice to meet requirements of a specific service in terms of functionality (security, mobility) or performance (latency, throughput) [8].

Inter-slice isolation is required to guarantee performance for each tenant and to preserve security, due to openness to third parties [9]. Such isolation can be deployed by: (i) using a different physical resource, (ii) separating a shared resource using virtualization, or (iii) sharing a resource between tenants enforcing an access control policy. A significant threat is that each tenant may share resources between its slices with different security parameters.

End-to-end isolation, notably for inter-domain (technology, administrative) slice segments, is an inherent property of network slices to facilitate service delivery from service providers to end-users [9]. Slices, therefore, combine resources belonging to distinct infrastructure providers and unify various network layers or technologies. This requires an end-to-end security management plane spanning device, edge, and cloud.

Finally, multiple stakeholders are involved, such as road authorities, municipalities, original equipment manufacturers (OEMs), network operators, and service providers. Vehicular use cases are usually *multi-tenant*. This requires strong multi-level access control for which there is no agreement today.

9.1.3 Software-Defined Vehicular Networking Security

In the following subsections the software-defined vehicular networking security is described. First in terms of principles and architecture, and then from the security benefits and threats perspective.

9.1.3.1 Principles and Architecture

SDVN adapts SDN to vehicular networks to make them more programmable and flexible [10]. The idea is to represent vehicles and elements of the road infrastructure as network switches in the data plane, while the SDN controller manages the vehicular network. A typical SDVN architecture is shown in Figure 9.3.

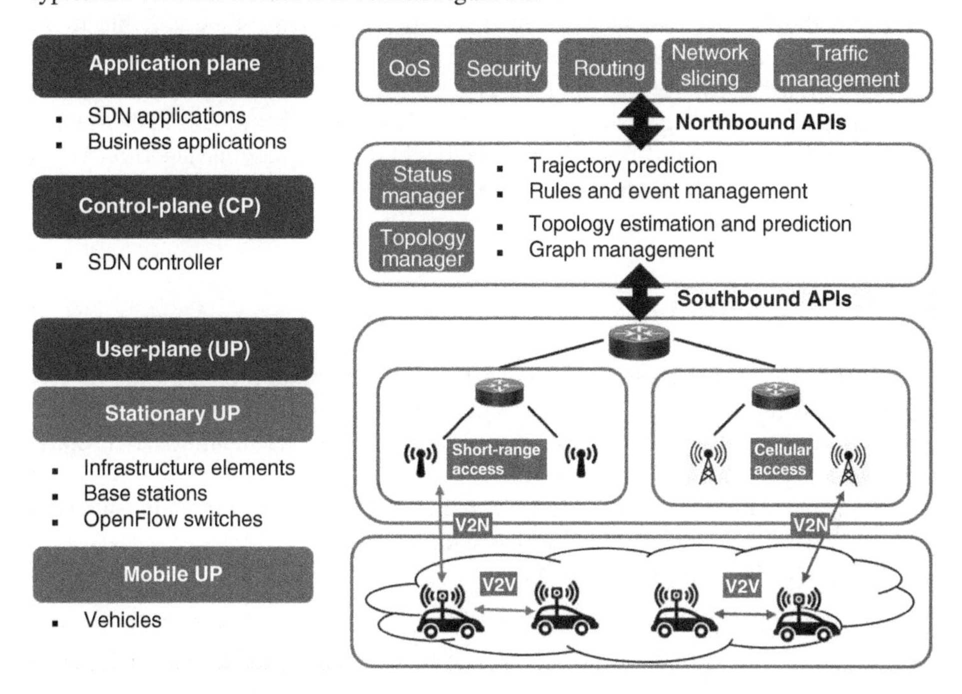

Figure 9.3 SDVN architecture.

The application plane contains network services and applications. The CP includes the SDN controller to manage the network in a logically centralized manner. It maintains the status of all vehicular network switches (including vehicle location, velocity, and network connectivity) and the current network topology state information. The *data plane* contains network resources: all vehicles, RSUs, and base stations are abstracted as SDN switches.

9.1.3.2 Security Benefits and Threats

SDVN contributes to smoother management of vehicular security and safety: it is a promising enabler for a wide range of automotive cybersecurity services, such as in-vehicle security - intrusion detection, firewalling, secure over-the-air (OTA) updates, security audits - 5G connectivity security - network isolation and slicing, network anomaly detection - and orchestration-level services for networks of vehicles - safety, data protection. SDVN-based monitoring also helps to bridge the safety-security gap by monitoring and reacting to both accidental and malicious faults threatening autonomous vehicle behavior. More broadly, architectures combining SDVN and edge technologies help to address automotive deployment challenges: context-awareness to meet real-time requirements, using fog cells to improve scalability and to manage handover between vehicles.

However, the propagation of misinformation from unauthorized entities can lead to serious accidents [11]. The SDN controller remains the central decision-making point. It should be strongly protected using a defense-in-depth approach. *Tightly coupled SDN layers* also facilitate the propagation of threats between layers. Application programming interfaces (APIs) between layers should thus be hardened and standardized. Finally, *vehicle mobility and openness in the lower data plane* amplify SDN threats to control and application planes. Both layers of the data plane should be secured, with the need for real-time authentication.

The impact may also be on *both security and safety*. Because the CAD ecosystem consists of tightly integrated systems of systems, failures in a subsystem may propagate to other subsystems and cascade, making recovery very difficult. Similarly, cyber threats to the vehicle decision-making logic remain critical: vulnerabilities of machine learning algorithms (e.g. to noise-injection attacks) can cause catastrophic safety outcomes in terms of safety if data integrity is compromised.

9.2 V2V/V2I Security

Both European Union and US security architectures follow similar principles and rely on public key infrastructures (PKIs) to deliver digital certificates to OBUs and RSUs, referred to as end entities (EEs) hereafter – US Security Credential Management System (SCMS) terminology.

Digital certificate formats for V2X are defined in [12] for the European Union and [13] for the US. The latest [12] standard is now based on [14].

Every EE is assigned two kinds of certificates:

- *A long-term certificate*, a.k.a. enrollment certificate (EC), which it receives during the enrollment phase. This long-term certificate is used not to sign V2X messages, but rather to sign requests sent to PKI authorities (especially requests to get and/or download pseudonym certificates).
- *Multiple short-term certificates*, a.k.a. pseudonym certificates or authorization tickets (ATs), usually 20 per week, which are used by EEs to sign messages. For privacy purposes, the EE regularly changes its identity, which requires changing its current pseudonym certificate; hence the provision of 20 pseudonym certificates per week.

Digital certificates are signed by a PKI authority, known as certificate authority (CA): enrollment certificates are signed by an enrollment authority (EA), a.k.a. enrollment certificate authority (ECA) or long-term certificate authority (LTCA). Pseudonym certificates are signed by a pseudonym certificate authority (PCA), a.k.a. authorization authority (AA). The mechanism is recursive, with EAs and PCAs using their certificates to sign enrollment and pseudonym certificates. The PKI architecture hence defines a tree of authorities: one parent authority signs the certificate of its child authorities. At the top is a root authority whose certificate is trusted by all.

A PKI also defines the protocol used by EEs to connect to different authorities and for authorities to exchange with each other. See [15, 16] for the European Union and [17] for the US.

A digital certificate specifies:

- *A verification public key*: This public key is used by the receiver of a V2X message to verify the signature of a message (if the verification is successful, it proves that the sender of the message owns the associated private key).
  Note: this is true for explicit certificates. In the US, enrollment and pseudonym certificates are *implicit* certificates: such certificates only specify a reconstruction value, instead of a verification public key and a (certificate) signature (hence an implicit certificate is shorter than an explicit one). The reconstruction value allows reconstructing a public key; if it allows successfully verifying a received message signature, then it also proves that the certificate was issued by the expected authority.
- *An (optional) encryption public key*: This public key can be used to encrypt data solely intended for the owner of that certificate (only the owner of the associated private key can decrypt the data).
- *A geographical validity* (e.g. one or multiple countries).
- *A time validity*.
- *Service-specific permissions* (SSPs) stating whether the certificate allows signing a given V2X message or given content inside a V2X message.

When a signed V2X message is received, the receiver must cryptographically verify the signature and also perform relevance (about time and/or position for instance) checks and consistency (against SSP) checks.

Both the European Union and US use elliptic curve–based cryptographic operations: and elliptic curve digital signature algorithm (ECDSA) for signing and an elliptic curve integrated encryption scheme (ECIES) for encrypting. The associated hashes and keys are either 256 bits or 384 bits long.

All public/private key pairs are required to be generated by a hardware security module (HSM), which ensures that private keys cannot be disclosed. As a consequence, private key-based signing and decrypting cryptographic operations are performed by the HSM.

9.2.1 Privacy

The European Commission draft Delegated Act [18] has a "Fundamental Rights" section that emphasizes that C-ITS services must comply with the EU law on the protection of personal data, in particular the General Data Protection Regulation (GDPR).

Personal data include those that identify a person, such as a driver, passenger, or vehicle owner, including the person's name, phone number or e-mail address, and vehicle license plate number or serial number. Personal data also include geographical location data, mileage, driving style, and other technical vehicle data.

The GDPR does not apply to genuinely anonymous data, provided that the anonymization mechanism cannot be reversible and that a person cannot be identified by cross-referencing multiple data sources. In that context, positioning data are challenging: for instance, someone making the same journey from home to work every day may be identified.

In that context, issues of which data are requested, whether they are stored, and, if they are, for how long need careful consideration. The *data minimization principle* must be enforced: for instance, an accurate position is not needed for a weather forecast application or map-distribution application.

Data must not be stored unless required to provide a service. In addition, data must not be shared with other entities unless explicit consent is given by the data's subject.

V2X protocols and messages that have been standardized so far support the adoption of pseudonyms: protocol identifiers and addresses, as well as security certificates, cannot be linked to a person or vehicle. Furthermore, standards require V2X devices for which privacy is required (such as privately owned vehicles or smartphones) to regularly change identifiers, addresses, and certificates, such as IPv6 addresses and/or sidelink layer 2 (i.e. medium access control – MAC) addresses for 3rd Generation Partnership Project (3GPP) based access technologies. The Delegated Act [18] about security policy proposes an algorithm for triggering such identity changes.

Utilization of pseudonyms and anonymization are not quite the same. To report misbehavior, an entity must exist that can link a pseudonym to its enrollment certificate (i.e. true identity) so that this enrollment can be revoked and no further pseudonym certificates allocated for and assigned to the misbehaving device: i.e. the adoption of pseudonyms needs be reversible by some authority of the PKI.

Note that Delegated Act [18] does not address *how* V2X devices, protocols, and applications may comply with the GDPR; it only states that they must comply. As a consequence, vehicle manufacturers may have to provide the ability for a driver to disable V2X. GDPR applicability to V2X should, therefore, be clarified; and for the sake of safety-related applications, regulation should allow a minimum set of information to always be sent by devices installed in vehicles or other road user equipment so that a minimum set of safety-critical services can operate.

For instance, connected and automated mobility (CAM) awareness messages, which are among the cooperative ITS (C-ITS) messages referenced in Delegated Act [18], are foreseen to be critical for many next generation (the so-called Day 2) safety applications. But their content has been met with reserve by the European Commission, and it is not clear whether sending CAM messages using short-range broadcast communication complies with GDPR, even though pseudonyms are used. The next-generation collective perception message (CPM) will most likely raise similar concerns.

On the other hand, other V2X services (such as lane merging, remote parking, etc.) may require explicit consent from the vehicle's owner or driver.

9.2.2 European Union Security Architecture

To obtain pseudonym certificates, an EE needs:

- *To be registered with the EA*. This is the "bootstrapping" procedure. For OBUs, it is expected that this registration will be performed by the car manufacturer, providing the vehicle's canonical identifier (and associated public key).

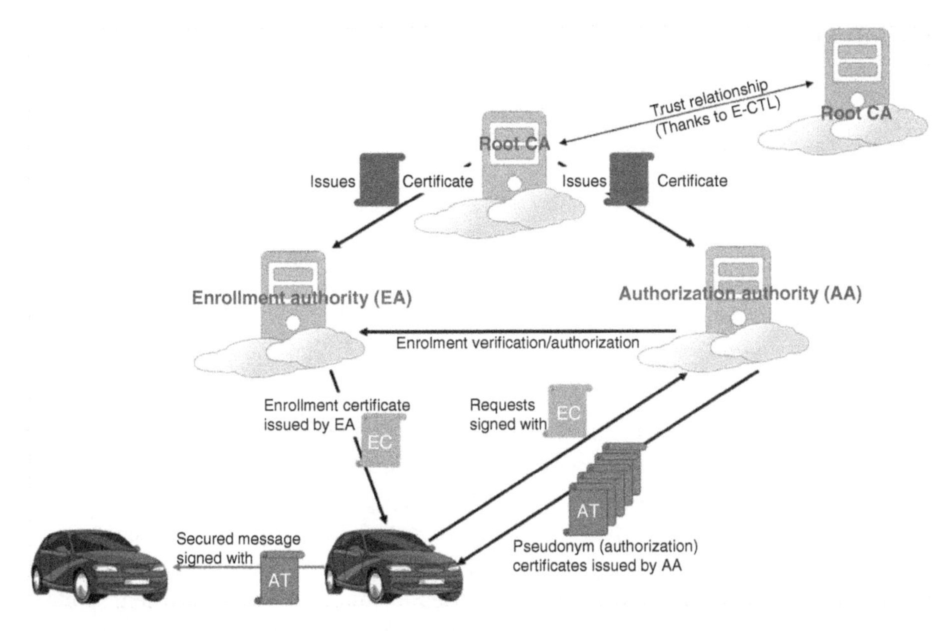

Figure 9.4 European Union security architecture.

- *To request an enrollment certificate to the EA* (Figure 9.4). The request is signed with the public key associated with the registered canonical identifier.
- *To request pseudonym certificates* (Figure 9.4): 20 per week, so that the OBU can keep signing messages even if it cannot connect to the PKI for some time. Such requests are signed using the enrollment certificate.

Note that the PKI protocol is designed in such a way that the AA, when answering the request for a pseudonym certificate (AT), does not know (and cannot track) the identity of the requesting EE. Hence the AA cannot tell whether two ATs have been assigned to the same EE. Similarly, the EA does not know which ATs have been generated for a given EE. Hence privacy is ensured for both EA and AA.

To sign a V2X message, an EE computes a signature and attaches its certificate to the message (the signature and certificate are part of the security envelope defined in [12]). For the sake of bandwidth, a specific rule applies to CAM messages (usually sent at 10 Hz by OBUs): the certificate is not attached to every message and can instead be replaced by its HashedId8 (i.e. 8 lower bytes of the certificate's hash). [12] defines a mechanism by which one OBU can request a missing certificate from other OBUs.

The PKI protocol supports two other critical functions:

- It allows EEs to regularly request certification revocation lists (CRLs). A CRL specifies a list of certificates' HashedId10 (i.e. 10 lower bytes of the certificate's hash) and allows identifying certificates that have been revoked (and thus messages signed by such certificates should not be accepted).
- It allows EEs to regularly request the last up-to-date certificate trust list (CTL) that provides PKI authorities' certificates and access points, e.g. unified resource locators (URLs).

A third important function (ongoing standardization) allows *misbehavior reporting* (Figure 9.4).

9.2.3 US Security Architecture

To obtain pseudonym certificates, an EE needs:

- *To request an enrollment certificate* (Figure 9.5). It is expected that this step will be triggered by the car manufacturer.
- *To trigger the generation of pseudonym certificates* (Figure 9.5). This request is issued once (in a normal situation), signed with the enrollment certificate, and requests the generation of pseudonym certificates batches.

If it accepts the request, the SCMS PKI, the registration authority (RA) and PCA together, generates three years' worth of per-week pseudonym certificates batches (each batch contains 20 certificates – more precisely, 20 encrypted and signed certificate responses).

The entire process is based on so-called *butterfly keys*. The initial request sent by the EE specifies two seed public keys (and some additional crypto materials) that are used by the PKI to generate two sets of keys. One set is used to encrypt every certificate response. Another set is used to generate the public verification key found in each pseudonym certificate. The two private keys associated with the two seed public keys are required.

- To reconstruct the private key, allowing the decryption of each certificate response.
- To reconstruct the private key associated with each pseudonym certificate (and therefore to sign messages with that certificate). Note that this description is simplified for the sake of this overview section. The critical point is that only the EE whose HSM stores

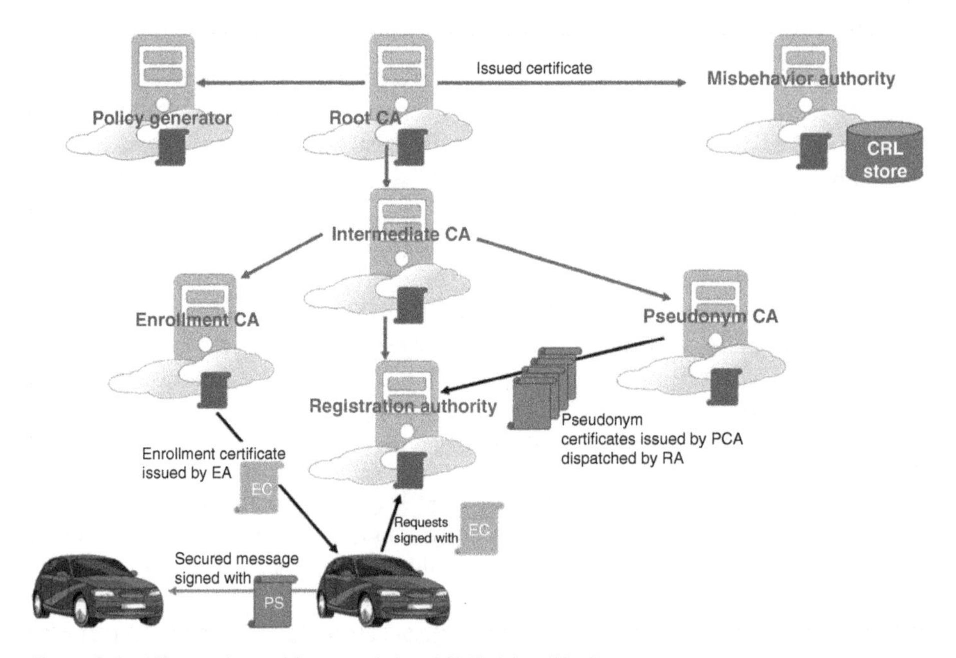

Figure 9.5 US security architecture (a.k.a. SCMS) (simplified).

the two seed private keys can decrypt the certificate responses and use the pseudonym certificates.
- To download per-week batches of pseudonym certificates. The EE can download as many batches as it wants – all three years' worth, if it has enough storage capacity.

Note that the PKI protocol is designed in such a way that the PCA cannot tell (and cannot track) the identity of the requesting EE. Hence the PCA cannot tell whether two pseudonym certificates have been assigned to the same EE. Similarly, the RA does not know which pseudonym certificates have been generated for a given EE. Hence privacy is ensured for both RA and PCA.

To sign a V2X message, an EE computes a signature and attaches its certificate to the message (the signature and certificate are part of the security envelope defined in [13]). For the sake of bandwidth, a specific rule applies to basic safety messages (BSMs) (usually sent at 10 Hz by OBUs): the certificate is not attached to every message and can instead be replaced by its HashedId8 (i.e. 8 lower bytes of the certificate's hash). [13] defines a mechanism, peer-to-peer certificate distribution (P2PCD), by which one OBU can request from other OBUs any missing certificate (Figure 9.5).

The PKI protocol supports two other critical functions:

- It allows EEs to regularly request CRLs. A CRL allows identifying certificates that have been revoked (and thus messages signed by such certificates should not be accepted).
  A CRL specifies either certificates' HashedId10 (i.e. 10 lower bytes of the certificate's hash) or their linkage value seeds. Linkage value seeds allow revoking a set of certificates at once. Two linkage authorities (two, so that one by itself cannot allow identifying all pseudonym certificates assigned to a given EE) are in charge of generating linkage values for generated pseudonym certificates. Disclosure of the two seeds (by the PKI authority) allows verifying whether a certificate's linkage value matches and has therefore been revoked.
- It allows EEs to regularly request the last up-to-date local certificate chain file (LCCF) that provides PKI authorities certificates and access points (e.g. URLs). It also allows EEs to regularly request the last up-to-date local policy file (LPF).

A third important function (ongoing standardization) allows misbehavior reporting to the SCMS misbehavior authority (MA).

9.3 Alternative Approaches

Security and privacy are two key requirements for V2X communication and applications, but the impact they can have on communication bandwidth and latency must be considered. As an example, some security operations may add too much latency and defeat use cases' latency requirements, such as signing all messages with an HSM, or universal subscriber identity module (USIM). Another example is the need for V2X messages to embed a signature and a full certificate that comply with the European Commission ITS security and certificates policies, to authenticate themselves and assert their permissions: doing so requires extra bandwidth.

An alternative approach for end-to-end security may be, whenever possible for a given application, to establish communication sessions between vehicles and application servers, or among a group of two or more vehicles. The benefit is that permission verification and signature only occur in initial messages. Once a session encryption key is agreed on, further message exchanges are only encrypted.

Such a scheme works fine for V2N (UE to V2X server application) and one-to-one V2V/V2I applications (note that V2I refers to both directions of communication between vehicle and road infrastructure). It does not work for one-to-many V2V or V2I applications (for instance, for exchanging awareness messages). One idea is to rely on a central entity in the back office to perform strong authentication and authorization checks, and to provide a session key (or multiple time-limited session keys) for a specific service and for the group of communicating UEs that want to join that service. Possession of the shared key from the back office is proof that the sending UE can be trusted. Note that this is (logically) the overall approach used by 3GPP for proximity service (ProSe) one-to-many security communication.

As an alternative, the central entity may be replaced by one UE that plays a specific role in the scope of one specific V2X application (such as the lead vehicle in a platoon). In both cases, it must be possible to retain the use of pseudonyms whenever required by the V2X service.

9.4 Conclusion

Many pilot projects in Europe, the US, and other countries have demonstrated the use of the V2V and V2I security protocols and public-key infrastructures defined by their respective standardization bodies. Note that V2N can be used to support PKI protocol exchanges between devices and authorities. However, a scalable, cost-effective, efficient solution may still require enhancements to the current standards. The same can be said for achieving privacy as required by the current legislation.

For V2N, given the broad set of security and isolation challenges for future smart vehicular networks, SDVN provides insight into the potential of machine learning and artificial intelligence [19]: a broad spectrum of approaches, techniques, and tools are increasingly receiving attention to tackle issues such as high levels of mobility, strong network dynamics, stringent and heterogeneous quality of service (QoS) requirements, security, and safety.

References

1 ONF. (2015). Principles and practices for securing software-defined networks. TR-511.
2 ENISA. (2019). Cyber security and resilience of smart cars.
3 Ren, K., Wang, Q., Wang, C. et al. (2019). The security of autonomous driving: threats, defenses, and future directions. *Proceedings of the IEEE* 108 (2): 357–372.
4 Upstream Security. (2020). Global automotive cybersecurity report.
5 Karnouskos, S. and Kerschbaum, F. (2018). Privacy and integrity considerations in hyper-connected autonomous vehicles. *Proceedings of the IEEE* 106 (1): 160–170.

6 Bazm, M., Lacoste, M., Sudholt, M., Menaud, J.M. (2017). Side-Channels Beyond the Cloud Edge: New Isolation Threats and Solutions. In: *Proceedings of IEEE Cyber Security in Networking Conference (CSNet).*
7 Bazm, M., Sautereau, T., Lacoste, M. et al. (2018). Cache-based side-channel attacks detection through intel cache monitoring technology and hardware performance counters. In: *Proceedings of 3rd IEEE International Conference on Fog and Mobile Edge Computing (FMEC).*
8 Ordonez-Lucena, J., Ameigeiras, P., Lopez, D. et al. (2017). Network slicing for 5G with SDN/NFV: concepts, architectures, and challenges. *IEEE Communications Magazine* 55 (5): 80–87.
9 Afolabi, I., Taleb, T., Samdanis, K. et al. (2018). Network slicing and softwarization: a survey on principles, enabling technologies and solutions. *IEEE Communications Surveys and Tutorials* 20 (3): 2429–2453.
10 Ku, I., Lu, Y., Gerla, M. et al. (2014). Towards software-defined VANET: architecture and services. In: *Proceedings of 13th Annual Mediterranean Ad Hoc Networking Workshop (MED-HOC-NET).*
11 Akhunzada, A. and Khan, M.K. (2017). Toward secure software defined vehicular networks: taxonomy, requirements, and open issues. *IEEE Communications Magazine* 55 (7): 110–118.
12 ETSI. (2017). Intelligent transport systems (ITS); security; security header and certificate formats. TS 103 097 v1.3.1.
13 IEEE. (2016). IEEE standard for wireless access in vehicular environments— security services for applications and management messages. Std 1609.2.
14 IEEE. (2017). IEEE standard for wireless access in vehicular environments— security services for applications and management messages amendment 1. Std 1609.2a.
15 ETSI. (2016). Intelligent transport systems (ITS); security; ITS communications security architecture and security management. TS 102 940 v1.2.1.
16 ETSI. (2018). Intelligent transport systems (ITS); security; trust and privacy management. TS 102 941 v1.2.1.
17 Crash Avoidance Metrics Partnership. (2016). Security credential management system proof-of-concept implementation. EE Requirements and Specifications Supporting SCMS Software Release 1.2.
18 European Commission. (2019). Commission Delegated Regulation (EU) of 13.3.2019 supplementing Directive 2010/40/EU of the European Parliament and of the Council with regard to the deployment and operational use of cooperative intelligent transport systems.
19 Liang, L., Ye, H., and Li, G.Y. (2019). Toward intelligent vehicular networks: a machine learning framework. *IEEE Internet of Things Journal* 6 (1): 124–135.

10

Status, Recommendations, and Outlook

Mikael Fallgren[1], Markus Dillinger[2], Toktam Mahmoodi[3], and Tommy Svensson[4]

[1] *Ericsson Research, Sweden*
[2] *Huawei German Research Center, Germany*
[3] *King's College London, UK*
[4] *Chalmers University of Technology, Sweden*

As discussed in Chapter 1, when entities share knowledge from their local surroundings, awareness among drivers and automated vehicles can increase immensely and lead to safe and efficient intelligent transport systems (ITS). The ability to connect and share information is the enabler for connected automated driving (CAD) services. Also, non-ITS related services for road users are evolving and require interoperability and trust between service providers. This book has addressed technical enablers for CAD within C-V2X in terms of business models, regulations, standardization, and technical components. However, further work is still needed in various aspects to fully implement the roadmap outlined in Chapter 1, Figure 1.1. The remainder of this chapter reflects the insights of the editors based on their acquired knowledge, including from the 5GCAR project.

In Section 10.1, we analyze and forecast further research and development (R&D) and standardization needs in C-V2X regarding broader aspects of CAD as well as for connected road users (CRU). In Section 10.2, we provide stakeholder recommendations to accelerate the development of the CAD ecosystem in terms of regulation, manufacturing, and operation. An outlook in Section 10.3 on challenges and opportunities ahead in C-V2X for CAD concludes the chapter.

10.1 Future Prospects of C-V2X and the CAD Ecosystem

Although broad aspects of C-V2X systems have been discussed in previous chapters, work is still required well beyond 5G (i.e. 6G and later 3GPP releases) in this new decade, before integration into a commercial CAD system can be accomplished. Such R&D can be continuously integrated into the standardization of mobile networks, since 3GPP ensures compatibility between releases, i.e. newer vehicles and road users in general can still communicate with older releases and vice versa.

Cellular V2X for Connected Automated Driving, First Edition.
Edited by Mikael Fallgren, Markus Dillinger, Toktam Mahmoodi, and Tommy Svensson.

In this section, we also outline the bigger picture of other aspects of CAD and other smart systems that can benefit CRU and thereby society. C-V2X is a core subsystem of the CAD system. Other subsystems for CAD are e.g. advanced physical infrastructure that supports cooperative functional safety and privacy aspects.

10.1.1 Future Needs for R&D and Standardization in C-V2X

The previous chapters have described a set of promising technology concepts for cellular vehicle-to-everything (C-V2X) to enable CAD, from business models, standardization, spectrum and channel modeling, C-V2X radio interface, network architecture, C-V2X application adaption, and positioning to security and privacy. In this section, we identify further R&D needs and missing items related to these topics.

Chapter 2, business models:

- More complete perspectives on sustainable business models related to how connectivity, telecom operators, road operators, original equipment manufacturers (OEMs), smart city operators, parking garages, insurance companies, etc. for new services can change the automotive, telecom, and other stakeholder value chains.
- Innovative business models beyond ITS, in particular mobility as a service, such as vehicle sharing, coordinated travel, and door-to-door transportation, and how they could benefit from connectivity.
- Further business analysis of the technical components of C-V2X and their significance.

Chapter 3, standardization and regulation:

- Coordinated international efforts among standardization bodies from wireless to the automotive industry, as well as to align the physical and digital infrastructure to support CAD services in vehicles.
- Harmonized regulation in different geographical regions to ensure cost-efficient, reliable, and uninterrupted service provision when e.g. crossing the border from one country to another.
- 3GPP solutions for advanced CAD services (such as maneuver alignment, group start, vehicle processing offloading to mobile networks, etc.) in coming years, as a technology to achieve the most challenging V2X requirements.

Chapter 4, spectrum and channel modeling:

- Assessment of the spectrum requirements below 1 GHz and from 1–7 GHz for the various V2X services in urban, suburban, and rural areas with related vehicle densities. For example, the 700 MHz band is attractive for C-V2X due to good coverage, but the related prices per hertz are higher.
- Identification of where to obtain the required spectrum and with what spectrum qualities (such as licensed or unlicensed spectrum). Short-range solutions, vehicle-to-vehicle (V2V), vehicle-to-infrastructure (V2I), vehicle-to-pedestrian (V2P), and long-range solutions, vehicle-to-network (V2N), have different spectrum demands.
- Many of the channel modeling results obtained have been discussed in 3GPP technical reports, but they need to be further assessed and extended toward new C-V2X spectrum and be included in 3GPP channel models.

Chapter 5, V2X radio interface:

- Consideration of vehicle user equipment (UE) in this work is based on standard UE. However, integrating 5G and beyond radio technology in vehicles, e.g. antenna design for mmWave bands, needs further studies.
- Design, evaluation, and performance analysis of sidelink for V2V, V2I, V2P communication and (partially) out-of-coverage communication.
- Design and analysis of ad hoc networking for providing alternative coverage in C-V2X in out-of-coverage regions.
- Vehicle UEs with distributed baseband processing and antenna systems for improved spectral efficiency and reliability, toward cell-free massive multiple input multiple output (MIMO) for C-V2X.
- Coexistence of C-V2X with other potential technologies, such as short-range communication like Bluetooth in the vehicle.

Chapter 6, network architecture:

- Design and evaluation of network slice for delivery of different services in CAD, in particular for services that need low latency and ultra-reliability.
- Study of softwarization, virtualization, and cloudification of infrastructure to allow further flexibility in delivering services with different requirements.
- Exploitation of multi-link connectivity for V2X as a key feature to support use cases with various requirements through optimized network-driven link selection and utilization. In addition, more work is required to understand challenges and impact of distributed architecture of the UE(s) in vehicles both at the transceiver and network level.
- Multi-operator support, including road operators, is necessary to enable seamless, low-latency communication for vehicles and road users, regardless of the mobile network operator (MNO) or road operator each vehicle is attached to. The challenge remains for distributed multi-access edge computing (MEC) over terminal network domains, access, core functions, and vehicles.

Chapter 7, V2X applications and their adaptations:

- Exchange of information between the network and applications is necessary for application adaptation and network optimization to enhance service delivery. However, many application protocols for automated driving and how they are interworking with the lower layers are yet to be developed and standardized. Additional work is needed to examine how to tailor the exchange of information for different use cases.
- CAD messages are exchanged through multi-domain transport infrastructures and between different actors. Each infrastructure has its own characteristics: for example, in terms of bandwidth and latency. However, it is not clear today which authorities should take responsibility for ensuring interoperability through multi-domain transport infrastructures and between actors.

Chapter 8, positioning:

- Obtaining sufficient relative and absolute positioning accuracy for all traffic scenarios and road conditions.

- Combining accurate positioning with mobility prediction not only of vulnerable road users (VRUs) but of all road users.
- System overhead for positioning and mobility prediction has not received much attention. Hence, optimizing e.g. demand reports, periodic reports, or both jointly is an important challenge that remains to be addressed.
- Combining multi-sensor techniques including global positioning systems (GPS), video, low-power accuracy positioning, and short-range and long-range communication.

Chapter 9, security and privacy:

- Further research is needed to understand how emerging CAD applications will comply with the legislation such as General Data Protection Regulation (GDPR) in Europe, e.g. in relation to cooperative awareness messages (CAMs), collective perception messages (CPMs), positioning, maneuver alignment, sensor sharing, and other issues such as implications of driver and owner consent.
- Many pilot projects in Europe, the US, and other countries have demonstrated the use of the V2V and V2I security protocols and public-key infrastructures defined by their respective standardization bodies. However, privacy aspects are not sufficiently addressed in the public discussion.
- Machine learning and artificial intelligence (AI) have shown potential for new security mechanisms by identifying hidden relations and learning mitigations. In combination with software-defined vehicular networking (SDVN), this opens up for new means to address the broad set of security and isolation challenges for future smart vehicular networks.

10.1.2 Broader Aspects of CAD and CRU Services

Research and innovation efforts in CAD have been significant in the last couple of years to develop, test, and validate technologies, systems, and applications to increase road safety and achieve better traffic efficiency. Although technological solutions are available, they have reached different levels of maturity; they need to be integrated into the ecosystem of automated transport. In addition, a significant step is required to go from advanced driver-assistance systems (ADAS) to higher levels of automation where connectivity helps with cost-efficiency, as well as additional safety and functionality to extend the operational range of sensors.

As introduced in Section 1.1.3, C-V2X supports not only CAD, but also many other CRU services such as infotainment, over-the-air (OTA) software updates, OEM clouds for proactive maintenance and diagnosis, weather forecasts, pollution monitoring, tolling for road usage, road maintenance, parking garage business support, services around parked vehicles (e.g. parking efficiency, charging the car, and protecting the car), insurance aspects, advertising, fleet management, car-sharing, and logistics toward integrated moving networks [1].

To fully explore these opportunities with C-V2X, several broader aspects need to be addressed. This section summarizes some of these remaining challenges for CAD (cf. [2] for further elaboration of some of those) and for CRU services. The key area of each of these remaining challenges is classified as: connectivity solutions *(C)*; device design

(D); system at large, e.g. involving network operators and OEMs *(S)*; or regulation and legislation *(RL)*:

- Cooperative functional safety involving more than one road user via radio, from a technical, trust, and business perspective. *(C)*
- More agile vehicle networks supporting cell-free massive MIMO for C-V2X. Multilink cooperation for improved spectral efficiency and reliability, involving evolved on-board and mobile network control transport networking. *(C)*
- Offloading sensor data processing and maneuver calculation, e.g. to the edge or another vehicle. *(C)*
- Further refining technical requirements and validation on CAD systems to enable upper-bound guarantees on key performance metrics, such as delay and jitter, with realistic implementation complexity. *(C)*
- Context-based 5G radio resource management to enable more efficient, reliable, and scalable massive CAD systems and CRU. *(C)*
- Harmonization of communication standards for CAD among mobile communication, networking, and automotive communities. *(C)*
- Physical integration of mmWave and visible light technologies related to both communication and sensing (e.g. radar and lidar) in connected vehicles. *(D)*
- In-vehicle (consisting of passengers networking and automotive components networking) wireless interface and radio access network concepts to maximize key performance indicators (KPIs). *(D)*
- Exploring how C-V2X can further enable more secure CAD systems as well as in-vehicle wireless networking. For example, lower-layer enhanced security mechanisms for the sidelink, and how to best leverage on the V2N security mechanisms for the sidelink as well as for in-vehicle networks. *(D)*
- Upgrading existing physical infrastructure and deploying related digital road infrastructure (radio, IT infrastructure along roads) for future vehicles. *(S)*
- Overall processing strategies for environmental perception. Collected large data must take into consideration the freshness of data, processing time, storage aspects, cleaning and categorizing the data, communication requirements, security, and privacy. *(S)*
- Exploiting AI for improving the road user environmental perception through object classification and enhanced decision strategies for various Society of Automotive Engineers (SAE) levels of automated vehicles maneuvers. *(S)*
- There are many open technical challenges to address when mixing human drivers and autonomous cars or robotic drivers, implying complex traffic management due to heterogeneous amounts of information and decision-making. Additional driver training, driver licensing, and modification of traffic rules will also be required. *(S)*
- On the way toward more advanced SAE levels, a question is how the system should interact with humans in the most clever way. Important research still needs to be carried out related to human-machine interaction, ranging from individual setup to fully autonomous under various legal and environmental constraints. *(S)*
- Developing, harmonizing, and standardizing message semantics, formats, and protocols (data governance) cross-borders, -OEMs, -MNOs, -stakeholders, and -domains. *(S)*

- Responsibilities of stakeholders need further clarification and harmonization to ensure liability and thereby clarify insurance terms. Evidence collection through technical functions, such as dash-cam monitoring and trajectory tracking supported by C-V2X, to assist arbitration. *(RL)*
- All the data-processing aspects relating to CAD must be compliant with data-protection policies. *(RL)*

In all these areas, there is also a need to develop tests and certification strategies that organize CAD testing with efforts and resources from many stakeholders. This can only be successful when all involved subscribe to common goals and aim for the same mobility targets, solutions, services, and use cases. When they do, a shared roadmap and agenda can be developed to tackle the various technical, regulatory, and business challenges and to serve as the basis for maximizing coherence and complementarities. To this end, it is essential to agree on the priority of use cases and details of all functionalities, technical and non-technical enablers, and other prerequisites to make progress with those use cases.

10.2 Recommendations to Stakeholders

To reach a successful deployment and adoption of 5G and beyond in the V2X domain, all stakeholders must be involved. First, the MNO is responsible for providing connectivity as a service with performance levels required for the different use cases. Second, OEMs are in charge of defining needs, equipping vehicles, and taking advantage of connectivity to foster the advent of the upper part of the application levels, like the vehicle control system of automated driving. Third, regulators have the key mission of providing a legal framework to motivate the infrastructure and service deployment needed for the success of 5G and beyond in the V2X domain. Fourth, other relevant stakeholders like telecom vendors and road operators are part of the equation, driving innovation and technology and also ensuring stable, continued service levels on the road.

CAD comprises not only a large set of challenges like complex automotive and telecom technology, but also human behavior, ethics, traffic management strategies, fair competition, and liability issues. For this reason, policies and regulations must be developed that allow the blending of mobility as a service with industrial policies, ecological and transformation to alternative energy, social and employment challenges, and the evolution of logistic chains into smart intermodal models. CAD will, therefore, play a vital role in the European or global transport policy, given its capacity in addressing important societal challenges, from safety and traffic efficiency to environmental issues. In particular, CAD addresses traffic safety by ensuring safe interaction of all road users in mixed traffic roads, especially with VRU. Likewise, acknowledging challenges related to cross-border interoperability, the public sector recognizes the need to foster mutual recognition policies and coordination for the development of cross-border and cross-sector experiences. The regulations should guarantee interoperability (e.g. semantics, message format, metadata, and positioning information) among all involved stakeholders. Additionally, common certification criteria and mutual recognition for the implementation of open road tests are vital to aid homogenization and reduce technical and bureaucratic barriers.

In the following, we give recommendations on the previous aspects for these important stakeholders and their interactions.

10.2.1 Mobile Network Operators

Recommendations for mobile network operators are presented in the following subsections. Various network-sharing alternatives are outlined, followed by new business models for connected vehicle services, and finally roaming and inter-operator cooperation.

10.2.1.1 Network-Sharing Alternatives

Different network-sharing alternatives will imply different return of investment timelines in C-V2X for ITS service providers. Depending on what national regulators allow in terms of network sharing, and how willing operators are to use the opportunities provided by network sharing, we will see a significant impact on the payback time of investments. As it seems unlikely that all users will use a single MNO on a single stretch of road, through either regulation or competitive pressure, network sharing will be essential to drive the infrastructure costs down and make connected and automated mobility a viable business case.

For MNOs, the arrival of 5G implies substantial telecommunication infrastructure investments. Network-sharing capabilities enabled by 5G are among the fundamental solutions to achieve a short return on this investment and might be beneficial for service continuity. Four network sharing possibilities have been identified: passive infrastructure sharing, active infrastructure sharing excluding spectrum sharing, active infrastructure sharing including spectrum sharing, and core network sharing. One of the new features of 5G is network slicing, which may be a good solution for sharing costs through network slice operators providing a dedicated slice to ITS/connected and automated mobility services.

For cost-efficient C-V2X services, MNOs should explore and analyze benefits of network sharing further than has been done before, depending on the regulatory framework.

10.2.1.2 New Business Models for Connected Vehicle Services

A recent telecom vendor report [3] indicated that in the smartphone market, consumers are willing to pay up to 20% more for 5G. There are also claims in the telecom market that business opportunities presented by connected homes and vehicles will not happen if the current model of paying for a bucket of gigabytes is maintained.

To illustrate a potentially profitable C-V2X business case, we will use OTA updating, described in Section 2.5, as an example. OTA is a service where a win-win approach should be possible for both parties – automotive and telecom stakeholders – but it needs a new business model definition in order to be viable. This service promises significant value both for end customers, in realizing new vehicle features after initial purchase, as well as for the OEM, which can add new features or solve other issues in the software without having to bring the vehicle in for costly garage visits. While customers might not be willing to pay for mandatory updates linked to safety, security, or bug fixes, they will be happy to pay for software updates that pertain to adding new features.

The new on-board electronic and electric architecture of next-generation vehicles will be able to manage a huge quantity of data. As an example, one OEM [4] has announced a processing power capacity of up to 4.5 terabytes per hour in its new electrical architecture. According to European Commissioner Thierry Breton, "*Autonomous cars will generate terabytes of data that can be used for innovative mobility-related services and for repair and maintenance. Innovation in this area requires that car data are shared, in a secure and well-framed way.*" In a connected vehicle paradigm, more data exchanged on-board will naturally mean an increase in data-offloading requirements in the environment. The willingness to share such data will influence the business opportunity for the telecom market in coming years.

10.2.1.3 Roaming and Inter-Operator Cooperation

International organizations like the European Commission must lead a digital single market (DSM) strategy to facilitate mobility for connected vehicles among different countries, reducing the cost and technical drawbacks of roaming (as has begun in Europe with the abolition of mobile roaming charges). There should be a clear and homogenous solution to allow all vehicles in the same location to receive the same complete information from a trusted source, which could be provided on the communication plane when possible or on the service plane, fulfilling the requirements such as throughput and latency.

Cooperation between operators, vertical partners, equipment vendors, horizontal technology enablers, and road operators will contribute to the sustained success of C-V2X deployments. To guarantee seamless service quality, inter-operator coordination requires integrating predictive quality of service (QoS), protocols used to route high-priority messages with low latency in combination with MEC-enabled distributed computing services, and 5G network-slicing technology to manage differentiated services. In the case of the C-V2X network slicing framework, all operators must agree to a basic set of functionalities for C-V2X slices through proper orchestration, while unlicensed bandwidth can be used for direct V2V and vehicle-to-infrastructure (V2I) communication in the absence of mobile network infrastructure. However, it is not yet clear whether an actor should coordinate communication between the licensed and unlicensed spectrum.

These significant technical and legal challenges must be solved, and ongoing and future C-V2X research projects should help to clarify the technical difficulties and develop corresponding solutions.

10.2.2 Original Equipment Manufacturers

Recommendations for OEMs are presented in the following subsections. Various aspects are further discussed related to connecting off-board sensors and vehicle-processing platforms supported by mobile networks and automotive standardization.

10.2.2.1 Connecting Off-Board Sensors

Automotive industry OEMs will adopt 5G and beyond as a consumer electronics standard, but the deployment timeline of this technology will depend on customer benefits and the added value of new use cases that are hardly possible with current 4G technology (e.g. SAE level 4). This is primarily due to the additional costs and complexity related to integrating new communication modules and testing them according to safety-critical automotive requirements.

The automotive industry should not link C-V2X adoption only to the arrival of autonomous driving, since connectivity is an addition to the assisted driving solutions, and it should be seen as another ADAS sensor. The business model and costs of autonomous drive (AD) are not clear yet, which makes it difficult to decide on 5G adoption. However, if C-V2X connectivity is thought of as another ADAS sensor, it can be analyzed to complement on-board sensors, helping human- or machine-driven vehicles, encouraging its adoption in terms of both time and volume. Off-board sensor connectivity requires an in-vehicle network to manage firewall delays in order to achieve timely data fusion for off-board and on-board sensors.

The financial cost of a 5G-enabled modem can be softened if a worldwide approach is fostered to adopting a single technology to be deployed in all regions for automotive applications. Fragmenting the market by selecting different solutions will increase the investment needed for the technology development, which in turn will delay 5G and beyond adoption. Moreover, the simultaneous adoption of different technologies will increase the cost and technical challenges: physical space, antenna installations, etc.

10.2.2.2 Vehicle Processing Platforms Supported by Networks

Since the arrival of connected vehicles, OEMs have defined a new paradigm for interacting with the off-board world through connectivity: the extended vehicle. The basic extended vehicle concept may start with same basic functionality for improving driver convenience, and it ends with the complete offloading of sensors and maneuver processing (advanced extended vehicle concept). This requires high-volume, low-latency connectivity. The vehicle can no longer be developed by the OEM as an isolated element, since vehicles are linked with the OEM cloud via connectivity. OEM applications have so far formed a central cloud to address customer needs, such as terminal precondition setting and electric vehicles (EV) charging control.

Besides the paradigm of integrating off-board (remote) sensors via C-V2X, we can also consider the design paradigm for vehicle platform processing offloading to the network, for example along 5G corridors. In fact, there are different views among OEMs, in particular between traditional companies and newcomers. Vehicle processing offloading for the advanced extended vehicle would require high volume and short-latency services that are not yet addressed in 5G.

With the arrival of 5G, two primary enablers must be considered: MEC and connectivity as an ADAS sensor. MEC provides new places where OEM applications can be located dynamically. The architecture of OEM applications should be ready to create, maintain, and destroy instances based on criteria such as traffic density, hazard events, weather conditions, available or required network performance, as well as cost.

On the other hand, as highlighted in this section, connectivity is an off-board sensor. This means ADAS applications can no longer be developed as strictly on-board: decision-making will be done on-board, but complementary information will come from outside the vehicle and provide non-line of sight information consolidated from various sources as well as new rules learned from other vehicles. In the longer term, on-board sensor processing and information for vehicle actuation must be offloaded to the network (e.g. MEC) to save vehicle energy and platform costs.

10.2.2.3 Automotive Standardization

To meet standardization requirements, it is essential to modify existing automotive standards by including cross-OEM data exchanges for cooperative driving. Many standards produced by International Organization for Standardization (ISO) TC204 WG14 describe current ADAS functions: for instance, ISO standard 15 622, which defines minimum performance requirements for ADAS, such as automated cruise control (ACC) based initially on a pure on-board sensor/actuators system. TC204 standards production has focused on cooperative ADAS such as cooperative adaptive cruise control (C-ACC) in ISO 20035, which will consider on-board and V2X information to perform more predictive enhanced ACC.

With regard to V2X communication standards, from 2006 to the present day, the European Telecommunications Standards Institute (ETSI) in Europe and the Institute of Electrical and Electronics Engineers (IEEE)/Society of Automotive Engineers (SAE) in the US have produced all the standards necessary to describe the ITS stack, independent of any kind of radio technology. These standards can be partly reused in the context of C-V2X technologies (LTE and 5G), but some adaptations are required in the access layers and stack management, in particular to control the sidelink and Uu (interface between the vehicle and mobile network) for optimal routing of messages related to road safety and traffic efficiency. However, new application protocols are needed to support advanced use cases: e.g. vehicles must individually negotiate and acknowledge actions.

As 5G and beyond technology will be used for communication by autonomous vehicles, automotive safety integrity level (ASIL) A or B software and hardware constraints will be taken into account for an important standard like ISO26262.

10.2.3 Regulators

The following subsections define future work areas for regulations in order to harmonize countries or regions following different traffic rules. In addition, regulatory automotive aspects for C-V2X are further detailed to enable CAD services along international roads. Regulatory involvement is also needed for monetization of road user data for commercial or societal purposes while ensuring personal data-protection rules. CAD spectrum demands and related data rate and delay performance will result in different deployment and coverage requirements given by regulations.

10.2.3.1 Deployment, Coverage, and Road Infrastructure

To enable the C-V2X ecosystem, a telco infrastructure must be deployed, and roads must have cellular coverage. Some countries have defined strong performance and deployment requirements for 5G frequency assignment to guarantee minimum throughput on all roads.

Traffic signals and information are already available on the internet in some parts of Europe, but this differs between countries. In most cases, road infrastructure is only partially connected to central control centers, and additional upgrades are required before more advanced safety services can be provided. Upgrades for traffic controllers available in the market include additional communication interfaces to allow the management, analysis, and provision of more advanced safety services to vehicles. Policies related to road infrastructure communication equipment will differ from one country to another, and road operators in each country may be able to choose technologies to distribute V2I safety messages.

The 5G Automotive Association (5GAA) has done a study on the economics of various ways to distribute traffic light information and has released a white paper [5] that shows clear benefits from using existing cellular networks for this purpose.

In early September 2019, the European Automobile Manufacturers Association (ACEA) published a 2019–2024 manifesto [6] describing the key elements of the mobility transformation, and a chapter is devoted to demanding that the EU make road transportation smart and convenient. The main recommendations are related to removing regulatory obstacles and enabling rapid deployment of digital communications infrastructure (V2X).

The required rollout investments, business models, and revenues – on which 5G and beyond V2X deployment is strongly dependent – are still uncertain. Without clear economic benefits, MNOs (including road infrastructure operators) will not be encouraged to start network deployments in areas where only ITS services are expected to be delivered. However, 5G networks will serve other CRU services and ITS services concurrently.

The first estimation of network deployment costs in terms of investment in capital expenditures (CAPEX) and operation expenditures (OPEX) and revenues of a 5G V2X rollout along a motorway has been provided by the 5G PPP Automotive working group [7]. This work can guide the 5G V2X community regarding required demand and service fees.

An ongoing Strategic Deployment Agenda (SDA) effort, including mobile networks (V2N) and roadside units (V2I) [8], is considering how C-V2X can support CAD in Europe and globally.

10.2.3.2 Simplifying and Harmonizing Regulation

A topic that has been high on the agenda of the European Commission is the Delegated Act [9] being drafted under the Intelligent Transport Systems Directive (Directive 2010/40/EU). This document sets the framework for conditioning the uptake of C-ITS and AD in Europe. The final version for public consultation was released in January 2019 and finally rejected in July 2019. Some industrial players had significant concerns because the Delegated Act was not committed to technology neutrality: it favored a specific, single-purpose Wi-Fi-based technology path (ITS-G5). In the latest European Radio Spectrum Committee announcement, dated July 2020 [10], it is expected that LTE-V2X will be included as additional short-range radio technology that can be used in the European ITS band. Meanwhile, at the Consumer Electronics Show (CES) in January 2019, Ford announced a C-V2X adoption beginning in 2022 in the USA. In Europe, Volkswagen has launched the latest Golf generation equipped with ITS-G5. These facts will have significant implications in terms of desired interoperability. In China, the government has decided on C-V2X adoption, with a roadmap proposed in terms of 5G deployment for an intelligent and connected vehicle (ICV) strategy: 13 national OEMs are planning to deploy this technology between the end of 2020 and the beginning of 2021 [11]. In addition, the FCC in the US has decided for C-V2X in November 2020 [21].

C-ITS services must comply with European Union laws regarding the protection of personal data, in particular the GDPR. This is a delicate topic and has not yet been completely solved; one example is CAM transmission. As a consequence, vehicle manufacturers may have to provide the ability for a driver to disable V2X. GDPR applicability to V2X should, therefore, be clarified; and for the sake of safety-related applications, regulations should allow a minimum set of information to always be sent by devices installed in vehicles or other road user equipment. The problem may also be a barrier for MEC deployment. In Europe, if CAM corridors are promoted by the European Commission, data from servers in different countries will move without restriction.

As another example, the security strategy proposed for broadcasting ITS messages based on pseudonyms is a significant market opportunity; but at the same time, the cost for each vehicle cannot be ignored [12]. Thus, this is likely to become another barrier to the adoption of sidelink communication on the OEM side.

10.2.3.3 Data Sharing and Monetization

GDPR has become a worldwide reference regulation for preserving people's privacy in the digital domain. It has provided a legal framework that addresses the concerns of many citizens regarding the use of their personal data by internet companies.

Internet giants have successfully mastered the monetization of data. OEMs (especially since the arrival of connected vehicles) and MNOs manage a huge quantity of customer data. These two sectors are strongly regulated domains to ensure privacy for customers. This has been recognized by the European Union, and activities are ongoing to modify this situation (see e.g. [13]) while keeping privacy requirements.

On the other hand, regulation provides room to enable profitable business models based on digital ITS data, always respecting data privacy. As an example, let's consider a recent pioneering study in Europe of the three main MNOs in Spain, demanded by the Spanish government [14]. The objective of the study was to analyze the primary mobility trends of Spanish citizens during a few days between July and December 2019 and thus learn valuable lessons to help improve public services. The data was anonymized to respect privacy protection regulations, such as GDPR requirements in Europe. This led to controversial discussions in the Spanish media. However, data is the foundation of artificial intelligence and machine learning, and if this source of information is not used, the European Union will lag behind other world regions and internet titans. A way forward is to anonymize all vehicle data (e.g. video cams in vehicles must remove personal data like faces before data is uploaded and processed in the network) for ITS purposes.

10.2.3.4 Spectrum Aspects

The European Union is holding 5G band auctions, as are other regions like South Korea and the USA. Important telecommunication markets such as Italy, Spain, and Germany have completed part of this process, with a significant quantity of spectrum made available.

Focusing on Europe, the re-farming of frequencies that are currently being used in 2G/3G and 4G is under consideration in many countries. The main candidates to be used for 5G are the 3G 2.1 and 2.6 GHz bands, since the allowed bandwidth in these bands is a good fit for 5G services. This band is harmonized in Europe and can potentially be re-farmed for use in 5G for supplemental downlinks. It is widely used for IMT applications due to its good coverage in outdoor scenarios and buildings [15]. This should be promoted by regulators and coordinated under the umbrella of the DSM framework.

5GAA has published a study of spectrum needs of intelligent transportation systems (ITS) [16], where the amount of bandwidth required for the introduction of day 1 and advanced ITS use cases like CAD is predicted to be 50 MHz < 1 GHz and 500 MHz from 1–7 GHz for long-range communications and 75 MHz in the 5.9 GHz band for short-range communications. Details of spectrum needs are provided in [16], and some are highlighted for selected use cases related to short- and long-range communication. The challenge will be to idendify where this spectrum can be assigned for V2X purposes.

The Global System for Mobile Communications Association (GSMA) released a report [17] centered in Europe, which concluded that higher prices are associated with more expensive, lower-quality mobile broadband and irrecoverable losses in consumer welfare. This is even more important for V2X communications, where cost will be a significant roadblock for 5G adoption.

The 700 MHz band presents opportunities for improving cellular network coverage, due to its propagation properties, and could be especially interesting for C-V2X use cases, as shown in [18]. This band will be made available in Europe from 2020 on. In the spectrum auctions so far, the prices per hertz of this band in Italy reached a notably high value. This will be a barrier for new use cases in this band, such as C-V2X, because operators will want a significant return on their investment in these frequencies. The example from Asia should be taken into account, where deployment has been prioritized over the price of frequency bands.

Some entities, like the General Services Administration (GSA) and Federal Communications Commission (FCC) in the US, are looking for new frequency bands beyond those just described. But even if they are finally harmonized and assigned, there will be a significant time to market for any of them.

Despite the expectations of the DSM for a common framework for frequency allocation, the truth is that there is no common European description for frequency auctions at a national level. Some countries have defined strong performance and deployment requirements for frequency assignments, while others have no requirements at all. Even though entities like the GSMA have provided auction best practices [19], the process has not been harmonized. This should be a priority for regulators.

On the unlicensed spectrum side, European adoption of the 5.9 GHz band remains in doubt. The rejection of the Delegated Act in July 2019 has led to some uncertainty, but a recent announcement [10] will also allow C-V2X in the EU ITS band to keep technology neutrality for short-range radios. Allowing two technologies must trigger coexistence discussions and agreed proposals. The European Association of Automotive Suppliers has published a position paper [20] on short-range technologies that speaks in favor of supporting the coexistence, interoperability, and compatibility of different technologies and against the segmentation and segregation of 5.9 GHz – a vision that is the opposite of the latest FCC decision in this regard [21].

10.2.4 Suppliers and Certification

Due to new C-V2X requirements, we need new network and radio features that can be provided to MNOs, vehicle OEMs, data center operators, and road operators. Other parties, like software providers for AI processing and high definition (HD) map providers, are also part of the ecosystem. Integrated CAD solutions require certification to ensure that they meet societal and safety demands. However, because regulations differ worldwide, we will see different market scenarios and involvement of stakeholders. In addition, all elements provided by equipment or software vendors must provide certified components for CAD end-to-end solutions. We forecast that no single body or agency today will be able to offer a global approach to the certification of C-V2X and involved vehicles. Thus, we foresee the need for a new CAD agency based on a partnership among all involved stakeholders, including regulators and the automotive and telecom industries.

10.3 Outlook

Safety and efficiency in the transportation system is an area that has attracted both the private and public sectors and will continue to do so in the future in this decade and beyond. The safety aspect primarily focuses on reducing fatal and severe accidents but aims to ultimately limit any type of injury. When it comes to efficiency, however, a plethora of issues are considered. Energy efficiency focuses on avoiding wasting resources and thereby limiting pollution while fulfilling users' need to get from point A to B. Another aspect of efficiency is reducing congestion and travel time in densely populated areas in cities.

Some developments and future trends from the automotive side are already achieving energy efficiency and reducing pollution: for instance, electric vehicles and their batteries will continue to evolve in the years to come. Widespread use of electric vehicles also play a part in the planning and management of the power grid. Another trend is an increased number of shared vehicles, which not only reduces overall vehicle density and congestion but also can be further enhanced by moving vehicles to parking spaces outside of dense areas. With the wider availability of vehicle and infrastructure data, global (optimal) solutions for better use of roads and parking could replace individual drivers' decisions. Such global solutions could then be communicated to drivers and be enforced through digitally connected infrastructure.

The ultimate performance improvement could be achieved through mobility as a service solution where users receive a customized solution for reaching a destination: for example, in the shortest time, with the least pollution impact, using various means of transportation. People could then effortlessly move from A to B while changing vehicles (e.g. road vehicles, trains, ships, and airplanes) as needed. Goods could be transported in a similar fashion: for example, on trucks and ships, with drones orchestrating and surveying them. Finally, with synergies and convergence trends, vehicles could become advanced moving nodes and part of the mobile communication infrastructure. Since these transportation services will take many years to deploy, we believe that beyond 5G and 6G, V2X as part of C-V2X will also significantly contribute to the CAD solutions required by OEMs, stakeholders, and society. In addition, CAD is to be expanded toward the broader scope of cooperative, connected, and automated mobility (CCAM), cf. Section 1.1.1.

Due to the pandemic crisis in 2020, reductions in industry investments have followed, and market analysis has indicated a strong revamp of investments. Likewise, vehicle automation investments have been forecast and the number of market players for CAD solutions after the pandemic crisis will be higher [22]. While pandemics like Covid-19 in 2020 could expedite both the industry move and societal acceptance toward the adoption of highly automated vehicles such as robot taxis, it could also raise new concerns for car-sharing platforms and use of public transport for a long time.

Overall, CAD enabled by C-V2X will be a paradigm shift in the way transportation is perceived today, and will bring important societal advantages in terms of safer roads, individual time spent commuting, and cleaner air, addressing increasing climate change concerns globally.

References

1 Fallgren, M., Dillinger, M., Alonso-Zarate, J. et al. (2018). Fifth-generation technologies for the connected car: capable systems for vehicle-to-anything communications. *IEEE Vehicular Technology Magazine* 13 (3): 28–38.
2 5GCAR. (2019). 5GCAR: executive summary, White paper, version 1.0.
3 Ericsson.(2019). 5G consumer potential.
4 Lienert, P. (2019). GM digital vehicle platform debuts, enables adoption of future technologies: Installed on newly-unveiled Cadillac CT5, with rollout to most GM vehicles globally by 2023.
5 5GAA. (2019). C-ITS vehicle to infrastructure service: how C-V2X technology completely changes the cost equation for road operators.
6 ACEA. (2019). Leading the mobility transformation: The future of the EU auto industry.
7 5G PPP Automotive Working Group. (2019). Business feasibility study for 5G V2X deployment.
8 5G PPP Automotive Working Group. (2019). 5G strategic deployment agenda for connected and automated mobility in Europe. Initial proposal.
9 Official Journal of the European Union. (2019). Regulation (EU) 2019/153204 of the European Parliament and of the council.
10 EU Radio Spectrum Committee. (2008). Draft Commission Implementing Decision on the harmonised use of radio spectrum in the 5 875-5 935 MHz frequency band for safety-related applications of intelligent transport systems (ITS) and repealing Decision 2008/671/EC.
11 Huawei. (2019). 13 auto makers jointly published commercial roadmap: launching mass production C-V2X car from 2020.
12 5GCAR. (2019). Automotive use cases and connectivity challenges, business models and spectrum related aspects. Deliverable D2.3.
13 Donkin, C. (2017). EC privacy law overhaul targets messaging services. Mobile World Live.
14 Maqueda, A. (2019). El INE seguirá la pista de los móviles de toda España durante ocho días. EL PAÍS.
15 Marks, P., Lavender, T., Wongsaroj, S., and Hogg, T. (2015). Global momentum and economic impact of the 1.4/1.5 GHz band for IMT. GSMA Spectrum.
16 5GAA. (2020). Working group standards and spectrum. Study of spectrum needs for safety related intelligent transportation systems – day 1 and advanced use cases. TR S-200137, Version 1.0.
17 Marsden, R. and Ihle, H-M. (2017). Effective spectrum pricing in Europe: Policies to support better quality and more affordable mobile services, GSMA Spectrum.

18 5GCAR. (2019). 5GCAR intermediate report on V2X business models and spectrum. Deliverable D2.2.

19 GSMA. (2019). Auction best practice: GSMA public policy position.

20 CLEPA. (2018). CLEPA position on short range V2V, V2P and V2I within C-ITS communication technologies (5.9 GHz frequency band).

21 W. Wiquist. (2020). FCC Modernizes 5.9 GHz Band for Wi-Fi and Auto Safety. *FCC News*.

22 Hausler, S., Heineke, K., Hensley, R. et al. (2020). The impact of COVID-19 on future mobility solutions. McKinsey.

Index

Cellular V2X for Connected Automated Driving, First Edition.
Edited by Mikael Fallgren, Markus Dillinger, Toktam Mahmoodi, and Tommy Svensson.

q

r

t